LIPID CHROMATOGRAPHIC ANALYSIS

Volume 1

LIPID CHROMATOGRAPHIC ANALYSIS

Edited by **GUIDO V. MARINETTI**

SCHOOL OF MEDICINE AND DENTISTRY
DEPARTMENT OF BIOCHEMISTRY
UNIVERSITY OF ROCHESTER
ROCHESTER, NEW YORK

Volume 1

1967

MARCEL DEKKER, INC., New York

PREFACE

The rapid advances in chromatographic methods for lipid analysis have been remarkable, particularly in the last decade. When this writer first entered the lipid field as a graduate student in 1951 he was immediately confronted in his research with the problem of separating, detecting, and analyzing on both a macro and micro scale the various major phosphatides from biological sources. Although column methods utilizing magnesium or aluminum oxide were available for fractionation of phosphatides, they did not have great resolving capacity. Moreover, paper chromatography of phosphatides was essentially nonexistent. Once a phosphatide was isolated there was no chromatographic method available for determining the purity of the material, in spite of the fact that paper chromatographic methods for amino acids and sugars were common procedures. It was obvious that the lipid field was lagging behind the protein and carbohydrate fields in chromatographic methodology.

It also was a severe hindrance to those working with isotopes and studying biosynthesis of phosphatides and glycerides not to have microchromatographic methods on hand. One had to be content with studying the incorporation of isotopes into total neutral lipids or total phosphatides as obtained, usually, by acetone precipitation.

The time was ready for a breakthrough in the lipid field, and a number of workers addressed themselves to this task. In 1952 Dr. George Rouser and this writer began a systematic attack on the problem of paper chromatography of phosphatides by first examining the solubility properties and ionic forms of these polar lipids and developing solvent systems which might effectively resolve these compounds on ordinary filter paper. This was a long and arduous endeavor which yielded some fundamental knowledge of the problems involved. One of the most aggravating aspects of this early work was the detection of the lipid components on chromatograms. The amino lipids offered no obstacle, since the ninhydrin reagent was available. The detection of choline-phosphatides was done with the phosphomolybdic-stannous chloride procedure, but it lacked sensitivity. What was needed was a general sensitive test for all lipids, since the inositol phosphatides, cardiolipin, glycolipids, and other lipids were not visualized on chromatograms by the two reagents just mentioned. A large number of dyes were tested as agents for detecting lipids. Rhodamine 6G turned out to be the most effective dye. It provided a general test for lipids on chromatograms and, moreover, gave characteristic fluorescent color tests with some lipids. Once a general test was available, the further development of paper chromatography of lipids was markedly facilitated. The next phase of the work, namely, the alteration of the stationary phase of the filter paper, was then inaugurated.

The early work with chromatography of phosphatides on ordinary filter paper made it clear that cellulose alone would not prove adequate, since it had very little loading capacity for lipids and bound some polar lipids too firmly. Reversed phase chromatography or filter paper impregnated with some suitable adsorbent was considered the most likely solution to this problem.

The use of silicic acid column chromatography for phosphatides was early realized in 1944 by D. Rathmann as part of her Ph.D. work at the University of Rochester. However, twelve years were to elapse before Lea, Rhodes, and Stoll revived interest in silicic acid for separating phosphatides. Very shortly thereafter these same workers in England and this writer in the United States independently developed solvent systems for resolving phosphatides on silicic-acid-impregnated filter paper. Lea, Rhodes, and Stoll used chloroform–methanol solvents, whereas this writer developed the diisobutyl-ketone–acetic acid–water system. Both of these solvent systems have been used widely since their inauguration in 1956–1957 and have also been successfully applied to thin-layer chromatography on silica-gel-coated glass plates. The researcher today has these techniques available for the separation and analysis of intact phosphatides and glycolipids on a microlevel.

Another major advance in phosphatide methodology was the separation and analysis of partial hydrolysis products of these lipids as developed primarily by Dr. R. M. C. Dawson in England. Together with the chromatography of intact lipids they offered important analytical tools to the lipid chemist and found wide application to a large number of problems.

The techniques mentioned above allowed one to analyze families of lipids such as lecithins, phosphatidylethanolamines, and neutral glycerides but did not distinguish the fatty acids of these lipids. The very timely development of gas-liquid chromatography, however, was to find its way directly in the lipid field, since fatty acid methyl esters were admirably suited to this analytical procedure. This opened a new dimension in lipid research because it allowed the study of the fatty acids of the various lipid classes.

The final stage in the analysis of lipids was now ready for exploration, and advances have already been made along these lines. What was needed to complete the picture was the separation of individual molecular species within a given class of lipids. For example, the separation of lecithins from a complex mixture of natural lipids was a fine achievement, and the analysis of the total fatty acids was also an important step forward. The question still to be resolved: how were these fatty acids distributed on individual lecithin molecules? This required the separation of the intact lecithin molecules which differed in their fatty acid composition. Of course, the same situation existed with the other phosphatides, neutral glycerides, cholesterol esters, and glycolipids. This final phase of resolution of different molecular species of intact lipids has already met with moderate success due in large part to the technique of argentation of the adsorbent. Thus silicic-acid-impregnated filter paper or glass plates coated with silicic acid can be treated with silver nitrate to produce systems which give a partial or sometimes nearly complete separation of the individual molecular species of a given lipid class. Consequently, it is now possible to separate a variety of cholesterol esters which differ in their fatty acid moieties. The lecithins, triglycerides, and other lipids have also been partially resolved into subgroups which differ in the total number of double bonds in their fatty acids, but the complete resolution of the lecithins and triglycerides will require more refined techniques, since one must deal not only with the number of double bonds in the fatty acids but also with the chain length of the fatty acids and with the positioning of the fatty acids in the molecule.

The future developments in chromatographic analysis of lipids will very likely be concerned with the refinement and improvement of existing techniques, but new breakthroughs can be anticipated. The gas chromatography of pyrolysis products of phosphatides and other lipids remains a virgin field at present despite its great promise. The coupling of gas chromatography with other physical methods such as mass spectrometry, infrared spectroscopy, optical rotatory dispersion, and nuclear magnetic resonance is beginning to be explored and, in the case of mass spectrometry, has already proved very successful.

When one considers these developments with the potential of automated chromatographic analysis, it becomes clear that although we have made great strides forward, we can gaze into the horizon and see that we have just begun to realize the full potential of modern technology.

This book, together with Volume 2, represents an attempt to compile the major chromatographic methods in the lipid field. It is hoped they will find wide application and serve to expedite research in this rapidly growing field.

G. V. M.

CONTRIBUTORS TO VOLUME 1

Wolfgang J. Baumann, *University of Minnesota, The Hormel Institute, Austin, Minnesota*

K. K. Carrol, *Collip Medical Research Laboratory, University of Western Ontario, London, Ontario, Canada*

R. M. C. Dawson, *Biochemistry Department, Agricultural Research Council, Institute of Animal Physiology, Babraham, Cambridge, England*

G. M. Gray, *Lister Institute of Preventive Medicine, London, England*

Morris Kates, *Division of Biosciences, National Research Council, Ottawa, Canada*

Gene Kritchevsky, *Department of Biochemistry, City of Hope Medical Center, Duarte, California*

A. Kuksis, *Banting and Best Department of Medical Research, University of Toronto, Toronto, Canada*

V. Mahadevan, *University of Minnesota, The Hormel Institute, Austin, Minnesota*

John M. McKibbin, *Department of Biochemistry, University of Alabama Medical Center, Birmingham, Alabama*

Helmut K. Mangold, *University of Minnesota, The Hormel Institute, Austin, Minnesota*

James F. Mead, *Department of Biological Chemistry, UCLA School of Medicine, Los Angeles, California*

O. Renkonen, *Department of Serology and Bacteriology, University of Helsinki, Helsinki, Finland*

George Rouser, *Department of Biochemistry, City of Hope Medical Center, Duarte, California*

Richard N. Roberts, *Electronics Laboratory, General Electric Company, Syracuse, New York*

B. Serdarevich, *Collip Medical Research Laboratory, University of Western Ontario, London, Ontario, Canada*

Vida Slawson, *Department of Biological Chemistry, UCLA School of Medicine, Los Angeles, California*

Robert A. Stein, *Department of Biological Chemistry, UCLA School of Medicine, Los Angeles, California*

Charles C. Sweeley, *Department of Biochemistry and Nutrition, Graduate School of Public Health, University of Pittsburgh, Pittsburgh, Pennsylvania*

Dennis E. Vance, *Department of Biochemistry and Nutrition, Graduate School of Public Health, University of Pittsburgh, Pittsburgh, Pennsylvania*

P. Varo, *Department of Serology and Bacteriology, University of Helsinki, Helsinki, Finland*

Benjamin Weiss, *Department of Biochemistry, New York State Psychiatric Institute and College of Physicians and Surgeons, Columbia University, New York, New York*

Akira Yamamoto, *Department of Biochemistry, City of Hope Medical Center, Duarte, California*

CONTENTS OF VOLUME 1

1

PAPER CHROMATOGRAPHY OF PHOSPHATIDES AND GLYCOLIPIDS ON SILICIC-ACID-IMPREGNATED FILTER PAPER

Morris Kates

DIVISION OF BIOSCIENCES
NATIONAL RESEARCH COUNCIL
OTTAWA, CANADA

I. Introduction

The successful application of paper chromatography to the problem of the resolution of natural mixtures of lipids has been achieved relatively recently. The first attempts, during the period 1951–1955, employed untreated or nonimpregnated paper and were only partially successful with the solvent systems developed (*1–8*). Better separations were later obtained by the use of acidic solvents containing ketones (*9–11*) and by employing two-dimensional chromatography (*11*).

However, these procedures were still only applicable to relatively simple mixtures and were limited by the small capacity of the paper and the resultant difficulties in detection and identification.

It became evident that modification of the stationary phase (hydrated cellulose) either by chemical treatment or by impregnation with various adsorbents might greatly improve the resolution and capacity of the papers. According to Marinetti et al. (*12,13*), silicic acid was first recognized as an excellent adsorbent for chromatography of phosphatides as a result of the studies of Rathmann in 1944 (*14*). This adsorbent was later used for the separation of phosphatides from nonphosphatides by column chromatography (*15,16*), and in 1955 Lea et al. (*17*), independently, developed a procedure using silicic acid columns for separation of relatively simple mixtures of phosphatides (e.g., from egg yolk). These authors also reported the separation of model mixtures of phosphatides on silicic-acid-impregnated paper using a mixture of chloroform and methanol 4:1, v/v, containing 2.5% water, as solvent. Independently, Marinetti et al. (*12,13*) developed their procedure for paper chromatography that effectively separated complex mixtures of tissue phosphatides. This procedure employed Whatman No. 1 paper impregnated with silicic acid, and the solvent found most effective was diisobutyl ketone–acetic acid–water 40:25:5, v/v. Marinetti has further refined and improved this procedure, a detailed description of which is given in his recent reviews (*18,19*).

A variety of other modified papers has also been investigated, including acetylated paper (*11,20*), phosphate-impregnated papers (*21*), formaldehyde-treated papers (*22,23*), tetralin-impregnated papers (*24*), and silicic-acid-impregnated glass-fiber paper (*25–28*). Each of these has certain advantages and also some disadvantages. Thus, for example, formaldehyde-treated paper, with butanol–acetic acid–water (4:1:5) as solvent, gives a good resolution of phospho-inositides and phosphatidyl serine but does not separate lecithin and phosphatidyl ethanolamine (*22,23*). Tetralin-impregnated papers have been used successfully for the separation of individual lecithins as their mercuric acetate addition compounds (*24*), and this procedure represents a start toward the resolution of a single phosphatide class into individual components having different fatty acid compositions. However, this method would not be useful for complex mixtures of different phosphatides.

By far the most successful and widely used technique has been the procedure of Marinetti (*18,19*) involving the use of silicic-acid-impregnated filter paper (*21,29–57*). Silicic-acid-impregnated glass-fiber paper has also been fairly widely used (*57–59*). Both types of impregnated paper give good resolution of components and allow for quantitative analysis of the phosphatides, but the impregnated filter paper is less fragile and easier to handle, more economical, and gives somewhat better resolution than the impregnated glass-fiber paper. Both papers also may be used for two-dimensional chromatography (*12,29*). A two-dimensional technique using formaldehyde treated, alumina-impregnated, or anion-exchange paper and silicic-acid-impregnated paper has been reported recently by Letters to give excellent separations (*60,60a*) ; the phosphatides are run in the first direction on formaldehyde-treated, alumina-impregnated or anion-exchange papers and then run in the second direction on silicic-acid-impregnated paper, using a strip-transfer technique.

In this laboratory we have used Marinetti's paper chromatographic procedure, modified slightly, to study the composition of plant (*50–52*) and bacterial lipids (*53–56*). The modification consists of using Whatman 3MM paper (a heavier and more rugged paper than Whatman No. 1) and a slightly more concentrated sodium silicate solution for impregnation of this paper. The chief advantages in using a heavier paper are greater ease of handling and higher load capacity.

This chapter will deal almost exclusively with chromatography on silicic-acid-impregnated paper and will describe in detail the techniques and procedures employed in preparation of the papers, running of the chromatograms, and detection and quantitative analysis of the separated components. The recent commercial availability of silicic-acid-loaded papers will undoubtedly result in an even more widespread application of paper chromatography to various biochemical problems of lipids [see Reference (*61*)]. However, it is important that the limitations of this technique be fully realized, and these will also be discussed here.

II. Procedures for Chromatography of Lipids on Silicic-Acid-Impregnated Paper

A. Impregnation of Paper with Silicic Acid

1. FILTER PAPER

Whatman 3MM paper is cut into strips 11.5 × 45 cm. A pencil line is drawn 2 cm from one end, this space being used for handling.

2. SODIUM SILICATE SOLUTION

310 g of silicic acid (Mallinckrodt, analytical reagent, 100 mesh) is added slowly with stirring (Teflon-coated magnetic stirrer) to 1 liter of a 7.2 N sodium hydroxide solution (288 g NaOH/liter). When the silicic acid has dissolved, the solution is allowed to cool to room temperature and diluted to 1500 ml with distilled water. This amount of solution is sufficient for impregnation of up to 48 papers.

3. PROCEDURE

The papers are dipped individually through the sodium silicate solution, in a Pyrex glass tray, and immediately hung to drain for 3–4 min; the drippings at the lower edge of the paper are removed by blotting with filter paper. The papers are then immersed in 2 liters of 6 N HCl (liter of conc. HCl diluted with an equal volume of water) in a Pyrex glass tray for 30 min, with intermittent rocking of the tray. Up to 12 papers may be placed in the acid bath at one time, taking the precautions described by Marinetti (*18,19*). The papers are then removed from the acid bath and washed under running tap water for 2–3 hr and finally in several changes of distilled water for a total period of 1–2 hr; the last wash water should be free of chloride ion as tested with silver nitrate. The papers are hung to dry overnight and placed in batches of 20 to 24 in a closed cylindrical jar (15 × 45 cm) with the lower ends immersed in 200 ml of chloroform–methanol 1:1 until the solvent front rises to within a few inches of the top (20–24 hr). This procedure serves to move traces of acid and lipids in the paper up to the solvent front. The papers are again hung to dry and finally pressed between glass plates for several hours, after which they may be used immediately. For storage, they are wrapped in filter paper and may be kept in a drawer for several months without affecting their chromatographic efficiency.

B. Application of Lipids to the Paper

The lipids are applied as spots on a starting line 5 cm above the line previously drawn, the spacing between the spots being not less than 1.5 cm; a maximum of seven spots can conveniently be applied. An aliquot of the lipid solution containing the total amount to be applied (4–5 μg lipid P per spot for total lipids of most tissues; 1–2 μg lipid P for individual phosphatides) is placed in a 15-ml conical centrifuge tube and evaporated just to dryness in a stream of nitrogen. The residue is immediately dissolved in chloroform to a concentration of 0.5–1 μg lipid P/μl, and the solution is brought down completely into the tip of the tube by centrifugation for a few seconds; 5–10 μl of this solution is then applied per spot by micropipette (preferably a 25-λ graduated pipette) using a pipette controller, under a gentle stream of nitrogen to give a small circular spot 5–6 mm in diameter. The bottom 2-cm strip is then cut off, and the chromatogram is run immediately.

C. Solvent Systems and Chromatography Chambers

The chromatograms are run ascending in the solvent system of Marinetti (*13,18,19*), diisobutyl ketone–acetic acid–water (DAW) 40:25:5, v/v, at 23–25°C, preferably in a temperature-controlled room, for 16–18 hr. For lipids rich in plasmalogens chromatography should be carried out at 0–5°C in diisobutyl ketone–acetic acid–water 40:20:3, diisobutyl ketone–*n*-butyl ether–acetic acid–water 20:20:20:3 (*13,19*), or diisobutyl ketone–acetic acid–water 50:20:4 (*35*). Other solvents giving good separations at room temperature are diisobutyl ketone–pyridine–water, 54:40:6 (*19*) or 50:37.5:5 (*35*), and diisobutyl ketone–methanol–water, 100:25:4 (*35*).

The chromatographic jar consists of an unlined Pyrex glass cylinder (15 × 45 cm) covered by a circular glass plate (15 cm in diameter) and sealed with masking tape; 200 ml of solvent is placed in the jar, and 18–20 hr preequilibration of the solvent is required initially. The chromatogram is suspended in the jar (by means of a suitable support) with its lower end immersed in the solvent to a depth of 1–2 cm. A simple convenient holder may be constructed from a piece of glass rod 7 mm × 13.5 cm, each end being inserted into a piece of rubber tubing 3 cm long. The over-all length of the holder is adjusted to fit tightly into the jar, the rubber ends serving to maintain the position of the holder without slipping. The upper end of the paper is rolled around the glass rod and secured with a common paper clip or a stainless-steel spring clip.

Chromatography may also be carried out in 2-quart wide-mouth Mason jars, using 21 × 20 cm papers rolled into a cylinder in the solvent system diisobutyl ketone–acetic acid–water 40:20:3 at room temperature; running time, 3.5–4.5 hr [see Marinetti (*18,19*) for further details].

D. Double Development

With complex lipid mixtures, such as lipids of photosynthetic tissues (see below), improved resolution of the components may be achieved by developing the chromatogram twice in the same solvent (*50*). After the first run, the chromatogram is dried for not more than 10–15 min, preferably in another cylindrical jar flushed with nitrogen, and then immediately developed again in the same solvent in an ascending direction. However, it should be noted that in spite of the precaution of drying the chromatogram in an atmosphere of nitrogen before the second run, peroxidation of some of the components may occur to some extent; the peroxidized components will then not move in the second run and will give rise to spurious spots.

E. Procedures for Detection and Identification of Lipids

The following stains have been found to be very useful for detection and identification of the lipid components separated on silicic-acid-impregnated chromatograms. However, it should be emphasized that identification by means of these stains is only tentative and must be confirmed by standard chemical procedures of isolation, analysis, and identification of hydrolysis products (see Chapters 4 and 14).

1. RHODAMINE 6G

This is the most versatile and useful of the lipid stains, inasmuch as all lipid components are revealed and the color of the fluorescent spot is often an indication of the chemical nature of the lipid. The use of this stain was first introduced by Marinetti (*12,13*), and the staining procedure has been described by him in detail (*18,19*). A 0.12% stock solution of Rhodamine 6G (color index, 752) (National Aniline Division, Allied Chemical and Dye Corp., New York) is made by dissolving 1.2 g in 1 liter of distilled water; this solution is stable indefinitely when kept in the dark. Before use, the stock solution is diluted 1:100 with distilled water to make a 0.0012% solution and transferred to a Pyrex glass tray. The paper chromatogram, after being dried in a hood for ½–1 hr, or less if the lipids are very susceptible to peroxidation, is immersed in the solution for 1–3 min; the excess dye solution is rinsed off with distilled water; and the wet chromatogram is viewed immediately under ultraviolet light (Mineralight Lamp, 366 mμ), the fluorescent spots being outlined with pencil. Acidic phosphatides and other acidic lipids give blue or purple fluorescent spots, whereas neutral lipids and neutral phosphatides give yellow or orange spots; however, neutral phosphatides and other lipids may give bluish-gray spots when peroxidized, and pigments such as chlorophyll and carotenoids also give intensely blue spots. On dry chromatograms most of the lipids appear as yellow fluorescent spots.

2. NINHYDRIN

This reagent is quite specific for phosphatides or lipids having a free amino group, and it can detect as little as 1 μg of amino phosphatide such as phosphatidyl ethanolamine or phosphatidyl serine. Marinetti (*18,19*) has described a most convenient and effective procedure for staining silicic-acid-impregnated chromatograms with Ninhydrin: The chromatograms are dried in a hood, preferably overnight, and sprayed with or dipped through a 0.25% solution of Ninhydrin in acetone–lutidine 9:1, v/v. After a few hours at room temperature, the amino lipids appear on the chromatograms as mauve spots. Color development is generally more rapid with papers that have been dipped through the reagent, but it can be accelerated with the sprayed papers by respraying. The Ninhydrin-positive spots are outlined in pencil, and the chromatogram may then be stained with Rhodamine in order to locate the other lipid components. It should be noted that this double-staining procedure gives better results with the Ninhydrin-sprayed papers; the dipped papers give streaky spots when overstained with Rhodamine.

3. STAINING REAGENTS FOR CHOLINE

Phosphomolybdic acid is commonly used to detect choline; but this reagent is not absolutely specific for choline-containing lipids, since unsaturated cephalins and other unsaturated lipids, e.g., mono- and digalactosyl dilinolenins (*50,51*), also react positively. However, fairly reliable results may be obtained by staining the chromatograms first with Ninhydrin to locate phosphatidyl ethanolamine and serine, a procedure used by Lea et al. (*17*). The Ninhydrin-stained chromatogram is then well washed with $1 N$ HCl and then with distilled water for 10 min and stained with phosphomolybdic acid by the procedure of Levine and Chargaff (*62*) as follows: The paper is immersed in a 1% solution of phosphomolybdic acid for 5–10 min and then washed three times in *n*-butanol, 5–10 min each time, and in running tap water for 1 hr to remove excess reagent. It is then passed through a dilute solution of $SnCl_2$ (stock solution of 40% $SnCl_2$ in conc. HCl diluted 1:100 before use), whereupon the choline-containing components (lecithin and sphingomyelin), and some highly unsaturated lipids (e.g., galactosyl diglycerides of leaves) appear as blue spots.

Choline-containing lipids may also be detected by spraying the chromatogram with the Dragendorff reagent prepared as follows (*63*): 1 g of potassium bismuth iodide is dissolved in 100 ml of methanol and 50 ml of glacial acetic acid, and the solution is diluted with two volumes of methanol before use. Choline-positive lipids appear as yellow spots. Dipicrylamine may also be used to test for choline (*11*): the dried chromatogram is immersed for 10 min in the reagent prepared by diluting 50 ml of a saturated solution of dipicrylamine in 10% Na_2CO_3 to 1 liter with water; the excess reagent is then removed by washing in running tap water, and the choline-positive lipids appear as yellow or orange spots.

4. PERIODATE—SCHIFF REAGENT

This stain is specific for lipids, such as glycolipids, cerebrosides, phosphatidyl glycerol, and phosphatidyl inositol, that contain vicinal hydroxyl groups. The method depends on the cleavage of vicinal hydroxyl groups with periodate and staining of the resulting aldehydolipids with Fuchsin bisulfite (Schiff reagent). The procedure used is an adaptation to silicic-acid-impregnated paper (*51*) of the method devised by Baddiley et al. (*64*) for unimpregnated paper. The dried chromatogram is passed through a 0.25% solution of sodium metaperiodate, hung for 15–20 min at room temperature, and then passed repeatedly through a 1% solution of sodium metabisulfite until the released iodine is completely decolorized. The chromatogram is then passed through a solution of Schiff reagent, whereupon the periodate-positive lipids appear within a few minutes as pink-mauve spots on a white background, which gradually becomes pink owing to reaction of the Schiff reagent with the paper. The chromatogram is then washed with distilled water, the positive spots are outlined in pencil, and the paper is stained with Rhodamine to locate the remaining lipids (all spots now fluoresce blue-purple). The Schiff reagent is prepared as follows: 1 g of Basic Fuchsin and 10 g of sodium metabisulfite are dissolved in 10 ml of conc. HCl and 100 ml of water, and the solution is treated with charcoal for 1 hr, filtered, and made up to 500 ml with distilled water. This stock solution is colorless and may be stored for several months. Before use it is diluted 1:2 with 1% sodium meta-bisulfite solution.

5. TEST FOR PLASMALOGENS

The procedure used is that of Hack and Ferrans (*33*), slightly modified: The dried chromatogram is immersed for 10 min in 500 ml of 0.05% sodium metabisulfite solution containing 5 ml of $0.05\,M$ $HgCl_2$ (1.35 g/100 ml of water) and 5 ml of Schiff reagent (stock solution prepared as described above). Free aldehydes (at the solvent front) immediately stain mauve, whereas plasmalogens require several minutes for hydrolysis of the vinyl ether linkage before appearing as mauve spots. To check for nonspecific staining with the Schiff reagent, the chromatogram should be immersed first in the diluted Schiff reagent without the $HgCl_2$, and any spots that appear are outlined in pencil; the $HgCl_2$ solution is added and the chromatogram is left for 10 min. When the plasmalogen-positive spots have developed, the paper is rinsed several times in 0.05% bisulfite solution and either blotted and dried between filter paper or stained with Rhodamine to locate the remaining lipid components.

Marinetti (*13,18,19*) prefers dinitrophenylhydrazine as reagent for detection of plasmalogens; his procedure is as follows: The dried chromatogram is washed in distilled water, dried, and immersed in a solution of 0.15% 2,4-dinitrophenylhydrazine in $3\,N$ HCl. After 1 to 2 min, the paper is washed four times for 10 min each in distilled water to remove excess reagent and is viewed under ultraviolet light (366 mμ). Plasmalogen-positive components appear as dark spots. Non-specific absorption of the reagent may result in the appearance of spots that are orange in visible light, but these do not appear as dark areas under UV light.

6. TEST FOR PHOSPHATE

Several variations of the Hanes and Isherwood test (*65*) have been used with silicic-acid-impregnated paper (*33,66;* see *18,19*). The procedure used by Hack and Ferrans (*33*) is as follows: The dried chromatogram is sprayed with a freshly prepared solution consisting of 2 ml conc. H_2SO_4, 12 ml of 2.5% ammonium molybdate solution, and 10 ml of 0.85% NaCl solution, heated for 1.5 min at 100°C, and then sprayed with stannous chloride solution (40% $SnCl_2$ in conc. HCl, diluted 1:10 before use) and dried at room temperature. Phosphatides appear as blue spots, but in the author's experience the background is often also blue, making it difficult to clearly distinguish the phosphatide spots. A more specific and reliable test for phosphatides is to employ ^{32}P-labeled lipids, using autoradiography to detect the phosphatide spots (see below).

Beiss (*63*) has described another procedure, in which the chromatogram is immersed in a solution of Zinzade's reagent prepared as follows: 6.85 g of $Na_2MoO_4 \cdot 2\ H_2O$ and 400 mg of hydrazine sulfate are dissolved in 100 ml of water; 250 ml of conc. H_2SO_4 is added; and the cooled solution is diluted with 600 ml of water. The phosphatides appear as blue spots after several minutes, and the excess reagent is then washed off with methanol.

7. OTHER STAINING REAGENTS

As general lipid stains, the following dyes have been used by various workers: Biebrich Scarlet (*3*), Nile Blue (*3*), Rhodamine B (*8*), Fast Acid Violet (*8*), protoporphyrin (*68*). Their use, however, has not found wide acceptance, and in general these dyes are less sensitive than Rhodamine 6G. Iodine has also been used as an effective and sensitive stain for unsaturated lipids by Marinetti (*18,19*). A tricomplex stain involving complex formation between phosphatides, uranyl ions, and Acid Fuchsin has been developed by Hooghwinkel et al. (*67*) and Bungenberg de Jong (*69*); zwitterionic phosphatides (lecithin, sphingomyelin, phosphatidyl ethanolamine) give red spots, and acidic phosphatides appear as green spots when the tricomplex stain is coupled with Brilliant Green.

8. AUTORADIOGRAPHY

The most reliable test for the presence of phosphatides and sulfolipids on chromatograms is the detection of ^{32}P and ^{35}S, respectively, in the separated spots. The chromatograms obtained with [32]P- or ^{35}S-labeled lipids (see below) are dried for several hours and stapled onto a sheet of Kodak no-screen X-ray film; the film is exposed for several hours to several days or weeks depending on the amount of radioactivity present per spot. If sufficient material was applied to the paper, the chromatogram may then be stained with Rhodamine 6G and superimposed on the autoradiograph to determine the position of the radioactive spots relative to the unlabeled components.

9. ENZYMATIC TESTS

The identification of phosphatides and glycolipids can often be confirmed by hydrolysis with specific enzymes and identification of the products formed. For phosphatides the following enzymes are most effective in this respect (see *18,19,33,47,70–73*).

a. *Phospholipase A*. This enzyme specifically hydrolyzes the β-fatty acid ester linkage in glyceryl phosphatides, including plasmalogens, cardiolipin, and phosphatidyl glycerol, to yield free fatty acid and the β-lysophosphatide (*71,72*). An ether solution of the substrate is incubated with a buffered (pH 7) aqueous solution of the enzyme (lyophilized venom of *Crotalis adamanteus, Naia naia,* or *Agkistrodon piscivorus*), containing calcium ion, for several hours (*33,47,70–72*). Chromatography of the chloroform-soluble products on silicic-acid-impregnated paper effectively separates the free fatty acids and the lysophosphatides (and/or lysoplasmalogens) from each other and from any unreacted phosphatides [*33,47;* see also examples given below (Figs. 10 and 11)].

b. *Phospholipase B*. This enzyme hydrolyzes both fatty acid ester linkages in lecithin and phosphatidyl ethanolamine to yield free fatty acids and the corresponding water-soluble glycerophosphate ester (*72,75*). An ultrasonified suspension of the substrate is incubated with an enzyme preparation from *Penicillium notatum* in acetate buffer at pH 4 for 2 hr (*74,75*). The chloroform-soluble products (free fatty acids) are analyzed by gas–liquid chromatography, and the water-soluble phosphate esters (e.g., glyceryl phosphoryl choline or ethanolamine) are identified by paper chromatography (see Chapter 4).

c. *Phospholipase C*. This enzyme catalyzes the hydrolysis of the digylceride–phosphate linkage in glyceryl phosphatides and of the ceramide–phosphate linkage in sphingomyelin [see (*72*)]. A solution of the substrate in ether–ethanol 98:2 is incubated with a Tris-buffered (pH 7) solution of the enzyme (*Clostridium perfringens* toxin or a *B. cereus* preparation) containing calcium ions for 1–2 hr (*47,70,72,76*). The chloroform-soluble products (diglycerides or cer-amides) are identified by chromatography on silicic-acid-impregnated paper (*18,19*), and the water-soluble phosphates are identified by paper chromatography (see Chapter 4).

d. *Phospholipase D*. This enzyme catalyzes the hydrolysis of glyc-eryl phosphatides to phosphatidic acid and the respective nitrogenous base [or other residue, such as glycerol from phosphatidyl glycerol (*47,70*)]. A suspension of the substrate in acetate buffer (pH 5.6), containing the enzyme (lyophilized preparation from brussels sprouts) and calcium ions, is shaken with ethyl ether and incubated for 3–4 hr (*70,72,73*). The chloroform-soluble phosphatidic acid is readily identified by silicic-acid-paper chromatography (*33,47*) and by de-acylation to glycerophosphate (*70,73*). The water-soluble products (free bases, glycerol, etc.) are identified by paper chromatography (see Chapters 4 and 14).

Enzymatic hydrolysis of cerebrosides may be achieved by a specific enzyme from brain or spleen that catalyzes the hydrolysis of the galactosidic linkage to form ceramide and free galactose (*72,77*). Galactosyl diglycerides are readily deacylated by specific enzymes in runner bean leaves (*78*) as follows. A suspension of mono- or diga-lactosyl diglyceride in phosphate buffer (pH 7 or 5.6, respectively) is incubated with a soluble extract of runner bean leaves for 2–3 hr. The chloroform-soluble products (free fatty acids) are readily separated from unreacted substrate and may be analyzed by gas–liquid chroma-tography. The water-soluble products, mono- and digalactosyl glycerol, as well as free galactose, are separated and identified by paper chroma-tography with pyridine–ethyl acetate–water 1:2.5:2.5, v/v, upper phase, as solvent (*78*). Whether this enzyme preparation will also attack other glycosyl diglycerides, such as the mannosyl diglyceride in *Micrococcus lysodeikticus* (*43,44*) or the glucosyl diglycerides in *Streptococcus faecalis* (*79,80*) has yet to be determined.

F. Quantitative Analysis of Phosphatides

Three methods have been used in this laboratory for quantitative analysis of phosphatide components separated as described above, depending on the amount of material available:

1. For a small sample, containing 5–6 μg of lipid P, sufficient for one spot on a chromatogram, a 3-cm-wide longitudinal strip is cut out of the developed and Rhodamine-stained chromatogram, taking care that all the separated spots lie well within this strip. The paper strip is cut up into rectangles each containing one of the separated spots as well as into two blank rectangles 2 cm wide, one above the solvent front and the other below the starting line; the width of each rectangle is measured in centimeters. Phosphorus is determined in each spot by a modification [cf. Letters (*60*)] of the micromethods of Allen (*81*) and Bartlett (*82*): The rectangles of paper are chopped into small pieces and transferred quantitatively to Pyrex digestion tubes (1.7×19 cm); 0.5 ml of 72% perchloric acid and one drop of 5% ammonium molybdate are added; and digestion is carried out by heating on a medium gas flame for 5–10 min. The ammonium molybdate acts as a catalyst for the oxidation of the paper, so that the digestion proceeds smoothly and rapidly. To the cooled digests are added a further 0.1 to 0.2 ml of 72% perchloric acid (to compensate for losses during digestion) and 6 ml of water, and after mixing on a Vortex mixer, the silicic acid is centrifuged down. A 5-ml aliquot of the clear supernatant is then removed, and the molybdenum blue color is developed by addition of 0.2 ml of 1% amidol solution (containing 20% sodium metabisulfite) and 0.2 ml of 5% ammonium molybdate and heating in boiling water for 7 min. After 20 min cooling, the optical densities are determined at 830 mμ in a Beckman Model B spectrophotometer. Standards containing 0.5 to 5 μg of P, as well as a reagent blank, are carried through the above procedure simultaneously; the calibration curve is linear over this range. The blank corrections for phosphorus in the paper are calculated as follows:

corrected μg P in component = μg P found
$$- \text{(average μg P in blanks} \times w/2)$$

where w is the width of each paper rectangle, the width of the blanks being 2 cm. The results are then expressed as a percentage of the total blank-corrected P in all the components.

2. When sufficient material is available, two other procedures may be used, depending on the type of analyses desired.

a. To determine both phosphorus content and fatty acid composition of each component, up to 20 spots, each containing 5–6 μg of lipid P, can be applied close together to a single chromatogram (e.g., see Fig. 1). After development, the paper is dried for 5–10 min in an

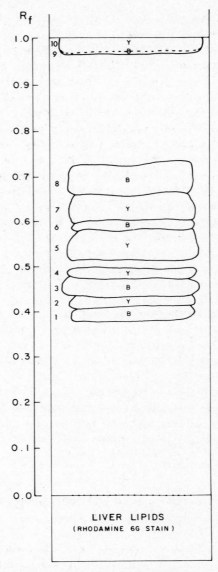

FIG. 1. Chromatogram of total rat liver lipids stained with Rhodamine 6G and viewed under ultraviolet light; 20 spots each containing 5 μg lipid P were applied to the paper. Solvent: diisobutyl ketone–acetic acid–water (DAW) 40:25:5, v/v. Abbreviations: B, blue; Y, yellow. Tentative identity of spots as given in Table 1. [After Beare and Kates (*88*).]

atmosphere of nitrogen and stained with Rhodamine 6G, the bands corresponding to each component are outlined in pencil, and the paper is partially dried under nitrogen for 15 min. The bands are cut out, their widths are measured, and they are chopped into small pieces, transferred to a 50-ml Erlenmeyer flask with side arm (*83*), and heated under reflux with 4 ml of 2.5% methanolic HCl for 1 to 2 hr. Water (1 ml) is added, and the mixture is extracted with several 5-ml portions of petroleum ether (b.p. 30–60°). The extracts are evaporated to dryness under nitrogen, and the residual fatty acid methyl esters (plus aldehyde dimethyl acetals if plasmalogens are present) are analyzed by gas–liquid chromatography. The methanol phases are centrifuged or filtered through glass wool into digestion tubes, using 5 ml of 90% methanol for washing purposes, concentrated to dryness at 50°C in an air jet, and analyzed for P by the method of Allen (*81*). Blanks prepared as described in (1) are carried through the procedure. The method is applicable to amounts of P ranging from 5 to 100 μg per component; for components containing less than 5 μg P, development of the molybdenum blue color should be carried out as described in (1).

b. When other analyses, such as deacylation and identification of the phosphate esters, are desired, one or more chromatograms are run as described in (2); the bands corresponding to each component are cut out, stapled to a strip of paper acting as a wick, and eluted by downward chromatography with chloroform–methanol–water 75:25:2, v/v, followed by methanol. The combined eluates of each component are evaporated to dryness; the residue is dissolved in chloroform to a known volume; and suitable aliquots are taken for analysis of P (*81*), fatty acids by GLC, and ester groups (*84*) or for deacylation and identification of the phosphate esters (*85,86*).

III. Separation of Phosphatides and Glycolipids from Animal, Plant, and Microbial Sources

The lipid compositions of animal, plant, and bacterial cells are generally quite distinct and readily distinguished from one another. Also, the procedure for extractions of lipids must be adapted to the particular tissue under investigation. In the following section, therefore, the results of lipid analyses as determined by chromatography on silicic-acid-impregnated paper, and also the lipid extraction procedure

used, will be given for a few representative animal, plant, and microbial cells.

A. Animal Lipids

1. LIPID EXTRACTION

Animal tissues may be conveniently extracted by a modification of
the Bligh and Dyer procedure (*87*); the procedure for rat liver is as
follows: to about 7 g of fresh liver is added 3 ml of water and 30 ml
of methanol–chloroform 2:1, v/v, and the mixture is blended in a
Lourdes homogenizer or a Sorvall Omnimixer (60 ml capacity) for 2
min. The homogenate is centrifuged, the supernatant is decanted, and
the residue is reextracted with 38 ml of methanol–chloroform–water
2:1:0.8 by homogenization for 2 min. After centrifugation, the com-

TABLE 1

CHROMATOGRAPHIC ANALYSIS OF RAT LIVER PHOSPHATIDES[a,b]

Spot	R_f value	Rho-damine 6G stain	Choline stain	Nin-hydrin stain	Plasma-logen stain	P, % of total	Tentative identity of components
1	0.38	Blue	—	Trace	—	Trace	Unidentified
2	0.42	Yellow	+	—	—	11	{ Lysolecithin
3	0.45	Blue	—	—	—		{ Phosphatidyl inositol
4	0.48	Yellow	+	Trace	Trace	5	Sphingomyelin (plus trace of lysophosphatidyl ethanolamine)
5	0.55	Yellow	+	—	+	56	Lecithin (diester and plasmalogen forms)
6	0.59	Blue	—	+	—	4	Phosphatidyl serine
7	0.63	Yellow	—	+	+	18	Phosphatidyl ethanolamine (diester and plasmalogen forms)
8	0.69	Blue	—	—	Trace	4	Diphosphatidyl glycerol (cardiolipin)
9	0.96	Blue	—	—	—	1	Phosphatidic acid
10	0.96–1.0	Yellow	—	—	—	—	Neutral lipids

[a] Data obtained from Reference (*88*).
[b] See Fig. 1.

bined supernatants are diluted with 20 ml each of chloroform and water and the phases are allowed to separate in a separatory funnel. The lower chloroform phase is withdrawn and concentrated in a rotary evaporator at 30–35°C (benzene is added to aid in removal of traces of water), and the residue is dissolved to a suitable volume (e.g., 10 ml) in chloroform. Aliquots of this solution are analyzed for lipid P (*81*).

2. CHROMATOGRAPHIC ANALYSIS OF LIPIDS

The rat liver lipids were chromatographed on silicic-acid-impregnated paper as described in Section II.F.2.a, 20 spots each containing 5 μg P being applied to the paper (*88*). The chromatogram obtained is shown in Fig. 1, and the staining behavior, quantitative analysis, and tentative identity of the components are given in Table 1.

TABLE 2

CHROMATOGRAPHIC ANALYSIS OF PHOSPHATIDES IN CALF KIDNEY CELLS[a,b]

Spot	R_f value	Rho-damine 6G stain	Choline stain	Nin-hydrin stain	Plasma-logen stain	P, % of total	Tentative identity of components
1	0.27	Blue	—	—	—	Trace	Unidentified
2	0.34	Blue	—	—	Trace	6	Phosphatidyl inositol
3	0.40	Yellow-orange	+	Trace	Trace	16	Sphingomyelin plus trace of lysophosphatidyl ethanolamine
4	0.48	Yellow	+	—	+	45	Lecithin (diester and plasmalogen forms)
5	0.53	Blue	—	+	+	8	Phosphatidyl serine (diester and plasmalogen forms)
6	0.62	Yellow	—	+	Weak	21	Phosphatidyl ethanolamine
7	0.69	Blue	—	—	—	3	Diphosphatidyl glycerol (cardiolipin)
8	0.88	Blue	—	—	—	1	Phosphatidic acid
9	0.95–1.0	Yellow	—	—	+	0	Neutral lipids

[a] Data obtained from Reference (*89*).
[b] See Fig. 2.

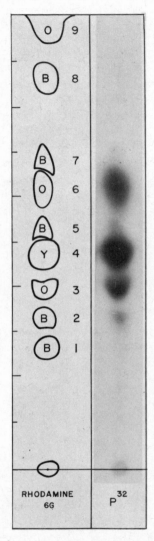

FIG. 2. Chromatogram and autoradiogram of ³²P-labeled lipids of calf kidney cells. Solvent; diisobutyl ketone–acetic acid–water, 40:25:5. Abbreviations: B, blue; Y, yellow; O, orange. Tentative identity of spots as given in Table 2. [After Kates *et al.* (*89*). Reproduced from *Biochim. Biophys. Acta,* **52,** 455 (1961), Fig. 3.]

Another example of the separation of lipids from animal tissues is shown in Fig. 2, which represents the chromatogram and autoradiogram of ^{32}P-labeled lipids of calf kidney cells cultured in vitro in the presence of ^{32}P-orthophosphate (*89*). The cells were extracted as described below for bacteria, and the total lipids were analyzed chromatographically as for the rat liver lipids. The staining behavior, quantitative composition, and identity of the phosphatide components are given in Table 2.

The chromatograms in Figs. 1 and 2 show the separation of the major components, lecithin, phosphatidyl ethanolamine, and sphingomyelin, and of the minor components, phosphatidyl serine, phosphatidyl inositol, and cardiolipin; but it should be noted that lysophosphatidyl ethanolamine is not resolved from sphingomyelin and that the diester and plasmalogen forms of phosphatidyl ethanolamine and of lecithin are not separated.

B. Plant Lipids

1. EXTRACTION OF LIPIDS

Photosynthetic tissues of higher plants contain phospholipases and galactolipid-hydrolyzing enzymes (*73,78*) that may result in degradation of the lipids during the extraction procedure unless precautions are taken to inactivate these enzymes. The following procedure has been found to be most effective in avoiding enzymatic hydrolysis of leaf lipids (*90*): 100 g of fresh leaves (spinach, runner bean, sugar beet, etc.) is fragmented under liquid nitrogen and immediately added portionwise to 300 ml of boiling isopropanol; the mixture is blended for 1–2 min in a Waring Blendor or Sorvall Omnimixer. The hot homogenate is filtered with suction, and the filter residue is washed with hot isopropanol (200 ml). The filter cake is then blended with 200 ml of chloroform–isopropanol 1:1; the homogenate is filtered; and the filter residue is washed with chloroform–isopropanol 1:1 and finally with chloroform. The combined filtrates are concentrated in vacuo; the residual lipids are taken up in chloroform (200 ml); and the solution is washed several times with water (or 1% sodium chloride solution). The chloroform solution is then diluted with benzene and concentrated to dryness in vacuo (30–35°C), and the lipid residue is immediately dissolved in chloroform (50 ml). Aliquots of this solution are taken for dry-weight determination, analysis of P, and analysis of total sugar if desired.

To obtain radioisotopically labeled lipids, the leaf petioles are first immersed in a solution of ^{32}P-orthophosphate, ^{35}S-sulfate, or ^{14}C-glucose and the leaves are allowed to photosynthesize for 1–2 hr in the light (2000 ft-candles) (*91*). Lipids of algae (*52,86*) or diatoms (*95*) may be labeled by culturing these organisms in the light in media containing orthophosphate-^{32}P, sulfate-^{35}S, bicarbonate-^{14}C, or acetate-^{14}C.

2. CHROMATOGRAPHY

The complex mixture of leaf lipids obtained cannot be completely resolved into individual components by silicic-acid-impregnated paper

TABLE 3

CHROMATOGRAPHIC ANALYSIS OF LIPIDS OF RUNNER BEAN LEAVES[a,b]

Spot	R_f value	Rhodamine 6G stain	Choline stain	Ninhydrin stain	Periodate–Schiff stain	P, % of total	Tentative identity of components
1	0.25	Blue	—	—	Weak	9	Phosphatidyl inositol
2	0.32	Blue	—	—	—	—	Sulfoquinovosyl diglyceride
3	0.41	Yellow	+[c]	—	+	—	Digalactosyl diglyceride
4	0.44	Pink	+	—	+	45	Lecithin plus cerebroside
5	0.51	Blue	—	—	+	22	Phosphatidyl glycerol
5′	0.59	Blue	—	Trace	—	6	Phosphatidyl serine
6	0.63	Pink	Weak[c]	+	—	17	Phosphatidyl ethanolamine
7	0.68	Blue	—	—	+	Trace	Diphosphatidyl glycerol plus unidentified glycolipid
8	0.75	Yellow	+[c]	—	+	—	Monogalactosyl diglyceride
9	0.81	Blue	—	—	—	1	Phosphatidic acid
10	0.90	Violet	—	—	—	—	Pigments and neutral lipids

[a] Data obtained from Reference (*50*).

[b] See Fig. 3.

[c] The positive choline test is due to the presence of highly unsaturated fatty acids in these components.

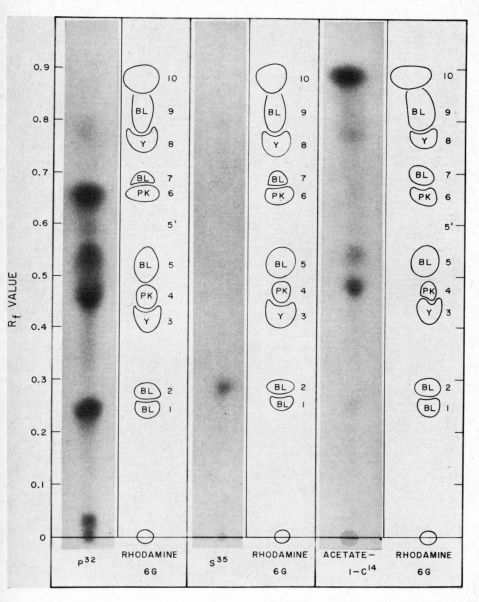

FIG. 3. Chromatogram and autoradiogram of ^{32}P-, ^{35}S-, and ^{14}C-labeled lipids of runner bean leaves; solvent, DAW 40:25:5. Chromatogram (double development) was stained with Rhodamine 6G (abbreviations: B, blue; Y, yellow; PK, pink). Identity of components as given in Table 3. [From Kates (*50*). Reproduced from *Biochim. Biophys. Acta*, **41**, 315 (1960), Fig. 2.]

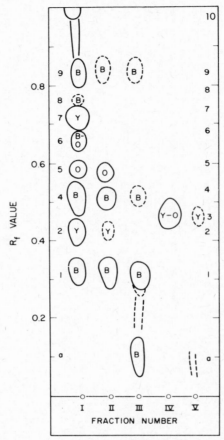

FIG. 4. Chromatogram of silicic acid column fractions of runner bean leaf lipids stained with Rhodamine 6G (B, blue; Y, yellow; O, orange); solvent, DAW 40:25:5. Fractions I and II were eluted with 2 and 3.5 column volumes, respectively, of chloroform–methanol 5:1, v/v; fractions III and IV were eluted with 2 and 4 column volumes, respectively, of chloroform–methanol 1:1, v/v; and fraction V was eluted with 2 column volumes of 100% methanol. Spots 1, 2, 4, 6, and 7 stained positively with the periodate–Schiff reagent; spot 5 was strongly Ninhydrin-positive; and spot 4 was weakly Ninhydrin-positive. Tentative identity of spots: 1, sulfoquinovosyl diglyceride plus phosphatidyl inositol; 2, digalactosyl diglyceride; 3, lecithin; 4, phosphatidyl glycerol + cerebroside + phosphatidyl serine (trace); 5, phosphatidyl ethanolamine; 6, diphosphatidyl glyceride + unidentified glycolipid; 7, monogalactosyl diglyceride; 8, unidentified; 9, phosphatidic acid; 10, pigments plus neutral lipids. [After Sastry and Kates (*51*) by permission of the American Chemical Society.]

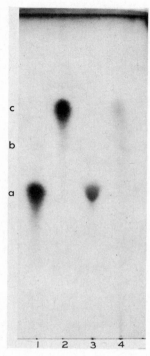

FIG. 5. Chromatogram of purified mono- and digalactosyl diglycerides stained with periodate–Schiff reagent; solvent, DAW 40:25:5. Material applied: 1, digalactosyl dilinolenin; 2, monogalactosyl dilinolenin; 3, hydrogenated digalactosyl dilinolenin; 4, hydrogenated monogalactosyl dilinolenin. Identity of components: a, digalactosyl diglyceride; b, phytosphingosyl glucocerebroside (contaminating trace); c, monogalactosyl diglyceride. (Sastry and Kates, unpublished results.)

chromatography, although the double-development procedure using ^{32}P, ^{35}S-, or ^{14}C-labeled lipids *(50)* does result in the resolution of at least ten components (Fig. 3, Table 3). However, a more useful procedure involves a preliminary fractionation on a column of silicic acid using chloroform with increasing proportions of methanol as eluting solvent. The fractions obtained are then more readily resolved by paper chromatography as shown in Fig. 4 *(51)*. It should be noted that certain components, e.g., phosphatidyl inositol and sulfoquinovosyl diglyceride, and phosphatidyl glycerol, phosphatidyl serine, and phyto-

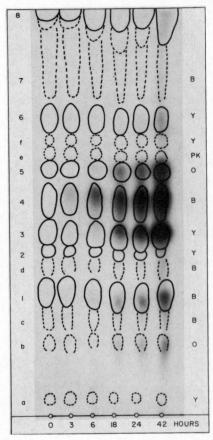

FIG. 6. Chromatogram and autoradiogram of ^{32}P-labeled lipids of *Chlorella vulgaris* exposed to ^{32}P-orthophosphate for various periods in the light; solvent, DAW 40:25:5; chromatogram stained with Rhodamine 6G (B, blue; Y, yellow; O, orange; PK, pink). Tentative identity of spots: a–d, unidentified; 1, sulfoquinovosyl diglyceride plus phosphatidyl inositol; 2, digalactosyl diglyceride; 3, lecithin; 4, phosphatidyl glycerol; 5, phosphatidyl ethanolamine; e and f, unidentified glycolipids; 6, monogalactosyl diglyceride (plus trace of diphosphatidyl glycerol); 7, lysophosphatidic acid; 8, pigments plus neutral lipids plus phosphatidic acid. [After Sastry and Kates *(52)*.]

sphingosyl glucocerebroside, are still not resolved; other procedures such as repeated column chromatography, acetone precipitation, and solvent fractionation are necessary to separate them (*51*). By a combination of such procedures it has been possible to isolate mono- and digalactosyl dilinolenin as fairly pure components. A chromatogram of these galactolipids (which reveals traces of contaminating phytosphingosyl glucocerebroside) is shown in Fig. 5.

An example of the use of paper chromatography to study the incorporation with time of ^{32}P-orthophosphate into the individual phosphatides of *Chlorella vulgaris* (*52*) is shown in Fig. 6. The cells were extracted as described below for microorganisms.

C. Microbial Lipids

1. LIPID EXTRACTION

Lipids of microorganisms are readily extracted by the procedure of Bligh and Dyer (*87*), except that it is not necessary to homogenize the cells, since they are easily broken by suspension in the extracting solvent. To 1 ml of cell suspension (50–100 mg cells per ml) in water (or salt solution for halophilic bacteria) is added 3.75 ml of methanol–chloroform 2:1, v/v; the mixture is shaken and left at room temperature for several hours with intermittent shaking. After centrifugation, the supernatant extract is decanted and the residue is resuspended in 4.75 ml of methanol–chloroform–water 2:1:0.8, v/v; the mixture is then shaken and centrifuged. To the combined supernatant extracts are added 2.5 ml each of chloroform and water, and the mixture is either centrifuged or allowed to settle in a separatory funnel. The lower chloroform phase is withdrawn, diluted with benzene (to aid in removal of traces of water), and brought to dryness in a rotary evaporator (30–35°C). The lipid residue is immediately dissolved in chloroform, and the solution is filtered through glass wool if necessary and made to a known volume. Aliquots are taken for analysis of total P (*81*) and for dry-weight determination.

2. CHROMATOGRAPHY

A chromatogram of the lipids of a mesophilic yeast, *Candida lipolytica* (grown at 25° and 10°C), and of three psychrophilic strains of

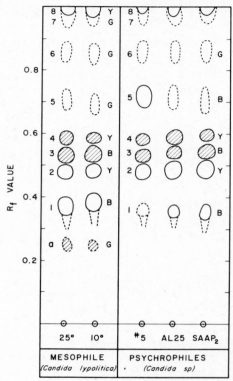

FIG. 7. Chromatogram of total lipids of mesophilic and psychrophilic yeasts (*Candida* sps.) stained with Rhodamine 6G (B, blue; Y, yellow, G, gray); solvent, DAW 40:25:5. Hatched spots are Ninhydrin-positive, and dashed lines indicate trace components. Identity of spots is given in Table 4. [After Kates and Baxter (*54*).]

Candida (*54*) is shown in Fig. 7. Staining behavior and identity of the components are given in Table 4 for *C. lipolytica* grown at 25°C [cf. results obtained by Letters (*60*)].

   Chromatographic data for a Gram-positive bacterium, *Bacillus cereus* (*53*), are given in Table 5 [cf. (*48,60*)], and data for a Gram-negative bacterium, *Serratia marcescens* (pigmented and nonpigmented strains) (*55*), are shown in Fig. 8 and Table 6. Comparison with the chromatograms of animal, plant, and yeast lipids (Figs. 2, 3, and 7) shows the striking absence of lecithin and the high proportions of

TABLE 4

CHROMATOGRAPHIC ANALYSIS OF LIPIDS OF *Candida lipolytica* GROWN AT 25°[a,b]

Spot	R_f value	Rhodamine 6G stain	Choline stain	Ninhydrin stain	P, % of total	Tentative identity of components
a	0.25	Gray	—	Weak	5	Unidentified
1	0.36	Blue	—	—	18	Phosphatidyl inositol[c]
2	0.48	Yellow	+	—	41	Lecithin
3	0.54	Blue	—	+	15	Phosphatidyl serine
4	0.59	Yellow	—	+	20	Phosphatidyl ethanolamine
5	0.70	Blue	—	—	1	Cardiolipin
6	0.85	Blue	—	—	Trace	Unidentified
7	0.95	Blue	—	—	Trace	Phosphatidic acid
8	0.98	Yellow	—	—	—	Neutral lipids

[a] Data from Reference (54).
[b] See Fig. 7.
[c] This component gave a weakly positive stain with the periodate–Schiff reagent; all other spots were negative.

TABLE 5

CHROMATOGRAPHIC ANALYSIS OF LIPIDS OF *Bacillus cereus*[a]

Spot	R_f value	Rhodamine 6G stain	Choline stain	Ninhydrin stain	Periodate–Schiff stain	P, % of total	Tentative identity of components
1	0.25	Yellow	Trace	—	—	3	Unidentified
2	0.32	Gray	—	Trace	—	2	Unidentified
3	0.40	Yellow	—	+	—	9	Phosphatidyl glycerol aminoacyl ester[b]
4	0.45	Yellow	Weak	—	—	5	Unidentified[c]
5	0.51	Blue	—	—	+	26	Phosphatidyl glycerol
6	0.57	Yellow	—	+	—	41	Phosphatidyl ethanolamine
7	0.60	Blue	—	—	—	14	Cardiolipin (?)
8	0.66	Gray	—	—	—	Trace	Unidentified
9	0.96	Yellow	—	—	—	—	Neutral lipids

[a] Data from Reference (53).
[b] This component has the same chromatographic and staining behavior as lyso-phosphatidyl ethanolamine, but its deacylation products are glyceryl phosphoryl glycerol and ornithine [see (48)].
[c] Although this spot gives a weakly positive reaction with the choline stain, it is probably not lecithin [see (48)].

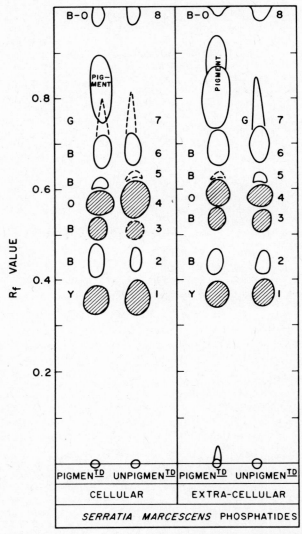

FIG. 8. Chromatogram of total lipids of pigmented and unpigmented cells of *Serratia marcescens* and their corresponding extracellular lipids; solvent, DAW 40:25:5; stained with Rhodamine 6G (B, blue; Y, yellow; O, orange). Hatched spots are Ninhydrin-positive; identity of spots is given in Table 6. [After Kates et al. (*55*).]

TABLE 6

CHROMATOGRAPHIC ANALYSIS OF LIPIDS OF *Serratia marcescens*[a,b]

Spot	R_f value	Rhodamine 6G stain	Ninhydrin stain	Periodate–Schiff stain	P, % of total	Tentative identity of components
1	0.36	Yellow	+	−	3	Phosphatidyl glycerol-O-aminoacyl ester[c]
2	0.45	Blue	−	+	4	Phosphatidyl glycerol
3	0.52	Blue	+	−	8	Phosphatidyl serine
4	0.59	Orange	+	−	63	Phosphatidyl ethanolamine
5	0.62	Blue	−	−	6	Cardiolipin
6	0.69	Blue	−	−	12	Unidentified[d]
7	0.76	Gray	−	−	3	Unidentified
8	0.98	Blue-orange	−	−	—	Nonphosphatides

[a] Data from Reference (*55*).

[b] See Fig. 8.

[c] Major amino acids detected: ornithine, glycine, alanine, leucine, and aspartic acid (*55*).

[d] Possibly phosphatidyl glycerophosphate.

phosphatidyl ethanolamine in these bacteria. Phosphatidyl glycerol is also a major component, occurring as such and as an O-aminoacyl ester ("lipoaminoacid"), a new phosphatide discovered in bacteria by Macfarlane (*46*).

The lipids of an extremely halophilic bacterium, *Halobacterium cutirubrum*, have been found to consist almost entirely of di-O-dihydrophytyl glycerol ether–derived phosphatides and glycolipids (*56,92*). A chromatogram of the total cellular lipids and of the lipids from the cell envelope and the intracytoplasmic membranes (*93*) is shown in Fig. 9. The composition of the two membrane fractions appears to be the same as that of the total cellular lipids, showing that most of the lipids are bound to membrane material (*93*).

The major phosphatide component (spot **6 + 7**, Fig. 9) accounts for 80–86% of the total lipid P, and its structure has been established (*56*) as **2,3-di-O-dihydrophytyl-glycerol-1-phosphoryl glycerophosphate** (phosphatidyl glycerophosphate, diether form). Spot 8 (Fig. 9) gives a positive reaction with the periodate–Schiff reagent and accounts for about 5% of the lipid P; it is most likely the di-O-dihydrophytyl

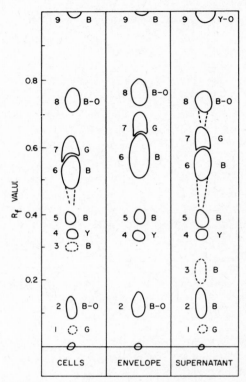

FIG. 9. Chromatogram of lipids of whole cells of *H. cutirubrum* and of envelope and supernatant (containing intracytoplasmic membrane fragments) fractions; solvent, DAW 40:25:5; chromatogram stained with Rhodamine 6G (B, blue; Y, yellow; G, gray). Spots 4 and 8 are periodate–Schiff positive; no Ninhydrin-positive spots were detected. Tentative identity of components: 1, 3, 5, un-identified; 2, glycolipid sulfate; 4, glycolipid; 6 + 7, phosphatidyl glycerophos-phate; 8, phosphatidyl glycerol; 9, pigments plus neutral lipids. All identified components are derivatives of di-*O*-dihydrophytyl glycerol. [After Kushner et al. (*93*).]

glyceryl ether analog of phosphatidyl glycerol (*92*). It is interesting to note that this diether analog of phosphatidyl glycerol has a much greater mobility (R_f 0.75) on silicic-acid-impregnated paper than the diester form (R_f about 0.5; see Tables 3, 5, and 6).

Spot 2, Fig. 9, is a major lipid component (accounting for about 40% of the lipids on a weight basis) that is periodate–Schiff negative but gives a strong positive in vitro test for sugars with the phenol–sulfuric acid reagent. It has been recently (*96*) isolated in pure form and shown to be a glycolipid sulfate tentatively assigned the structure 2,3-di-*O*-dihydrophytyl glycerol-1-(glucosyl mannosyl galactosyl sulfate). Spot 4 (Fig. 9) is periodate–Schiff positive and is presumed to be a diether glycolipid; mild acid hydrolysis of spot 2 gives rise to a spot having R_f value and staining behavior identical with those of spot 4, suggesting that the latter is a desulfated derivative of the glycolipid sulfate (spot 2).

It is interesting to note that the glycolipid (spot 4), having three sugar residues, has an R_f value only slightly less than that of digalactosyl diglyceride (see Figs. 3 and 4). This is unexpected, since the presence of one more sugar residue should reduce its mobility much below that of digalactosyl diglyceride [see discussion by Marinetti (*19*)]. However, the presence of highly branched chain groups (dihydrophytyl groups), as well as ether linkages, might account for the unexpectedly high mobility of this glycolipid. In this connection it may be mentioned that the diether lecithin, (dioctadecyl)-L-α-lecithin, has only a slightly higher R_f value than the diester analog, (dioctadecanoyl)-L-α-lecithin (*94*), suggesting that ether groups are only partly responsible for the high mobility of the *H. cutirubrum* lipids and that the major factor may be the presence of the dihydrophytyl groups.

IV. Chromatography on Commercial Silicic-Acid-Impregnated Paper

Commercial silica-gel-loaded filter paper has recently been made available by Schleicher and Schüll (Kieselgel papier, Schleicher and Schüll 289) and by H. Reeve Angel and Co. Inc. (Whatman SG-81 silica-gel-loaded paper). The Schleicher and Schüll paper has been investigated by Beiss (*63*), who developed a two-directional solvent system (tetrahydrofurane–chloroform–diisobutyl ketone–acetic acid–water 45:10:5:6:6) and chloroform–diisobutyl ketone–methyl isobutyl ketone–methyl ethyl ketone–formic acid–water 110:30:26:20:30:3, which gave excellent separations of plant lipids into 23 components.

The Whatman SG-81 paper has been investigated by Marinetti (*61*), who found it to be uniformly impregnated and to have properties, such as load capacity and resolution, similar to those of the paper

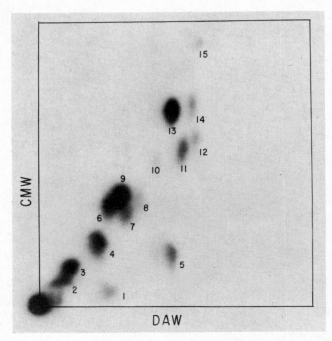

FIG. 10. Autoradiogram of ³²P-labeled pea root lipids and their lysoderivatives obtained by phospholipase A hydrolysis; chromatogram was run on Whatman SG-81 paper first in CMW then in DAW. Tentative identity of spots: 1, ?; 2, lysophosphatidyl inositol (lyso-PI); 3, lysolecithin (lyso-LEC); 4, phosphatidyl inositol (PI); 5, ?; 6, lysophosphatidyl ethanolamine (lyso-PE); 7, lysophosphatidyl glycerol ? (lyso-PG); 8, lysophosphatidic acid (lyso-PA); 9, lecithin (LEC); 10, lysocardiolipin (lyso-DPG); 11, phosphatidyl glycerol (PG); 12, ?; 13, phosphatidyl ethanolamine (PE); 14, phosphatidic acid (PA); 15, cardiolipin (DPG). The following changes were effected by phopholipase A treatment: 5 → 1 (?; 5 moves like phosphatidyl serine but is Ninhydrin-negative); 4 → 2 (PI → lyso-PI); 9 → 3 (LEC → lyso-LEC); 11(12) → 7 (?); 13 → 6 (PE → lyso-PE); 14 → 8 (PA → lyso-PA); 15 → 10 (DPG → lyso-DPG). (Marinetti, personal communication.)

prepared according to his procedure (*18*). Excellent separations were obtained with natural mixtures of lipids from various sources using chloroform–methanol–water 65:25:4 or diisobutyl ketone–acetic acid–water 40:25:5 (containing 0.05% butylated hydroxytoluene as antioxidant) as solvent systems, both in one direction (ascending) and

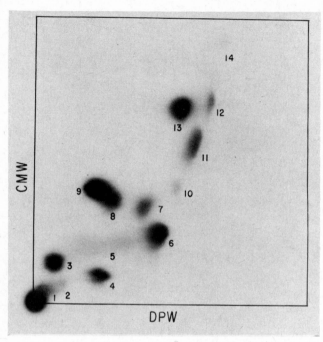

FIG. 11. Autoradiogram of ^{32}P-labeled pea root lipids and lysoderivatives, as in Fig. 10, except that the chromatogram was run first in CMW then in DPW. Tentative identity of spots (abbreviations as in Fig. 10): 1, 2, ?; 3, lyso-LEC; 4, lyso-PI; 5, ?; 6, PI; 7, lyso-PG; 8, lyso-PE; 9, LEC; 10, lyso-DPG; 11, PG; 12, PA ?; 13, PE; 14, DPG. The following changes occurred after phospholipase A treatment: 5 → 1,2 (?); 6 → 4 (PI → lyso-PI); 9 → 3 (LEC → lyso-LEC); 11 → 7 (PG → lyso-PG); 12 → ■ (PA → lyso-PA); 13 → 8 (PE → lyso-? PE); 14 → 10 (DPG → lyso-DPG). (Marinetti, personal communication.)

two-directionally (*61*). Recently, Marinetti (personal communication) introduced the use of another pair of solvents for two-directional chromatography, chloroform–methanol–water 65:25:4 and diisobutyl ketone–pyridine–water 25:25:4, to be used in conjunction with the previously mentioned pair for more complete resolution of complex mixtures.

Examples of the use of commercial silica-gel-loaded paper with these solvent systems (Figs. 10 and 11) were kindly provided by Dr. G. V. Marinetti, University of Rochester. In Figs. 10 and 11 are shown two-dimensional chromatograms of ^{32}P-labeled pea root lipids and corresponding lyso derivatives obtained as follows: Dormant pea seeds were allowed to take up water containing ^{32}P-orthophosphate, and after germination the roots were removed and extracted with chloroform–methanol 1:1, v/v. An aliquot of the total lipids in ethyl ether was subjected to the action of phospholipase A (*Naia naia*) until the reaction was complete. The enzyme was inactivated by addition of methanol, and the sample was evaporated to dryness. Another aliquot of the untreated pea lipids was added and the mixture (20 μl containing 5 μg lipid P) was chromatographed on a 20 × 20 cm sheet of Whatman SG-81 silica-gel-loaded paper, first in chloroform–methanol–water (CMW) 65:25:4, v/v, for 2.5 to 3 hr and, after drying for 20 min, either in diisobutyl ketone–acetic acid–water (DAW) 40:25:5, v/v, (Fig. 10) for 4 to 4.5 hr or in diisobutyl ketone–pyridine–water (DPW) 25:25:4, v/v, (Fig. 11) for 3.5 to 4 hr. All chromatographic systems contained 0.05% butylated hydroxytoluene as antioxidant to prevent autooxidation and subsequent streaking of labile lipids such as phosphatidyl serine.

Considering the extreme complexity of the mixture, the separations obtained are excellent and the spots are discrete and show little streaking. In comparing the CMW–DAW (neutral–acidic) system with the CMW–DPW (neutral–basic system) it is evident that phosphatidyl inositol separates better from lecithin in the latter system (compare spot 4, Fig. 10, with spot 6, Fig. 11). Also, lysolecithin is better resolved from lysophosphatidyl inositol in the CMW–DPW system (compare spots 2 and 3, Fig. 10, with spots 3 and 4, Fig. 11), as are also the components around the lecithin spot (cf. spots 6 to 9, Fig. 10, with 7 to 10, Fig. 11). However, the CMW–DAW system gives a better resolution of spots 10 to 12, Fig. 10, than the CMW–DPW system in which these components move as one spot (spot 11, Fig. 11); and spot 5 in Fig. 10 moves much better than it does in Fig. 11.

Wuthier (*97*) has also investigated the Whatman SG-81 paper and has developed several pairs of solvents for two-directional chromatography, optimum results being obtained with chloroform–methanol–diisobutyl ketone–acetic acid–water 45:15:30:20:4 in the first direction and chloroform–methanol–diisobutyl ketone–pyridine–water 30:25:25:35:8 in the second. All the major phosphatides and their lyso derivatives were completely separated, and as many as 22 lipid components from a variety of tissue extracts could be resolved. A higher loading of lipids could be used, making it possible to determine the quantitative composition of the phosphatides by estimation of lipid P in each spot.

Commercial silica-gel-loaded paper used in conjunction with the above solvent systems has obvious advantages, such as ease of manipulation of the paper, rapidity of the chromatography, and high degree of resolution, which should make it extremely useful for analysis of complex natural lipid mixtures. The availability of this commercially impregnated paper has removed a major deterrent to the widespread use of silicic-acid-impregnated paper chromatography, namely, the preparation of the paper.

V. Conclusions

It is important to realize that the mobility of a lipid on silicic-acid-impregnated paper in a given solvent is not an absolute value but will depend on the manner of impregnation of the paper, the type of chromatographic jar used, temperature, humidity, amount of material applied to the paper, and other factors, a detailed discussion of which has been presented by Marinetti (*18,19*). Nevertheless, for a particular natural mixture of lipids, the order of movement of the components is remarkably constant regardless of absolute R_f values. Thus, the R_f values given in Tables 1 to 6 should be regarded as indicating the relative mobilities, or relative order of positions, of the components on the chromatogram. The only exceptions to this rule that have been encountered here are the occasional inversion of the sulfolipid and phosphatidyl inositol spots in leaf lipids (see Fig. 3) and the inversion of the lysolecithin and phosphatidyl inositol spots (see Fig. 1). Also, phosphatidyl glycerol may occasionally appear just above the phosphatidyl ethanolamine spot.

It should also be emphasized that identification of a lipid component by the staining reactions listed in Section II.E and by its relative mobility can only be considered as tentative and requiring further verification by standard chemical procedures such as isolation, analysis, and identification of hydrolysis products. Some examples of possible ambiguities in identification are evident from the data given here. Thus, lysophosphatidyl ethanolamine and phosphatidyl aminoacyl glycerol (see Tables 1 and 5) have the same staining properties with Rhodamine, Ninhydrin, etc., and very similar relative mobilities. They obviously cannot be identified by these means, but they can easily be distinguished by examination of their deacylation products. Other examples are lecithin and phytosphingosyl glucocerebroside (Table 3), which are not resolved on chromatograms of total leaf lipids and stain yellow with Rhodamine. They are, however, readily distinguished by the positive periodate–Schiff reaction of the latter and can be separated by silicic acid chromatography or by saponification, which hydrolyzes the lecithin but has no effect on the cerebroside (*51*).

Apart from these limitations, the technique of silicic-acid-paper chromatography has proved to be extremely useful for preliminary survey of previously unexamined natural lipid mixtures, for monitoring column chromatography and enzymatic hydrolysis of lipids, and for biosynthetic studies with radioisotopically labeled lipids.

Thin-layer chromatography of phosphatides and glycolipids is discussed by Renkonen and Varo in Chapter 2. The choice of these techniques depends on the nature of the problem and on the idiosyncrasies of the individual.

REFERENCES

1. T. H. Bevan, G. I. Gregory, T. Malkin, and A. G. Poole, *J. Chem. Soc.,* **1951**, 841.

2. E. Hecht and C. Mink, *Biochim. Biophys. Acta,* **8**, 641 (1952).

3. M. H. Hack, *Biochem. J.,* **54**, 602 (1953).

4. F. M. Huennekins, D. J. Hanahan, and M. Uziel, *J. Biol. Chem.,* **206**, 443 (1954).

5. D. Amelung and P. Böhm, *Z. Physiol. Chem.,* **298**, 199 (1954).

6. G. Rouser, G. V. Marinetti, and J. F. Berry, *Federation Proc.,* **13**, 286 (1954).

7. G. V. Marinetti and E. Stotz, *J. Am. Chem. Soc.,* **77**, 6668 (1955).

8. G. Rouser, G. V. Marinetti, R. F. Witter, J. F. Berry, and E. Stotz, *J. Biol. Chem.,* **223**, 485 (1956).

9. R. F. Witter, G. V. Marinetti, A. Morrison, and L. Heiklin, *Arch. Biochem. Biophys.,* **68**, 15 (1957).

10. R. F. Witter, G. V. Marinetti, L. Heiklin, and M. A. Cottone, *Anal. Chem.,* **30**, 1624 (1958).

11. O. Armbruster and U. Beiss, *Z. Naturforsch.,* **13b**, 79 (1958).

12. G. V. Marinetti and E. Stotz, *Biochim. Biophys. Acta,* **21**, 168 (1956).

13. G. V. Marinetti, J. Erbland, and J. Kochen, *Federation Proc.,* **16**, 837 (1957).

14. D. M. Rathmann, Ph.D. thesis, University of Rochester, Rochester, N.Y., 1944.

15. B. Borgstrom, *Acta Physiol. Scand.,* **25**, 101 (1952).

16. D. L. Fillerup and J. F. Mead, *Proc. Soc. Exptl. Biol. Med.,* **83**, 574 (1953).

17. C. H. Lea, D. N. Rhodes, and R. D. Stoll, *Biochem. J.,* **60**, 353 (1955).

18. G. V. Marinetti, *J. Lipid Res.,* **3**, 1 (1962).

19. G. V. Marinetti, in *New Biochemical Separations* (A. T. James and L. J. Morris, eds.), Van Nostrand, Princeton, N.J., 1964, pp. 339–377.

20. O. Armbruster and U. Beiss, *Naturwiss., 44,* 420 (1957).

21. G. Rouser, J. O'Brien, and D. Heller, *J. Am. Oil Chemists' Soc., 38,* 14 (1961).

22. L. Hörhammer, H. Wagner, and G. Richter, *Biochem. Z., 331,* 155 (1959).

23. L. Hörhammer and G. Richter, *Biochem. Z., 332,* 186 (1959).

24. Y. Inouye and M. Noda, *Arch. Biochem. Biophys., 76,* 271 (1958).

25. J. W. Diekert and R. Reiser, *Science, 120,* 678 (1954).

26. J. W. Diekert and R. Reiser, *Federation Proc., 14,* 202 (1955).

27. J. W. Diekert and R. Reiser, *J. Am. Oil Chemists' Soc., 33,* 535 (1956).

28. M. Brown, D. A. Yeadon, L. A. Goldblatt, and J. W. Diekert, *Anal. Chem., 29,* 30 (1957).

29. G. V. Marinetti, R. F. Witter, and E. Stotz, *J. Biol. Chem., 226,* 475 (1957).

30. G. V. Marinetti, J. Erbland, J. Kochen, and E. Stotz, *J. Biol. Chem., 233,* 740 (1958).

31. G. V. Marinetti, M. Albrecht, T. Ford, and E. Stotz, *Biochim. Biophys. Acta 36,* 4 (1959).

32. G. V. Marinetti and E. Stotz, *Biochim. Biophys. Acta, 37,* 571 (1960).

33. M. H. Hack and V. J. Ferrans, *Z. Physiol. Chem., 315,* 157 (1959).

34. M. H. Hack and V. J. Ferrans, *Circulation Res., 8,* 738 (1960).

35. M. H. Hack, *J. Chromatog., 5,* 531 (1961).

36. M. H. Hack, A. E. Gussin, and M. E. Lowe, *Comp. Biochem. Physiol., 5,* 217 (1962).

37. M. H. Hack, R. G. Yaeger, and T. D. McCaffery, *Comp. Biochem. Physiol., 6,* 247 (1962).

38. M. H. Hack and F. M. Helmy, *Comp. Biochem. Physiol., 16,* 311 (1965).

39. M. P. Cormier, P. Jouan, and L. Girre, *Bull. Soc. Chim. Biol., 41,* 1037 (1959).

40. L. E. Hokin and M. R. Hokin, *J. Biol. Chem., 233,* 805 (1958).

41. E. Baer, D. Buchnea, and T. Grof, *Can. J. Biochem. Physiol., 38,* 853 (1960).

42. M. G. Macfarlane, *Biochem. J., 78,* 44 (1961).

43. M. G. Macfarlane, *Biochem. J., 79,* 4P (1961).

44. M. G. Macfarlane, *Biochem. J., 80,* 45P (1961).

45. M. G. Macfarlane, *Biochem. J., 82,* 40P (1962).

46. M. G. Macfarlane, *Nature, 196,* 136 (1962).

47. F. Haverkate, U. M. T. Houtsmuller, and L. L. M. Van Deenen, *Biochim. Biophys. Acta, 63,* 547 (1962).

48. U. M. T. Houtsmuller and L. L. M. Van Deenen, *Koninkl. Ned. Akad. Wetenschap., Proc., 66B,* 236 (1963).

49. F. Haverkate and L. L. M. Van Deenen, *Koninkl. Ned. Akad. Wetenschap., Proc., 68B,* 141 (1965).

50. M. Kates, *Biochim. Biophys. Acta, 41,* 315 (1960).

51. P. S. Sastry and M. Kates, *Biochemistry, 3,* 1271 (1964).

52. P. S. Sastry and M. Kates, *Can. J. Biochem., 43,* 1445 (1965).

53. M. Kates, D. J. Kushner, and A. T. James, *Can. J. Biochem. Physiol., 40,* 83 (1962).

54. M. Kates and R. M. Baxter, *Can. J. Biochem. Physiol.,* **40,** 1213 (1962).

55. M. Kates, G. A. Adams, and S. M. Martin, *Can. J. Biochem.,* **42,** 461 (1964).

56. M. Kates, L. S. Yengoyan, and P. S. Sastry, *Biochim. Biophys. Acta,* **98,** 252 (1965).

57. B. W. Agranoff, R. M. Bradley, and R. O. Brady, *J. Biol. Chem.,* **233,** 1077 (1958).

58. J. E. Muldrey, O. Miller, and J. G. Hamilton, *J. Lipid Res.,* **1,** 48 (1959).

59. W. E. Cornatzer, W. A. Sandstrom, and J. H. Reiter, *Biochim. Biophys. Acta,* **57,** 568 (1962).

60. R. Letters, *Biochem. J.,* **93,** 313 (1964).

60a. R. Letters and B. Brown, *Biochim. Biophys. Acta,* **116,** 482 (1966).

61. G. V. Marinetti, *J. Lipid Res.,* **6,** 315 (1965).

62. C. Levine and E. Chargaff, *J. Biol. Chem.,* **192,** 465 (1951).

63. U. Beiss, *J. Chromatog.,* **13,** 104 (1964).

64. J. Baddiley, J. G. Buchanan, R. E. Handschumacher, and J. F. Prescott, *J. Chem. Soc.,* **1956,** 2818.

65. C. S. Hanes and F. A. Isherwood, *Nature,* **164,** 1107 (1949).

66. K. Saito and S. Akashi, *J. Biochem. (Tokyo),* **44,** 511 (1957).

67. G. J. M. Hooghwinkel, H. Hoogeveen, M. J. Lexmond, and H. G. Bungenberg de Jong, *Koninkl. Ned. Akad. Wetenschap., Proc.,* **62B,** 222 (1959).

68. L. L. Sulya and R. R. Smith, *Biochem. Biophys. Res. Commun.,* **2,** 59 (1960).

69. H. G. Bungenberg de Jong, *Koninkl. Ned. Akad. Wetenschap., Proc.,* **64B,** 467 (1961).

70. F. Haverkate and L. L. M. Van Deenen, *Biochim. Biophys. Acta,* **106,** 78 (1965).

71. L. L. M. Van Deenen and G. H. de Haas, *Advan. Lipid Res.,* **2,** 167 (1964).

72. M. Kates, in *Lipide Metabolism* (K. Bloch, ed.), Wiley, New York, 1960, pp. 165–237.

73. M. Kates, *Can. J. Biochem. Physiol.,* **34,** 967 (1956).

74. M. Kates, J. R. Madeley, and J. L. Beare, *Biochim. Biophys. Acta,* **106,** 630 (1965).

75. J. L. Beare and M. Kates, *Can. J. Biochem.,* **45,** 101 (1967).

76. D. J. Hanahan and R. Vercamer, *J. Am. Chem. Soc.,* **76,** 1804 (1954).

77. A. K. Hajra, D. M. Bowen, Y. Kishimoto, and N. S. Radin, *J. Lipid Res.,* **7,** 379 (1966).

78. P. S. Sastry and M. Kates, *Biochemistry,* **3,** 1280 (1964).

79. M. L. Vorbeck and G. V. Marinetti, *J. Lipid Res.,* **6,** 3 (1965).

80. M. L. Vorbeck and G. V. Marinetti, *Biochemistry,* **4,** 296 (1965).

81. R. J. L. Allen, *Biochem. J.,* **34,** 858 (1940).

82. G. R. Bartlett, *J. Biol. Chem.,* **234,** 466 (1959).

83. M. Kates, *J. Lipid Res.,* **5,** 132 (1964).

84. F. Snyder and N. Stephens, *Biochim. Biophys. Acta,* **34,** 244 (1959).

85. R. M. C. Dawson, *Biochem. J.,* **75,** 45 (1960).

86. R. A. Ferrari and A. A. Benson, *Arch. Biochem. Biophys.,* **93,** 185 (1961).

87. E. G. Bligh and W. J. Dyer, *Can. J. Biochem. Physiol.,* **37,** 911 (1959).

88. J. L. Beare and M. Kates, *Can. J. Biochem.,* **42,** 1477 (1964).

89. M. Kates, A. C. Allison, D. A. J. Tyrrell, and A. T. James, *Biochim. Biophys. Acta,* **52,** 455 (1961).

90. M. Kates and F. M. Eberhardt, *Can. J. Botany,* **35,** 895 (1957).

91. F. M. Eberhardt and M. Kates, *Can. J. Botany,* **35,** 907 (1957).

92. M. Kates, B. Palameta, C. N. Joo, D. J. Kushner, and N. E. Gibbons, *Biochemistry,* **5,** 4092 (1966).

93. D. J. Kushner, S. T. Bayley, J. Boring, M. Kates, and N. E. Gibbons, *Can. J. Microbiol.,* **10,** 483 (1964).

94. N. Z. Stanacev, E. Baer, and M. Kates, *J. Biol. Chem.,* **239,** 410 (1964).

95. M. Kates and B. E. Volcani, *Biochim. Biophys. Acta,* **116,** 264 (1966).

96. M. Kates, M. B. Perry, and G. A. Adams, *Biochim. Biophys. Acta,* **137,** 214 (1967).

97. R. E. Wuthier, *J. Lipid Res.,* **7,** 544 (1966).

2

THIN-LAYER CHROMATOGRAPHY
OF PHOSPHATIDES AND GLYCOLIPIDS

O. Renkonen and P. Varo

DEPARTMENT OF SEROLOGY AND BACTERIOLOGY
UNIVERSITY OF HELSINKI
HELSINKI, FINLAND

I. Introduction

Thin-layer chromatography (TLC) is one of the main analytical tools of current lipid research. It is an extremely simple technique, requiring very little equipment and minimal amounts of manipulative skill. Its great advantages include sensitivity and rapidity. Chromatography on silica-gel-loaded paper or silicic-acid-impregnated paper has also found wide application and is discussed by Kates in Chapter 1.

The versatility of TLC becomes apparent when one realizes how many different types of problems are studied with it. The fractionation of complex natural lipid mixtures is one of its principal applications, but other uses exist. The purity of lipid preparations can be assayed with TLC, and simple lipid components can be identified with it. Fairly well-established regularities between structure and chromatographic mobility sometimes permit good guesses as regards the structure of unknown lipids even prior to the actual isolation. TLC is an excellent method for monitoring extractions and preparative column separations, and also all kinds of reaction mixtures.

This chapter is intended to draw attention to some aspects of TLC as applied to the more common phosphatides and glycolipids found in nature. A brief description of the isolation and preliminary fractionation of these lipids is also included, since these operations normally precede the TLC analysis. It might be expected that much of the subject matter selected would describe microbial lipids, since the authors work in a bacteriological laboratory. Nevertheless, with a few exceptions, the rather unique microbial lipids are not discussed at all. Only the types of polar lipids that appear very widely distributed among the living organisms are discussed; the microbial glycolipids as well as the plant glycosides are excluded.

The first task in lipid TLC, separation of the classes of natural lipids, is exceptional in that each class contains many different molecular species. Therefore, the primary task in polar lipid TLC is twofold: to separate the classes and, simultaneously to "shepherd" all molecular species within each class so that they remain together. In other words, selectivity toward class differences and nonselectivity toward species differences are required in the first stage; the molecular species should be separated from each other first in the final stages of the analysis. This ambivalence naturally enhances the difficulties, and the charm, of lipid separations quite remarkably.

II. Isolation and Preliminary Fractionation of Phosphatides and Glycolipids

Extraction of the natural lipids is probably most often carried out with chloroform–methanol mixtures; the work of Folch et al. (*1*) that elaborates this has become almost classical in recent lipid chemistry. An excellent paper by Bligh and Dyer (*2*) should also be consulted when Folch extractions with chloroform–methanol are planned. Such extractions give solutions of all types of lipids from the most nonpolar hydrocarbons to the most polar gangliosides and phytoglycolipids. This spectrum of different polarities is so wide that for purely practical reasons it is advantageous to divide the lipids into a few more homogeneous subgroups for further fractionation. Typical subgroups are "neutral lipids" and "polar lipids." The former include simple glycerides, sterols and their esters, free fatty acids, and hydrocarbons; the latter, phosphatides and glycolipids. The best-known polar lipids are of animal origin, and they can relatively easily be further divided into gangliosides and other polar lipids.

A. Separation of Neutral Lipids and Polar Lipids

This separation used to be carried out by precipitating the polar lipids with acetone, but nowadays it is more effectively performed with three different methods utilizing rubber dialysis, solvent partition, or adsorption chromatography.

The hydrocarbon dialysis through rubber described by van Beers et al. (*3*) is simply carried out by dissolving the lipids in petroleum ether, introducing the solution into a rubber sack, and dialyzing against petroleum ether. The neutral lipids will pass the rubber membrane, whereas the polar lipids will not. This simple method is easily applicable to relatively large samples of lipids, and it is useful when the polar lipid fraction is very small in comparison to the neutral lipid fraction. The neutral lipids passing the rubber membrane include the glycerides, probably even monoglycerides, and cholesterol and its esters. It is quite important that free fatty acids also penetrate the membrane (*4*). In this respect the dialysis procedure is superior to the partition methods that distribute the free acids equally between the neutral and polar fractions. The nondialyzable lipids include phosphatides and glycolipids. The dialysis method has been successfully used in many laboratories and with several different types of lipid mixtures.

The solvent partition system described by Galanos and Kapoulas (*5*) consists of petroleum ether and 87% aqueous ethanol. In this system the polar lipids prefer the ethanol layer, whereas the neutral lipids are enriched in the hydrocarbon phase. The authors describe a simple procedure for the separation of the neutral and polar fractions. Two separatory funnels containing 45 ml of the hydrocarbon phase are used. To the first one are introduced up to 10 g of lipids under study and 15 ml of the lower phase; the mixture is shaken well; and the layers are separated. The lower layer is transferred to the second funnel, and 15 ml of fresh lower phase is introduced into the first funnel. Both funnels are now shaken; and after the layers are separated, the lower layer of the second funnel is withdrawn, that of the first is transferred into the second, and fresh ethanol phase is introduced to the first funnel. In this way altogether eight portions of the lower layer are used to extract the two hydrocarbon phases. The result is that the hydrocarbon phase contains the neutral lipids and the combined lower layers contain practically the total amount of the polar lipids contaminated only with minute amounts of the neutral lipids. For instance a 2995-mg sample of total milk lipids that contained 1050 μg lipid phosphorus (i.e., about 26 mg phosphatides) gave a neutral fraction that weighed 2961 mg and was free of phosphorus; the ethanol phases contained 1046 μg phosphorus.

Solvent partition between heptane and 95% aqueous MeOH was used by Carter et al. (*6*) for fractionation of wheat flour lipids. The heptane phase contained mainly triglycerides plus sterols and sterol esters, and the methanol phase contained monogalactosyl and digalactosyl glycerides.

The obvious advantages of these partition procedures include simplicity and the possibility of treating large samples. The disadvantages include emulsion formation, which may cause severe trouble. An additional drawback is that free fatty acid is reported to distribute fairly evenly between the two phases (*7*); in this respect the dialysis method appears more suitable.

Attention should also be paid to the cations combined with the acidic lipids. A dramatic example of their role was described by Carter and Weber (*8*), who found that, when Na salt of phosphatidyl inositol was partitioned between ether and water, 94% of the phosphatide was recovered from the aqueous layer. However, with Ca salt of this lipid 75% was found in the ether layer under the same experimental conditions. It thus appears that the conversion of the lipid extracts into pure Na form or K form with chelating resin columns (*8,9*) should become a general practice before any fractionation steps are undertaken.

Chromatographic procedures on all adsorbents used in lipid separations form an easy method for separating the neutral and polar fractions. Silicic acid in particular is a very suitable adsorbent for this type of separation, because both the elimination of the neutral lipids *and* the subsequent recovery of the polar lipids are easily achieved. An advantage of the chromatographic approach is that the polar lipids and neutral lipids can be separated into several classes simultaneously. The drawback of the chromatographic procedure is that relatively large column and solvent volumes are needed with mixtures containing a large amount of neutral lipids and very little polar lipid.

B. Separation of Gangliosides from Other Lipids

Gangliosides are sphingoglycolipids containing sialic acid, which are present in the brain, especially in the gray matter, in the spleen, and in the erythrocytes. Their carbohydrate part is so large that they are freely soluble in water, but they are "true" lipids in that they also dissolve in chloroform–methanol mixtures. Thus they are extracted together with the other lipids by chloroform–methanol. Their elimination from the bulk of other lipids takes place during the partition step included in the procedure of Folch et al. (*1*): When 0.2 part of water or a suitable salt solution is added to the chloroform–methanol extract, a biphasic system results and all other lipids (neutral and polar) favor the lower, chloroform layer, whereas the gangliosides together with nonlipid material are enriched in the upper aqueous layer.

Svennerholm (*10*) has described a very carefully controlled procedure for the isolation of gangliosides, which is based on the Folch partition. Acetone-dried or lyophilized material is extracted in a Soxhlet apparatus first with chloroform–methanol 2:1 for 8 hr and then with chloroform–methanol 1:2 for 16 hr. Paper chromatography revealed that the second solvent was necessary for the extraction of the more polar gangliosides. To separate the gangliosides and the other brain lipids, the chloroform–methanol 2:1 extract is shaken vigorously with one-sixth volume of 0.1% NaCl solution. After standing for 12 hr, the chloroform phase is transferred to another separatory funnel containing one-third volume† of fresh upper phase, chloroform–methanol–water 1:10:10. The lower phase is extracted in all 5–7 times with fresh upper phase. The combined upper layers are then evaporated in vacuum until foaming ensues. The residual extract is then dialyzed for at least 3 days against running water and then evaporated to dryness and extracted with boiling chloroform–methanol 2:1. The resulting extract contains crude gangliosides. The recovery of the more polar gangliosides from the chloroform–methanol 1:2 extract is simpler, because there is very little other lipid in this extract. Thus the partition step can be omitted and the nonlipids are eliminated simply by dialysis and chloroform–methanol extraction of the dry nondialyzable material.

The yield of ganglioside in the partition step can be estimated by analyzing hexosamine and sialic acid. Svennerholm reported that his ganglioside fractions contained 90–95% of the total hexosamine and 95–98% of the total sialic acid present in his original extracts. However, the yields with the ganglioside typical of Tay–Sachs disease are not more than 80%. These findings allow some crude ideas about the partition coefficients of different glycolipids to be formed. The main gangliosides of normal brain contain ceramide with two fatty chains and an oligosaccharide with five to seven monosaccharide units, whereas the Tay-Sachs ganglioside contains a normal ceramide with two fatty chains, but it has only four monosaccharides in its carbohydrate part.

The purity of the gangliosides that are obtained by the partition procedure is not very high, the chief contaminants include cephalins, sulfatides, nonlipid material, and smaller amounts of cerebrosides, lecithin, and sphingomyelins. A new possibility for removal of nonlipids both from the gangliosides and other lipids lies in the use of Sephadex columns (11,12).

C. Separation of Phosphatides and Glycolipids of Comparable Polarity

The procedures described above lead to three main subgroups of lipids; neutral, polar, and very polar. The neutral fraction contains no phosphatides or glycolipids, the very polar group almost exclusively

† Calculated from the original volume of the extract.

glycolipids, but the middle group both phosphatides and glycolipids. Thus the question that arises is whether these two types can be separated by any simple method.

One procedure that is of considerable use is based on the principle that most animal glycolipids are derivatives of sphingosine and consist of alkali-resistant amide and glycosidic linkages, whereas most phosphatides are glycerolipids and contain alkali-labile ester groups. Therefore it is quite convenient to subject the polar lipids to mild alkaline methanolysis whereby the glycerolipids are largely converted into methyl esters of fatty acids and water-soluble products, whereas the sphingoglycolipids and the sphingomyelins survive this treatment in intact form. The water-soluble products are then eliminated by Folch partition or dialysis, and the methylesters, e.g., by hydrocarbon dialysis, leaving the glycolipids contaminated only with sphingomyelins. These phosphatides can be eliminated in many cases by chromatography on Florisil columns or by combined use of silicic acid and alumina. The disadvantage of the alkaline methanolysis technique is that, besides glycerophosphatides, glyceroglycolipids also are lost.

For the separation of the glyceroglycolipids from the phosphatides the recent chromatographic procedure of Vorbeck and Marinetti (*13*) seems of considerable promise. These authors found that acetone elutes only neutral lipids and glyceroglycolipids from silicic acid columns. Since chloroform elutes only the neutral lipids, it was possible to devise an elution scheme for some bacterial lipids in which first the neutral lipids were eluted from the column with chloroform, next the neutral glyceroglycolipids were eluted with chloroform–acetone 1:1 and pure acetone, and finally the phosphatides were eluted with the usual chloroform–methanol mixtures.

Radin and associates have described another chromatographic procedure that can be used for similar separations (*14,15*). It is based on the use of Florisil (see Chapter 6), a commercial preparation of magnesium silicate, which adsorbs the phosphatides under suitable conditions, whereas the cerebrosides are not retained. This method was later improved by Rouser et al. (*16*), who observed that Florisil is most useful as a very dry adsorption column. They developed a method for the elimination of traces of water by adding 2,2-dimethoxypropane to all solvents used in the chromatography; this agent reacts with water to yield methanol and acetone. Under these conditions both cerebrosides and sulfatides run ahead of all brain phosphatides. Also, the monogalactosyl diglyceride, digalactosyl diglyceride, and the plant sulfolipid run ahead of phosphatides (17). Some observations suggest that glycolipids having still larger carbohydrate moiety can be eluted from superdry Florisil ahead of lecithin and sphingomyelin (18).

Of considerable interest are some observations that suggest that the water content may have profound influence upon the relative mobility of glycolipids and phosphatides even on silica gel. G. Nichols has presented TLC figures showing that on silica gel G with anhydrous chloroform–methanol, with or without acetic acid, the plant galacto-lipids run much faster than phosphatidyl ethanolamine, whereas in chloroform–methanol–water 65:25:3 the spots of digalactolipid and phosphatidyl ethanolamine fused together (19). Rouser et al. (20) have reported that brain cerebroside sulfates move on silicic acid in anhydrous chloroform–methanol faster than phosphatidyl ethanol-amines, whereas in water-containing systems they move more slowly than phosphatidyl ethanolamines, sometimes even more slowly than phosphatidyl cholines. The authors emphasize that small amounts of water in silicic acid systems influence the relative migration rates of acidic lipids versus neutral lipids, but it seems also that the migration rates of carbohydrate-containing lipids versus typical phosphatides are affected.

Another possiblity for the separation of the phosphatides and glycolipids lies in direct chromatography on silicic acid, assay of the fractions for carbohydrate and phosphorus, and rechromatography of mixed fractions on borate-containing silicic acid (21) or on DEAE-cellulose or alumina.

There might be a possibility of separating the nonionic glycolipids from the phosphatides also by dialysis in organic solvents in the presence of suitable organic cations to form "nonmicellar" salts with the phosphatides (22).

III. Separation of Different Classes of Phosphatides and Glycolipids

A. TLC of Animal Phosphatides Using Silica Gel G and Neutral Solvents

Among the adsorbents employed in polar lipid TLC, silicic acid, or silica gel, is by far the most important. Plaster of paris is often added to this material; it binds the adsorbent to the glass plate during its hydration. Mixtures of this kind are commonly designated as silica gel G. Adsorbents that either contained no plaster of paris or contained magnesium silicate instead have been used to some extent, but most of the essential observations about the TLC of polar lipids that we possess today have been obtained with silica gel G.

The basic procedures of TLC of the polar lipids originate from the work of Schlemmer (*23*) and Wagner et al. (*24*). These two laboratories introduced the use of silica gel G and the solvent system of chloroform–methanol–water, which had proved so successful with silicic-acid-impregnated papers in the hands of Lea et al. (*25*). The plates are prepared by mixing 25 g silica gel G (E. Merck AG, Darmstadt, Germany) and 50 ml water; for instance, the Desaga applicator (Desaga GmbH, Heidelberg, Germany) is used to apply the slurry to the plates usually in the thickness of 0.25 mm. The adsorbent is allowed to solidify for a while, and the plates are finally activated at 110°C for 1 hr.

The lipids are applied to the plates by delivering about 0.5–1% solutions with microsyringes (Hamilton Company, Whittier, California). Usually the amount of the mixtures applied is such that the major components are present in quantities of 5–50 μg. The lipids are often introduced in chloroform–methanol solutions, but chromatographic movement during the application results easily in formation of rings of the lipids at the origin instead of compact starting spots. The introduction of lipids should therefore be carried out in nonpolar solvents, e.g., in petroleum ether whenever feasible. It is important that the surface of the plates should not be unduly disturbed during the application of the lipids. To eliminate autoxidation of unsaturated lipids during TLC, all manipulations may be carried out under nitrogen; a commercial preparation box (obtainable from Desaga Gmbh, Heidelberg, Germany) allows the delivery of lipids to the plates to be carried out in this way.

Ascending development is carried out in a tank 21 × 21 × 9 cm (obtainable from Desaga Gmbh, Heidelberg, Germany). In our laboratory solvent-soaked filter liners are not used in the tanks; instead, the closed tanks are well shaken before inserting the plates. We believe that straight solvent fronts are best obtained by inserting the plates in vertical position, in almost complete contact with the tank wall. The development should often be carried out in cold rooms, especially when multiple developments are necessary and the lipids stay relatively long in contact with the adsorbent. The low temperature slows down the rate of possible hydrolytic reactions. To protect the lipids against autoxidation, small amounts of an antioxidant, 4-methyl-2,6,di-t-butylphenol (BHT), are often added to solvents used for TLC (*25a,25b*).

In the detection of the lipids the charring agents such as sulfuric, perchloric, and phosphoric acids, with or without added oxidative agents, are of particular importance in analytical TLC; in fact, the possibility of using these corrosive reagents is one of the prime advantages of TLC over paper chromatography. These reagents may detect as little as 1 μg of lipid. We like very much a new corrosive spray developed by Borowski and Ziminski (*26*), that consists of 20% aqueous ammonium bisulfate; it gives as good charring as sulfuric acid on the heated plates, but it is less corrosive and much more agreeable to use than the strong sulfuric acid solutions. This spray may be especially useful for laboratories where hoods may be lacking. So far, however, we have used it only for qualitative work.

Other universal detecting agents for the polar lipids are dyes and fluorescent agents such as Bromthymol Blue, Rhodamine 6G, and 2′,7′-dichlorfluorescein. These agents are widely used, especially in preparative work, because they are harmless to the lipids and can be removed from the preparations later.

Iodine too, either as vapor or as chloroform solution, is a good agent for detection of lipids; but as the unsaturated fatty chains react with it (*93*), the chief use of iodine is in analytical work. Whenever specific reagents are needed to aid in the identification of the spots, iodine is very advantageously used first as a general detecting agent; it can later be eliminated from the plates simply by subliming in vacuo, and afterward the specific sprays can be applied.

Since a number of common phosphatides and glycolipids have similar mobilities on silica gel G, there certainly is a need to differentiate at least between these two large families of lipids. This should be easy, since there are several good carbohydrate reagents for glycolipids and at least one promising reagent for phosphatides (*27*). Another interesting specific spray is the benzidine reagent of Bishel and Austin (*28*), which reacts with amide groups and thus differentiates sphingolipids and glycerolipids.

With neutral chloroform–methanol–water mixtures on silica gel G the neutral lipids, i.e., sterols and their esters, simple glycerides, and free fatty acids, move with the solvent front ahead of the common phosphatides; then in order of decreasing mobility come cardiolipin, phosphatidyl ethanolamine, phosphatidyl choline, sphingomyelin, and lysolecithin.

The separation between the neutral lipids and the more rapid phosphatides is often not quite satisfactory. In such cases the technique of multiple development is of great help. In this technique the plate is first developed in the usual way with the chloroform–methanol–water mixture whereby the solvent is allowed to rise 10 cm, then the plate is dried for 10 min, and finally it is rerun in the same direction with a less polar solvent, for instance, with hexane–ether 4:1. The latter solvent, which is allowed to rise 15 cm, hardly moves the polar lipids at all, but it carries the unpolar ones away and distributes them in separate spots on the fresh 5-cm path above the most rapid phosphatides. This technique of multiple development is useful in treating mixtures containing lipids of widely different polarities, but in addition it is of great potential value for very difficult separations. Most people familiar with chromatography will probably agree that systems of relatively slow movement of the components are the most effective ones. With such systems good separations can be achieved by using multiple development three or four times with the same solvent. Morris (*28a*) has presented an excellent demonstration of this principle by separating some positional isomers of methyl oleate (Fig. 1). The paper of Hooghwinkel et al. (*29*) on cerebroside fractionation, which is discussed below, is another example of this efficient method.

Native plasmalogens and 1-alkyl-2-acyl analogs of phosphatides run together with the corresponding phosphatidyl compounds; this was observed, for example, with pure samples of the choline lipids isolated from ox heart lecithin and also with glyceryl phosphoryl ethanolamine lipids of ox brain (*30*). However, there is a slight tendency of the plasmalogens and alkyl phosphatides to run ahead of their phosphatidyl analogs, because this is the case with the dephosphorylated form, the diglyceride and diglyceride acetates (*31*). We have tried to fractionate the ox heart lecithins, but we have never obtained any clearly reproducible separation between the plasmalogen and phosphatidyl form. However, the dimethyl esters of plasmalogenic and ordinary phosphatidic acid are separable on silica gel G with ether as solvent (*31a*). Horrocks and Ansell have separated 1-alkyl glyceryl phosphoryl ethanolamine (R_f 0.34) from the 1,2-acetal of glyceryl phosphoryl ethanolamine (R_f 0.46) that is formed from the alkenyl

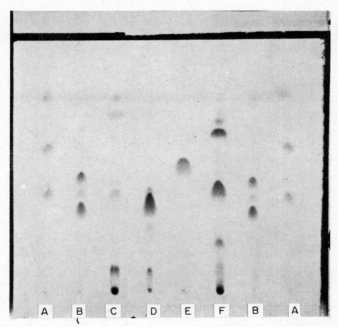

FIG. 1. Thin-layer chromatogram of some fatty acid methyl esters on silica gel G impregnated with silver nitrate (10%, w/w). The plate was developed three times with toluene at −15°C, and spots were located by spraying with chlorosulfonic acid–acetic acid 1:2 and charring. Samples: A, methyl elaidate (upper) and methyl oleate (lower); B, methyl *cis*-vaccenate (Δ^{11}, upper), methyl oleate (Δ^9, middle), and methyl petroselinate (Δ^6, lower); C, rat liver mixed esters (small proportion of Δ^{11}–18:1 not visible on reproduction); D, parsley seed oil mixed esters (separation of Δ^9– and Δ^6–18:1 isomers); E, methyl stearolate; F, *Exocarpus cupressiformis* seed oil mixed methyl esters. [From L. Morris, *J. Lipid Res.*, **7**, 717 (1966).]

derivate by cyclization under certain acidic conditions (*31b*). They used chloroform–methanol–water 60:30:6.

The sphingomyelins often show two spots on TLC plates run with neutral chloroform–methanol–water mixtures; Wood and Holton (*32*) studied these two spots of human serum sphingomyelins more closely. The two spots obtained with chloroform–methanol–water 80:35:5 as solvent are shown in Fig. 2. The material was separated on preparative scale, and the component fatty acids were analyzed with GLC. The more rapidly moving component SM (U) contained principally C-22

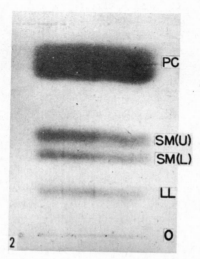

FIG. 2. Separation of the two human plasma sphingomyelins on a silica gel G plate. Developing solvent: chloroform–methanol–water 80:35:5. Detection of spots by charring. PC = phosphatidyl choline; SM (U) = sphingomyelin, upper fraction (less polar); SM (L) = sphingomyelin, lower fraction (more polar); LL = lysophosphatidyl choline. [From P. D. S. Wood and S. Holton, *Proc. Soc. Exptl. Biol. Med.,* **115**, 990 (1964).]

and C-24 acids, whereas the slower component SM (L) contained mainly palmitic acid. No differences were observed in the sphingosine bases of the two fractions. This experiment is important, because it proves that silica gel G also effects certain separations within given lipid classes; in this case the differences in the chain length led to fractionation.

In glycerophosphatides the differences in chain length are less than in sphingolipids and, what is more important, the longer chains usually contain several double bonds, which in turn tend to retard the mobility. Thus the glycerophosphatides are not usually separated as clearly as sphingomyelins into different molecular species on silica gel G. However, we have recently separated lysolecithins resulting from phospholipase A hydrolysis of Baltic herring lecithin. Silica gel G and chloroform–methanol–water 65:25:4 led to a clear separation of two fractions, the more rapid one consisting mainly of docosahexaenoic ester of glyceryl phosphoryl choline, whereas the slower one was a mixture of corresponding palmitic and oleic esters (*33*). This separation was also reproduced on silica gel G columns. The two types of positional isomers, α and β-form, of lysophosphatides appear not to be separated on silica gel G plates (*34,35*), but silica seems to cause some migration of the acyl residue of the β forms (*35a*).

On the whole, the neutral chloroform–methanol–water systems and silica gel G work well as long as there are no acidic phosphatides such as phosphatidyl serine or phosphatidyl inositol in the mixture to be analyzed, but for the acidic lipids this type of system is rather unsatisfactory. For instance, phosphatidyl serine and phosphatidyl inositol streak badly and are not separated, and their R_f values vary with concentration to such a point that their analysis becomes impossible. These shortcomings clearly reveal a need for improved methods; but nevertheless the combination of moist chloroform–methanol and silica gel G has survived as a useful system for all neutral phosphatides, and it has proved to be an excellent one for the neutral glycolipids also.

B. Refined TLC Systems for Animal Phosphatides

Phosphatidyl serine was the major animal phosphatide causing trouble with neutral silica gel G systems. Several laboratories attacked this problem simultaneously and successfully and with partly different methods. Horrocks (*36*) and Skidmore and Entenman (*37*), as well as Müldner et al. (*38*) and also Redman and Keenan (*38a*), devised ammonia-containing solvent systems to be used with silica gel G. This type of solvent gives compact spots from phosphatidyl serine; this lipid is also well separated from phosphatidyl ethanolamine. Horrocks, for instance, reported an R_f value of 0.47 for phosphatidyl ethanolamine and a value of 0.11 for phosphatidyl serine with chloroform–methanol–aqueous ammonia 75:25:4.

Another approach was reported by Skipski et al. (*34*), who first studied the incorporation of acetic acid into the neutral mixture of chloroform–methanol–water. But the streaking of phosphatidyl serine could not be eliminated, and the position of the streak depended on the amount of the lipid. Therefore, these authors resorted to modified silica gel G plates; the plates were made "basic" by preparing the adsorbent slurry in 0.01 M Na_2CO_3 instead of water. In fact, a silicic acid–silicate type of adsorbent elaborated earlier by Rouser et al. (*39*) had similar properties. This adsorbent proved adequate in many respects; with chloroform–methanol–acetic acid–water 50:25:8:4 as solvent, phosphatidyl serine was well separated from phosphatidyl ethanolamine, and its position was independent of the load (*34*).

The ammonia-containing solvents devised by Skidmore and Entenman (*37*) resolved phosphatidyl inositol from phosphatidyl ethanolamine and phosphatidyl serine on silica gel G, but phosphatidyl choline, its lysoform, and sphingomyelin had R_f values in the same region. This difficulty led the authors to study two-dimensional chromatography of the phosphatides (*37*).

Another possibility of separating phosphatidyl inositol from the other animal lipids was reported by Skipski's group (*40*). They adopted the use of silica gel without calcium sulfate as binding agent.† When this type of adsorbent and 0.001 M Na_2CO_3 were used to prepare the plates, an excellent separation of phosphatidyl ethanolamine, phosphatidyl serine, and phosphatidyl inositol was obtained with acetic-acid-containing solvents; the choline phosphatides are not disturbing, as they move much more slowly. In Fig. 3 is shown an example of the separations obtained with this system. Quite recently, Skipski et al. have devised complementary systems capable of separating acidic phosphatides such as cardiolipin, phosphatidic acid, and phosphatidyl glycerol from each other and also from phosphatidyl ethanolamines and some common sphingoglycolipids (*40a*). These systems consist of two-step undimensional development on neutral and basic Ca_2SO_4-free plates; the first development is actually a "prewashing" of the chromatograms that moves the disturbing spots away from the acidic phosphatides.

A third approach for the problem presented by the "cephalin group" of lipids was reported by Hofmann (*41*), who obtained good resolution of phosphatidyl ethanolamine, phosphatidyl serine, and phosphatidyl inositol on 5% NH_4NO_3-containing silica gel G plates. He used as a solvent chloroform–propionic acid–*n*-propanol–water 10:10:10:4. This system also separated the choline phosphatides from each other, but lecithin shows the same mobility as phosphatidyl inositol. We have used a slightly modified version of Hofmann's system; our solvent mixture simply contains the solvents in ratio 2:2:3:1. This has proved

† There are at least three manufacturers selling such silica gel, which yields layers of sufficient mechanical strength without added gypsum. They are Camag, Muttenz, Switzerland; Merck, Darmstadt, Germany; and Woelm, Eschwege, Germany.

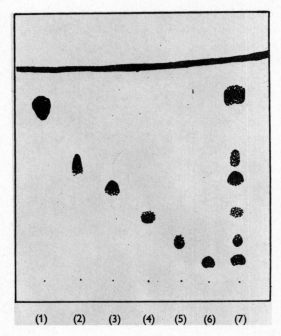

FIG. 3. Separation of reference phosphatides on pure silica gel. Developing solvent, chloroform–methanol–acetic acid–water 25:15:4:2. Detection of spots by charring. 1, phosphatidyl ethanolamine; 2, phosphatidyl serine; 3, phosphatidyl inositol; 4, phosphatidyl choline; 5, sphingomyelin; 6, lysophosphatidyl choline; 7, mixture of 1–6. [From V. P. Skipski et al., *Biochem. J.*, **90**, 374 (1964).]

a very reliable system for the separations within the cephalin group. Lysophosphatidyl inositol also migrates well in this system.

The two-dimensional approach of Skidmore and Entenman (*37*) has been extended by Rouser et al. (*20*), who applied rather different solvents in the two runs; chloroform–methanol–water in the first direction and butanol–acetic acid–water in the second. They also used magnesium silicate instead of gypsum as binding agent in the adsorbent; this prevents tailing of the acidic lipids in the first run. The resulting pattern, Fig. 4, shows good separation of the principal phosphatides of brain, and in addition allows for the simultaneous separation of the main glycolipids, cerebrosides, sulfatides, and gangliosides.

Abramson and Blecher (*42*) have reported a combination of am-

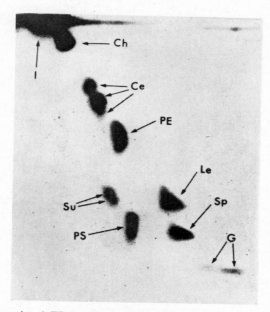

FIG. 4. Two-dimensional TLC of 200-μg beef brain lipids on silica gel that contains magnesium silicate as binder. First solvent (vertical) chloroform–methanol–water 65:25:4; second solvent (horizontal) butanol–acetic acid–water 60:20:20. Detection by charring. The substances are Ch, cholesterol; Ce, cerebrosides; PE, phosphatidyl ethanolamine; Le, lecithin; Sp, sphingomyelin; Su, sulfatide; PS, phosphatidyl serine; and G, gangliosides. [From G. Rouser et al., *J. Am. Oil Chemists' Soc.*, **42**, 215 (1965).]

monia- and acetic-acid-containing mixtures of chloroform–methanol–water for two-dimensional TLC on basic silica gel G. This system eliminated the need for magnesium silicate, which is necessary in the system of Rouser et al. (*20*).

C. TLC of Sphingoglycolipids

There is considerable confusion in the present nomenclature of these types of natural lipids; the vocabulary almost resembles that of Humpty Dumpty, who said, "When I use a word, it means just what I choose it to mean." We employ the word cerebroside to mean a ceramide monohexoside regardless of the type of hexose, sphingosine base, or fatty acid present; words like ceramide dihexoside and trihexoside are used for the characterization of the more complex lipids of this family regardless of the presence or absence of amino sugars in the carbohydrate moiety; the term "sulfatide" is used to denote sulfate esters of cerebrosides only, although even more complex sulfate-containing sphingolipids are known (*43*); the term "ganglioside," finally, is used only when sialic acid is present in the sphingolipid.

The sphingoglycolipids in the animal kingdom have been known for a long time—brain in particular is a rich source of these lipids, but in recent years sphingoglycolipids have also been found in plants (*44,45*) as well as in yeasts (*46*). The general structure of the molecules in the different sources is the same, but there are small structural differences in the central sphingosine moiety, and the most complex sphingolipids of plant and microbial origin, the phytoglycolipids (*46,47*), differ considerably from the gangliosides (*48*), which are their counterpart in animal kingdom.

The major sphingoglycolipids of brain are the cerebrosides and sulfatides. Since these lipids are present in brain in concentrations comparable to those of the major phosphatides, their behavior on TLC is well known. In the neutral system of chloroform–methanol–water 65:25:4 and silica gel G, elaborated by Wagner et al. (*24*), the brain cerebrosides occupy a position ahead of phosphatidyl ethanolamine but behind cardiolipin; the sulfatides are slower and appear mostly in the region between phosphatidyl ethanolamine and phosphatidyl choline. The two-dimensional technique of Rouser et al. (*20*) allows a good separation of these glycolipids along with the phosphatides of brain. The new systems of Skipski et al. (*40a*) are specifically devised for study of these glycolipids when they appear together with acidic phosphatides like cardiolipin, phosphatidic acid, and phosphatidyl glycerol. Horrocks (*36*), as well as O'Brien et al. (*49*), has used ammonia-containing solvents to eliminate the tailing of the sulfatides, but it appears that the problem of tailing is relatively unimportant with the sulfatides.

A good example of use of TLC in monitoring enzymatic reactions was presented by Mehl and Jatzkewitz (*50*), who isolated cerebroside sulfatase from kidney. How simply the various reaction mixtures were analyzed with TLC for sulfatides and cerebrosides is shown in Fig. 5.

Brain cerebrosides are ceramide galactosides, but other cerebrosides are known too. Thus Gaucher spleen, for instance, is known to contain

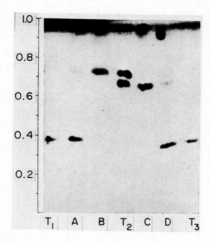

FIG. 5. TLC revealing enzymatic hydrolysis of sulfatides into cerebrosides by an enzymatic preparation of kidney. Silica gel G plates; development with chloroform–methanol–water 14:6:1; detection by Kaegi–Miescher reagent. The substances are T_1, kerasin sulfate; T_2, kerasin and cerebron; T_3, cerebron sulfate; B, formation of kerasin from kerasin sulfate; C, formation of cerebron from cerebron sulfate; A, control of B with enzyme inhibitor; D, control of C with enzyme inhibitor. [From E. Mehl and H. Jatzkewitz, *Z. Physiol. Chem.*, **339**, 260 (1964).]

ceramide glucosides (*51*). These two types of cerebrosides are only partially separated on unmodified silica gel, but they were effectively separated on borate-impregnated plates as shown by Young and Kanfer (*52*). The plates were prepared by adding 33 ml of saturated sodium tetraborate and 66 ml water to 50 g of silica gel G; the development was carried out with chloroform–methanol–water 65:25:4. This technique of glycol complexing will undoubtedly become increasingly important in several ways in lipid chemistry. Morris (*21*) has reviewed the method as applied to di-, tri-, and tetrahydroxy fatty acids; he has also presented the details of borate incorporation into the plates. Horrocks (*36*) has shown how the relative rate of migration of glycolipids versus phosphatides can be varied with borate; and Thomas et al. (*53*) have separated 1-monoglycerides from 2-monoglycerides on these plates.

Just as are the sphingomyelins, the brain cerebrosides and sulfatides are usually seen as two or even three separate spots on TLC. These lipids are known to contain both very long and medium chain fatty acids, but besides this, there are other differences in the component fatty acids, namely, the normal unsubstituted acids and the hydroxy acids. Both these differences evidently affect the mobility. The lipid spots must be extracted and subjected to fatty acid analysis in order to identify their type. This problem was studied by Hooghwinkel et al. (29) in a very instructive paper. These authors analyzed beef spinal cord cerebrosides on silica gel G plates by using a solvent system chloroform–methanol–90% aqueous formic acid 70:18:12; the technique of multiple development was used. As shown in Fig. 6, the cerebrosides had separated into two spots after one run (A), but in (B), where four successive runs had been carried out, the cerebro-

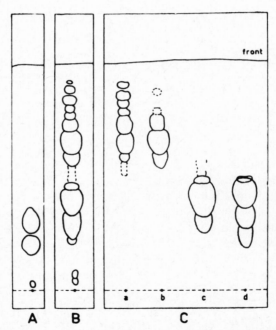

FIG. 6. Schematic presentation of TLC of beef spinal cord cerebrosides. Silica gel G plates; development with chloroform–methanol–90% formic acid 70:18:12; detection with bromthymol blue. A, separation of total cerebrosides after one ascent of the solvent; B, the same after four ascents of the solvent; C, column fractions of cerebrosides (see Table 1) separated by four ascents of the solvent.
[From G. J. M. Hooghwinkel et al., *Rec. Trav. Chim.*, **83**, 576 (1964).]

sides were separated in 12 or 13 different spots. To elaborate these findings, the authors subjected the cerebrosides to column chromatography on a large silicic acid column and collected four successive fractions a, b, c, and d. These fractions were subjected to TLC (Fig. 6,C); also, they were methanolyzed and the methyl esters were analyzed by GLC. The results are shown in Table 1. It appears that *both* the

TABLE 1

FATTY ACID COMPOSITION OF THE CEREBROSIDE FRACTIONS SHOWN IN FIG. 6,C

Fatty acid[a]	a	b	c	d
20:0		8.6		
22:0	4.6	14.4		
22:1	4.4	20.0		
23:0	2.4	4.9		
23:1	1.7	1.2		
24:0	5.8	8.3		
24:1	37.5	31.9		
25:0	16.9	2.1		
25:1	1.3			
26:0		3.9		
26:1	25.5			
18 h: 0			3.7	36.3
20 h: 0			9.3	7.4
20 h: 1				5.2
22 h: 0		2.8	5.3	
22 h: 1				7.5
23 h: 0			4.2	4.4
24 h: 0			8.9	14.5
24 h: 1			65.5	22.6
Unidentified			3.2	2.1

[a] h = hydroxy acid.

chain length of the acid and the hydroxyl group affect the mobility of the cerebrosides; on the other hand, unsaturation does not play a major role in the subfractionation of these lipids, since there are maximally two double bonds in the base and only one in the acid component.

Allen and co-workers (*54*) have shown that spinach cerebroside runs with the same rate as the hydroxy-acid-containing cerebrosides of beef brain on silica gel G. This suggests that phytosphingosine or dehydro-phytosphingosine are present, but not hydroxy acids.

Next in the series of complex sphingolipids are the ceramide hexo-sides containing more than one monosaccharide per molecule. These substances are very interesting, not least because of their hapten properties, discovered by Rapport et al. (55). The TLC behavior of these lipids has been studied notably by the Svennerholms (56) and Gray (57).

The Svennerholms (56) studied neutral ceramide hexosides isolated from very large samples of human serum. These lipids were first separated into less polar (IA) and more polar (IIA) fractions, which were then subjected to TLC on silica gel G by using chloroform–methanol–water 65:25:4. An analytical run of these two fractions is shown in Fig. 7. The four components visible in the IA material were separated on preparative scale (TLC) and analyzed. The two rapid spots (component A) turned out to be glucose-containing ceramide monohexosides, i.e., cerebrosides. The faster-moving of these two spots contained normal fatty acids, and the slower band contained α-hydroxy fatty acids. The slower pair of IA material (component B) represented ceramide dihexosides containing both galactose and glucose. Even here the faster component contained normal fatty acids and the slower α-hydroxy fatty acids. Mild acid hydrolysis of component B gave component A and galactose. Therefore this lipid is very similar to cytolipin H of Rapport (58), which also is ceramide lactoside. The different components of IIA material shown in Fig. 7 were also iso-lated, and the faster pair of spots (component C) was identified as ceramide trihexosides containing galactose and glucose in molar ratio 2:1. TLC revealed that after weak acid hydrolysis this material was converted into ceramide monohexosides and dihexosides. The separa-tion of component C in two spots was again shown to depend on normal fatty acids in the faster and hydroxy acids in the slower com-ponent. The slowest spot of IIA material, Fig. 7, component D, was finally identified as ceramide tetrahexoside containing glucose, galac-tose, and galactosamine, most likely in molar ratios 1:2:1. This ma-terial was seen only as one spot on the TLC plate; and in accordance with this, only normal fatty acids were found after methanolysis. After acid hydrolysis the lipid gave products having same R_f values as components A, B, and C with normal acids.

Gray (57) has studied the ceramide hexosides present in different

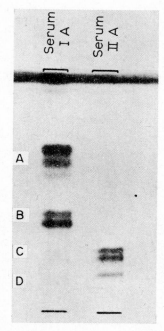

FIG. 7. TLC of neutral glycolipids of serum. Silica gel G plates; development with chloroform–methanol–water 65:25:4; detection by charring. Fraction IA, component A and component B (both appear as double spots). Fraction IIA, component C (double spots) and component D (single spot). [From E. Svennerholm and L. Svennerholm, *Biochim. Biophys. Acta,* **70,** 432 (1963).]

ascites tumors by using the conditions of the Svennerholms. His findings agree well with those of the Gothenburg group in that he could separate ceramide glucosides, ceramide lactosides, ceramide digalactosyl glucosides, and *N*-acetyl galactosaminyl digalactosyl glucosides; in all cases the characteristic double spots were observed. Gray's figures suggest that ceramide trihexosides that contain normal hexoses run slightly faster than trihexosides of two hexoses and one galactosamine. Two slow glycolipids present in some of the ascites tumors probably contained a larger carbohydrate moiety than tetrasaccharide; another possibility would be the presence of sialic acid. An interesting observation was that, although all the tumors contained ceramide dihexoside, only one of these revealed immunochemical similarity with cytolipin H, which is ceramide lactoside (58). In one case a dihexoside was isolated and partially characterized. Although it was shown to be a ceramide galactosyl glucoside, it appeared to have no cytolipin H activity. This suggests, perhaps a little surprisingly, that the family of ceramide hexosides may be a rather large group of different lipids.

The work of Shapiro et al. (57a,57b) and the recent papers of Flowers (57c,57d) are an example of consistent use of TLC in chemical synthesis of cerebrosides, sulfatides, and ceramide dihexosides. For instance, Flowers (57d) has separated the two isomeric sulfatides with the sulfate group on either 3 or 6 position of the galactose; the 6-sulfate migrated significantly faster.

It is important to note that the polarity of ceramide hexosides is such that they can be chromatographed with the same system as the common phosphatides; moreover, they occupy the same region on the plates as the phosphatides. Therefore the use of specific staining procedures is to be recommended. The animal ceramide hexosides appear very often as a series of double spots on TLC plates; this is so characteristic that it helps in their identification. The glycerophosphatides and the gangliosides do not appear to share this property, but the sphingomyelins do.

One of the major gangliosides (G$_I$) of human and beef brain has the following structure according to Kuhn and Wiegandt (59):

gal(1 → 3) galNAc(1 → 4) gal(1 → 4) glu(1 → 1) ceramide†

$$\begin{pmatrix} 3 \\ \uparrow \\ 2 \end{pmatrix}$$

NANA

The other major gangliosides of brain can be derived from this lipid by introducing a second molecule of N-acetylneuraminic acid on the terminal galactose (G$_{II}$) or on the first NANA molecule (G$_{III}$) or on both positions (G$_{IV}$). Minor gangliosides in which the terminal galactose and galactosamine are lacking and one or two NANA molecules are linked to the "middle" galactose are known (48). Thus it is obvious that one of the major differences between the gangliosides and the ceramide hexosides described above lies in the presence of sialic acid in the gangliosides.

TLC of gangliosides is widely carried out by using the system

† Abbreviations used: gal = galactose, galNAc = N-acetyl galactosamine; glu = glucose; NANA = N-acetylneuraminic acid.

chloroform–methanol–water 60:35:8 with silica gel G described by Wagner et al. *(24)*; another good system is propanol–water (7:3) with silica gel G described by Kuhn and Wiegandt *(59)*. TLC separations of the major brain gangliosides obtained with the latter system are shown in Fig. 8. G_0 in Fig. 8 refers to a monosialoganglioside that is present in small amount normally but increases very much in Tay–Sachs disease; this lipid lacks the terminal galactose of G_I; G_V could be a tetrasialoganglioside or a compound identical with the new trisialoganglioside reported by Penick and McCluer *(60)*.

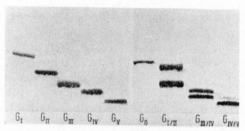

FIG. 8. Separation of gangliosides on silica gel G with propanol–water 7:3 as solvent. Running time 5 hrs, detection of spots with Ehrlich reagent. [From R. Kuhn and H. Wiegandt, *Chem. Ber.,* **96,** 866 (1963).]

As the gangliosides are acidic compounds, some basic TLC systems have also been described for their separation; thus Wherret and Cumings *(61)* substituted 2.5 N ammonia for the water of the Wagner system; Weicker et al. *(62)* again described a propanol–water system of the Kuhn type in which there was ammonia; and Klenk and Gielen *(63)* have used *n*-butanol–pyridine–water 3:2:1 on silica gel G. Acidic systems are not used, since the gangliosides are labile under acidic conditions.

The great polarity of the complex gangliosides gives them a very slow migration rate. This had led many laboratories to the use of multiple development or other techniques. Ledeen *(48)* used plates of double length (20 × 40 cm) and chloroform–methanol–2.5 N ammonia as solvent; two successive long runs with this solvent, approximately seven hours each, resulted in good resolution of the major gangliosides (Fig. 9). Eichberg et al. *(64)* used an ascending system that allowed evaporation of the solvent from the top of the plate.

Another solution is the descending system of Korey and Gonatas (65), in which the solvent is applied to the top of the plate by means of filter paper.

It appears that types of gangliosides other than those directly related to the main molecular types of brain may occur. Thus Kuhn and Wiegandt (66) have isolated from beef erythrocytes a ganglioside

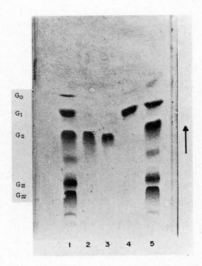

FIG. 9. TLC of gangliosides. Silica gel G plates of double length (40 cm); multiple development by ascending technique with chloroform–methanol–2.5 N ammonia 60:40:9; detection with resorcinol. 1 and 5, ganglioside mixture of normal human brain; 2, ganglioside of structure NANA(2 → 8) NANA(2 → 3) gal(1 → 4) glu(1 → 1) cer; 3, ganglioside G_{II}; 4, ganglioside G_I. [From R. Ledeen, *J. Am. Oil Chemists' Soc.*, **43**, 57 (1966).]

that contains glucosamine and N-glycolylneuraminic acid instead of galactosamine and N-acetylneuraminic acid. With propanol–water 7:3 this lipid moves on silica gel G plates slightly behind the common monosialoganglioside G_I of brain; its R_f value is slightly less than that of G_I.

Thin-layer chromatography has been of great use in many laboratories in studies on partial hydrolysis of the different gangliosides. Kuhn and Wiegandt (59) established that sialidase splits the gangliosides G_{II}–G_V to G_I. G_V gives first G_{IV} and then G_{III}, which is converted to G_I without any G_{II} appearing in the hydrolysate. Pure G_{II}, on the other hand, is also hydrolyzed to G_I. These observations suggest that the gangliosides G_V and G_{IV} are derivatives of G_{III} and that G_{III} and G_{II} are different types of disialogangliosides.

$$G_{II} \rightarrow G_I \leftarrow G_{III} \leftarrow G_{IV} \leftarrow G_V$$

Svennerholm (*67*) studied mild acid hydrolysis of the gangliosides with the aid of TLC. G_{II}† and G_{III} gave G_I and ceramide-*N*-tetrose, and G_{IV} gave about equal amounts G_{III} and G_{II}. This showed that there is some relation also between G_{IV} and G_{II}.

It has already become apparent that the gangliosides are such a complex group of lipids that two solvent systems are occasionally needed to identify a given pure component. Thus, a disialoganglioside of the structure: with chloroform–methanol–ammonia ceramide-glu-gal-NANA-NANA‡ runs together with G_{II} (Fig. 9), whereas with propanol–water it runs together with G_I. Accordingly, McCluer's group (*68*) has recently advocated the use of four different solvents, two of which are neutral and two ammoniacal, for TLC of gangliosides.

D. TLC of Polar Plant Lipids

The phosphatides and glycolipids of plants are significantly different from those of animal origin; the plant extracts appear to contain the same types of glycerolipids as the animal tissues, but in addition the plants contain large amounts of monogalactosyl diglyceride, digalactosyl diglyceride, sulfolipid, and phosphatidyl glycerol. These compounds, together with different sterol glycosides and sphingoglycolipids, make the extracts of plant polar lipids often quite complex.

Probably the first really successful TLC separation of the polar plant lipids was reported by Nichols (*19*), who studied several different solvents on silica gel G. He used first the neutral chloroform–methanol–water system and its ammoniacal modification familiar from the studies of animal tissues; but he also applied to the TLC work the acidic solvent system developed by Marinetti and co-workers (*69*) for paper chromatography of phosphatides on silicic-acid-impregnated

---

† The symbols of Kuhn and Wiegandt are used.

‡ glu = glucose; gal = galactose; NANA = *N*-acetylneuraminic acid.

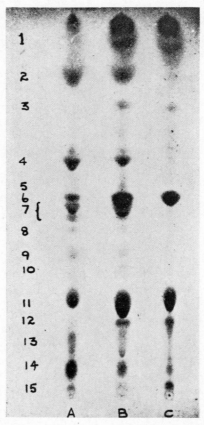

FIG. 10. Separation of phosphatides and glycolipids of lettuce stalk (A), whole lettuce leaf (B), and lettuce leaf chloroplasts (C). Key to numbered components: 1, pigments and neutral lipids; 2, sterols; 3, pigments; 4, sterol glycoside; 5, phosphatidic acid; 6, monogalactolipid; 7, sterol glycoside; 8, unknown; 9, phosphatidyl ethanolamine; 10, phosphatidyl glycerol; 11, digalactolipid; 12, lipid A; 13, phosphatidyl choline; 14, phosphatidyl inositol, and lipid B; 15, unknown. [From B. W. Nichols, *Biochim. Biophys. Acta,* **70,** 417 (1963).]

papers. This mixture, diisobutyl ketone–acetic acid–water 40:25:3.7, gave an excellent separation of at least 15 plant lipids on silica gel G (Fig. 10). Later Nichols described a two-dimensional TLC technique (7) in which a basic and acidic solvent are used in succession (Fig.

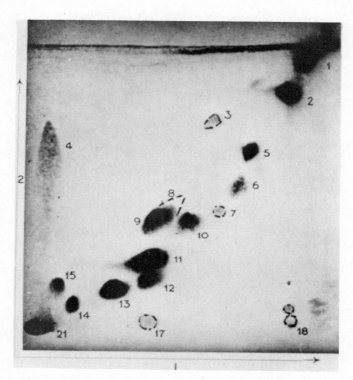

FIG. 11. Two-dimensional chromatography of lettuce leaf lipids. Mobile phase: first direction, chloroform–methanol–7 N ammonium hydroxide 65:30:4; second direction, chloroform–methanol–acetic acid–water 170:25:25:6. Spots detected by charring. Identity of components: 1, neutral lipid and pigment; 2, monogalactosyl glycerol lipid; 3, polyglycerophosphatide; 4, phosphatidic acid; 5, sterol glycoside; 6, cerebroside; 7, unidentified; 8, unidentified; 9, phosphatidyl ethanolamine; 10, phosphatidyl glycerol; 11, digalactosyl glycerol lipid; 12, sulfolipid; 13, phosphatidyl choline; 14, phosphatidyl inositol; 15, unidentified; 17 and 18, decomposition products of galactosyl–glycerol lipids; 21, unidentified. [From B. W. Nichols, in *New Biochemical Separations* (A. T. James and L. J. Morris, eds.), Van Nostrand, Princeton, N.J., 1964, p. 334.]

11). A similar system was also described by Lepage (70), who combined the neutral Lea–Wagner solvent with the acidic Marinetti–Nichols mixture into a two-dimensional system that gave good resolution of potato tuber extracts. The most striking difference between the use of neutral and ammoniacal chloroform–methanol–water mixtures is in the position of phosphatidic acid. With the neutral solvent this lipid migrates with a R_f value of 0.74 (70), whereas with the ammoniacal solvent it is almost stationary (7). Haverkate and van Deenen (71) have also used the two-dimensional technique in their analysis of spinach lipids. These authors isolated phosphatidyl glycerol and studied its properties very carefully. An example of their work

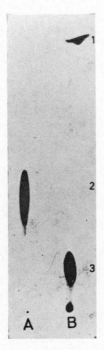

FIG. 12. TLC of enzymatic hydrolysis of phosphatidyl glycerol with phospholipase A. Silica gel G plate; development with chloroform–methanol–acetic acid 70:20:2. The spots are 1, fatty acids; 2, phosphatidyl glycerol; and 3, lysophosphatidyl glycerol. [From F. Haverkate and L. L. M. van Deenen, *Biochim. Biophys. Acta,* **106,** 78 (1965).]

FIG. 13. Subfractionation of intact phosphatidyl glycerol of spinach leaves (A) on silica gel G that contains silver nitrate. B, phosphatidyl glycerol containing fully saturated fatty acids. Solvent system: chloroform–ethanol–water 65:30:3.5 [From F. Haverkate and L. L. M. van Deenen, *Biochim. Biophys. Acta,* **106,** 78 (1965).]

is shown in Fig. 12—the hydrolysis of phosphatidyl glycerol by phospholipase A as revealed by TLC.

In view of the large number of terpenoid substances found in plants, it is quite understandable that in most reports on plant lipid TLC there are relatively many unknown or only partially identified spots. For instance, the group of cardiac glycosides present in *Strophanthus* species forms a very complex family of plant "polar lipids." But also the more common polar lipids present problems not yet completely solved. For instance, the acidic compounds, phosphatidic acid, phosphatidyl glycerol, phosphatidyl glycerophosphate, and diphosphatidyl glycerol (= cardiolipin), together with the corresponding lysoforms, form a complex group of substances. Their separation has been studied notably by Faure et al. (*72,73*) in connection with cardiolipin breakdown.

One possible reason for the confusion in this area might be found in phosphatidyl transferase activity detected by Benson et al. (*74*) in plant vascular tissues. These authors found that, besides hydrolysis, phospholipase D of plants catalyzed alcoholysis of lecithin and other phosphatidyl lipids. The danger of this transfer reaction, which leads to phosphatidyl methanol, phosphatidyl ethanol, and analogous acidic lipids, appears to be present in several currently used extraction procedures. Another difficulty that might cause a lot of trouble with these acidic lipids is the question of the cations. As already discussed above, there are dramatic differences in the partition coefficients of the sodium and calcium salts of phosphatidyl inositol. This explains why the chromatographic mobility of phosphatidyl inositol is affected by the cations. The behavior of phosphatidyl serine and phosphatidyl inositol diphosphate also depends upon the cations present (*9,75,76*), and it seems reasonable to expect this to be the case with the phosphatidic acids and the polyglycerophosphatides also.

IV. Analysis of Molecular Species within Polar Lipid Classes

The samples of different polar lipid classes as obtained from natural sources always contain several different fatty chains; hence they represent complicated mixtures of several molecular species. For instance, a lecithin preparation containing only two fatty acids A and B could be composed of four different lecithin species, AA, AB, BA, and BB.

Quite obviously, the analysis of the composition of any such mixture requires further fractionation work; the molecular species within each class of polar lipids must be separated from each other. As regards the neutral lipids, this problem has been rather successfully studied, but with the polar lipids the progress has been slow until quite recently.

A. TLC of Molecular Species of Intact Polar Lipids

The separation of sphingomyelins, cerebrosides, and lysolecithins into different subfractions has been discussed above. In these examples

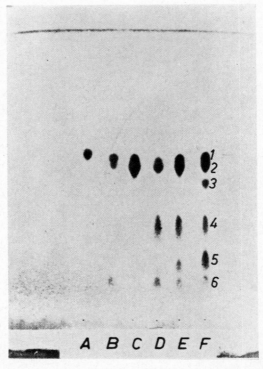

FIG. 14. TLC of lecithins on silica gel H (E. Merck) impregnated with silver nitrate and heated for 5 hr at 175°C. Development, chloroform–methanol–water 65:25:4; detection by charring. A, hydrogenated egg lecithin; B, egg lecithin; C, rabbit liver lecithin; D, rat liver lecithin; E, pig liver lecithin; F, bovine liver lecithin. [From G. A. E. Arvidson, *J. Lipid Res.*, **6**, 574 (1965).]

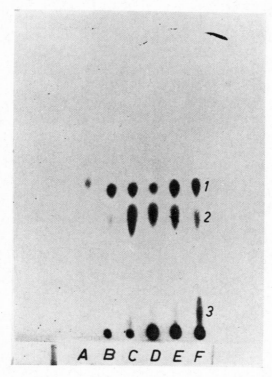

FIG. 15. TLC of lecithins on silica gel H (E. Merck) impregnated with silver nitrate and heated for 24 hr at 180°C. Solvent system, samples, and detection method as in Fig. 14. [From G. A. E. Arvidson, *J. Lipid Res.*, **6**, 574 (1965).]

the separation obtained depended on hydroxyl substituents or differences in the length of the lipid chains.

Other attempts to fractionate molecular species of polar lipids have utilized differences in the unsaturation of the lipid chains. Inouye and Noda *(78)* separated mercuric acetate adducts of intact egg lecithins with reversed-phase partition. But it is by no means necessary to resort only to the slightly cumbersome partition technique when using mercuration. For instance, Blank et al. *(78a)* reported quite recently the separation of fully saturated lecithins from all other lecithin species by using adsorption techniques, i.e., ordinary silica gel G plates and chloroform–methanol–water 70:30:4 after mercuration. Argenta-

TABLE 2

FATTY ACID COMPOSITION OF THE LECITHIN FRACTIONS SHOWN IN FIGS. 14 AND 15

Pig liver

Fatty acid	Total	1[a]	2	4	5	6
16:0	24.8	30.9	23.7	16.7	20.2	25.5
16:1	1.3	1.0	0.8	1.2	1.7	2.0
18:0	28.8	22.1	25.7	36.5	30.3	24.2
18:1	14.7	46.0	6.4	1.6	2.1	3.3
18:2	13.3	—	43.5	—	2.4	—
18:3	Trace	—	—	—	—	—
20:3[b]	1.5	—	—	—	—	—
20:4	10.9	—	—	44.0	4.5	—
20:5[b]	1.2	—	—	—	9.1	—
22:5[b]	2.2	—	—	—	29.6	—
22:6	1.3	—	—	—	—	45.1
Unidentified	—	—	—	—	—	—

Bovine liver

Fatty acid	Total	1[a]	2	3	4	5	6
16:0	16.1	23.4	12.7	5.3	11.7	12.6	15.6
16:1	2.4	1.8	1.9	1.1	1.3	1.6	2.0
18:0	35.3	27.0	32.8	39.9	42.3	38.1	38.2
18:1	16.6	47.8	6.5	7.8	2.8	3.1	3.5
18:2	9.2	—	46.1	6.4	—	—	—
18:3	1.3	—	—	—	3.1	—	—
20:3[b]	4.4	—	—	39.6	—	—	—
20:4	5.9	—	—	—	32.1	3.4	—
20:5[b]	1.2	—	—	—	—	—	6.0
22:5[b]	5.8	—	—	—	—	41.2	10.1
22:6	0.9	—	—	—	6.6	—	24.6
Unidentified	1.0	—	—	—	—	—	—

[a] Numbers 1–6 refer to spot number in Figs. 14 and 15.
[b] Tentative identification.

tion chromatography (*21,28a*) has proved even more practical than the mercuration methods in separations of lipophilic materials according to number, type, and location of their unsaturated centers. Kaufmann et al. (*79*) were the first to report separation of lecithins on silver nitrate containing silica gel G. They claimed the separation of egg and soya lecithins into several components with a solvent mixture containing chloroform, ether, and acetic acid. Some difficulties appear to be inherent in this, because Pelick et al. (*80*) and Morris (*28a*) could not reproduce Kaufmann's work. However, Haverkate and van Deenen (*71*) were partially successful with intact phosphatidyl glycerol (Fig. 13), and Arvidson (*81*) reported quite successful procedures for the separation of intact lecithins on silver-nitrate-containing plates. He prepared the plates by mixing 40 g silica gel H,† 12 g AgNO₃, and 112 ml of water; the plates of 0.35 mm thickness were dried for 24 hr at 20°C and then activated by heating for 5 hr at 175°C or 24 hr at 180°C. The developing solvent was chloroform–methanol–water 65:25:4. The separations obtained between different molecular species of lecithin are shown in Figs. 14 and 15. The less activated plates separated well the components 3–6, which contained from three to six double bonds, whereas the components 1–2, which were monoenes and dienes, respectively, were fractionated on the more active plates. For instance, pig and bovine liver lecithins (lanes E and F in Figs. 14 and 15) were separated into six different spots. The fatty acid analysis of these fractions (Table 2) revealed that approximately equal amounts of saturated and unsaturated acids were present in each spot and that the number of double bonds in the unsaturated chains increased with decreasing mobility of the lecithins.

B. TLC of Molecular Species of Chemically Modified Polar Lipids

Privett and Blank (*82*) have reported a method of analysis that allows the separation and quantitative estimation of the following four subtypes of lecithins:‡

$$S-\begin{matrix} -S \\ -OPN \end{matrix} \qquad U-\begin{matrix} -S \\ -OPN \end{matrix} \qquad S-\begin{matrix} -U \\ -OPN \end{matrix} \qquad U-\begin{matrix} -U \\ -OPN \end{matrix}$$

(1)   (2)   (3)   (4)

† E. Merck, Darmstadt, Germany. This adsorbent does not contain gypsum.
‡ S represents saturated and U unsaturated fatty acids.

In their procedure the lecithin mixture is subjected to reductive ozonolysis whereby the saturated molecules (**1**) remain intact and the unsaturated ones yield modified lecithins ("aldehyde cores") in which the unsaturated chains, still fixed on glycerol, are converted to pieces whose length corresponds to the carbon chain between the carbonyl group and first double bond of the chain. The "outer" end of the ozonolyzed chain is converted to a formyl group in the process. The three different aldehydic cores resulting from the lecithin types (**2**), (**3**) and (**4**) can be separated from each other as well as from the intact saturated lecithins by reversed-phase TLC. For this purpose the plates are prepared in the normal way from silica gel G and then impregnated with the silicone (Dow Corning 200 fluid; viscosity 10 cs) by immersing them into a 5% solution of the silicone in ether. Some laboratories impregnate the plates for reversed phase TLC by a chromatographic technique, which means simply that the silica gel G plates are put into TLC chambers and the appropriate solution of silicone oil is allowed to ascend onto the plates. The actual developing solvent used by Privett and Blank was acetic acid–water 85:15 saturated or almost saturated with the silicone. The developed plates were charred with sulfuric acid and subjected to densitometry; typical results are shown in Fig. 16. Wurster and Copenhaver (*82a*) have applied this process successfully to synthetic model compounds and adrenal lecithins.

A totally different approach has been studied in our laboratory. We convert the glycerophosphatides into diglyceride acetates and fractionate these molecules with all common methods of triglyceride chemistry (*83–85*). This approach has some drawbacks in comparison to the procedures of Arvidson and Privett—the polar part of the molecules that is useful in many biochemical studies is lost—but several significant advantages are inherent in our method.

The approach was based on the idea that the phosphatides have intermolecularionic forces in native form, but after elimination of the polar part of the molecules these forces are removed. This change has remarkable implications in the fractionation. If two lecithin species A and B are to be separated in a TLC system where there are forces that tend to retain specifically the A molecules, the B molecules will also be retained if the two species form "aggregates." In other words, the specificity of the system is thus more or less lost because of the strong intermolecular forces. If, on the other hand, two species of diglyceride acetates A′ and B′ are to be separated in a similar system

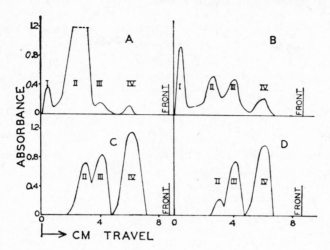

FIG. 16. Densitometer curves of saturated lecithins (I) and three "aldehydic cores" obtained from different types of unsaturated lecithins. Reversed phase, liquid–liquid TLC; stationary phase, silicone; mobile phase, acetic acid–water 85:15. A, egg; B, beef spinal cord; C, soybean; D, wheat germ lecithins. II, "core" from α-saturated-β-unsaturated; III, core from α-unsaturated-β-saturated; IV, core from α-unsaturated-β-unsaturated lecithin. [From O. S. Privett and M. L. Blank, *J. Am. Oil Chemists' Soc.,* **40,** 70 (1963).]

and A′ is to be retained, then B′ will not be retained if the intermolecular forces are so weak that no aggregates are formed. In this case the specificity of the system will not be lost.

This rather naive way of thinking has indeed led us to some remarkable TLC separations. We have, for instance, separated the three subclasses of diglyceride acetates which obtained from phosphatidyl compounds (**5**), corresponding plasmalogens (**6**), and corresponding alkyl-acyl phosphatides (**7**).

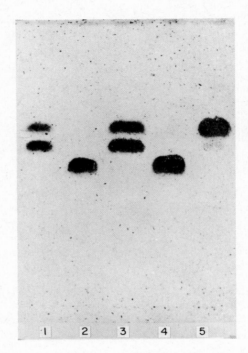

FIG. 17. Separation of three different subtypes of diglyceride acetates on silica gel G. Dual development. First run (7 cm) with hexane–ether 1:1; second run (15 cm, in the same direction) with toluene. Detection by charring. 1, Diglyceride acetate derived from choline plasmalogen (upper spot) and from alkyl acyl lecithin (lower spot). 2, Ordinary 1,2-diglyceride acetate derived from phosphatidyl choline. 3, same as 1; 4, same as 2; 5, almost pure plasmalogenic diglyceride acetate. [From O. Renkonen, S. Liusvaara, and A. Miettinen, *Ann. Med. Exptl. Biol. Fenniae (Helsinki)*, **43**, 200 (1965).]

The three types of diglyceride acetates can be separated on silica gel G by a technique of multiple development (*31*). With hexane–ether mixtures as solvent the diglyceride acetates derived from (**6**) and (**7**) run together, moving well ahead of the diglyceride acetates of type (**5**). A second run with an aromatic solvent, for instance toluene, separates type (**6**) and (**7**) diglyceride acetates, the plasmalogenic derivative (type **6**) running faster than the alkyl-acyl derivative [type (**7**)]. As this aromatic solvent does not separate type (**5**) and type (**7**) diglyceride acetates too well, a technique of multiple development was needed. When the plates are first developed with hexane–ether 1:1 and then with toluene, the separations obtained are as shown in Fig. 17.

Each of the three types of diglyceride acetates has been fractionated on silver nitrate–silica gel G (*84,85,85a*). With benzene–chloroform 9:1 as solvent, the saturated monoenoic and dienoic diglyceride acetates are well separated from each other and also from the more unsaturated molecules. The latter are separated with chloroform–methanol 97:3 into an unresolved mixture of the less unsaturated molecules and well-resolved spots of trienes, tetraenes, pentaenes, and hexaenes (*88*). Quite recently we have isolated highly unsaturated diglyceride acetates derived from Baltic herring lecithins (*86*). These lecithins revealed docosahexaenoic acid in both C-1 and C-2 of glycerol, and a combination of two docosahexaenoic acids in one lecithin molecule appeared possible. The study of the corresponding diglyceride acetates revealed that such lecithins indeed exist. The solvent that separates diglyceride acetates of six to twelve double bonds on silver nitrate–silica gel G is chloroform–methanol–water 65:25:4, the "magic mixture" for polar lipid TLC. With this solvent the diglyceride acetates of 0–3 double bonds migrate with the solvent front, those of 5–7 double bonds have an R_f value in the range of 0.7, and the very unsaturated diglyceride acetates of 10–12 double bonds migrate with R_f values of about 0.25 (Fig. 18).

The selectivity of silver-nitrate-impregnated plates is sufficient to separate some isomers. In the dienoic diglyceride acetates, for instance, the two double bonds can occupy different parts of the chains and very many positional isomers are possible. However, the most commonly met dienoic molecular species of diglyceride acetates contain either two oleic acid chains or one palmitic and one linoleic acid. These two types of dienoic isomers run together on silver nitrate–silica gel G with benzene–chloroform 9:1 as solvent, whereas aliphatic solvents, like hexane–ether 1:1, separate them quite clearly (*85*). In Fig. 19 is shown how the diolein acetate type of lipid runs faster in this solvent than the isomeric diene of palmitoyl–linoleoyl type. Similar separations have recently been observed between trienes too; 1-oleoyl-2-linoleoyl acetate runs faster than diglyceride acetates containing a saturated and a trienoic fatty chain (*86*). It seems that double bonds that are close "conformational neighbors" cause less retardation than is usual on silver-nitrate-containing plates.

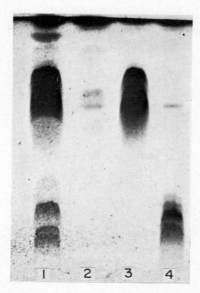

FIG. 18. TLC of highly unsaturated diglyceride acetates on silica gel G impregnated with silver nitrate. Development with chloroform–methanol–water; detection by charring. 1, Diglyceride acetates of total lecithins from Baltic herring; 2, relatively unsaturated (0–3 double bonds) diglyceride acetates derived from 1; 3, diglyceride acetates of medium unsaturation (5–7 double bonds) derived from 1; 4, diglyceride acetates of high unsaturation (10–12 double bonds) derived from 1.

Isomeric triglycerides of type OSS and SOS are separable, but analogous synthetic diglyceride acetates OSAc and SOAc have practically identical mobility on silver-nitrate-containing plates (*86a*). This shows that the double bond in the middle chain is "hindered" in SOS but "not hindered" in SOAc; the acetyl chain is simply too short to "cover" the double bond of the neighboring chain.

In addition to the examples cited above, the proper use of aromatic and aliphatic solvents in TLC of diglyceride acetates (and other lipids) may have further importance, since with aliphatic solvents the unsaturated compounds are retained on silica gel G plates in comparison with the saturated analogs, but in our experience this retardation is eliminated to a great extent when aromatic solvents are used (*86*). When lipid classes are separated from each other, it is naturally important to "shepherd" all molecules of any one class so that they

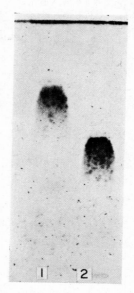

FIG. 19. TLC of dienoic diglyceride acetates on silica gel G impregnated with silver nitrate. Developing solvent hexane–ether 1:1. 1, Diglyceride acetates containing two *cis*-monoenoic acids; 2, diglyceride acetates containing one saturated and one *cis-cis*-dienoic acid. [From O. Renkonen, *Biochim. Biophys. Acta,* **125,** 288 (1966).]

stick together. In this shepherding the obvious interaction between the aromatics and the polyenes can and should be profited from.

The difference of chain length causes some separation of molecular species of diglyceride acetates on the silver-nitrate-containing plates in the same way as on the usual silica gel G plates. This is a drawback when TLC is used as a preparative method and a representative sample would be needed. We have therefore returned to the use of silver nitrate–silica gel columns, which allow easy monitoring of the fractions by GLC of the methyl esters.

Having been only partly successful in the separation of molecular species of intact phosphatidyl glycerols, Haverkate and van Deenen *(71)* converted these lipids into diglycerides that could be separated into different molecular species with great efficiency. Van Golde et al. *(87)* and, more recently, van Golde and van Deenen *(87a)* have described the analysis of liver lecithins after conversion to diglycerides.

These papers show that the separation of diglyceride species according to the degree of unsaturation is possible much in the same way as that of diglyceride acetates tritylated derivatives. Lands and Hart, too, have quite recently achieved very similar separations after conversion of the lecithins into diglycerides followed by acylation with heptadecanoyl anhydride (87b). Dyatlovitskaya et al. (87c) separated tritylated diglycerides with similar results. The quantification in this case becomes particularly easy through the UV absorption of the trityl group.

In general, TLC on silver nitrate–silica gel G does not yield pure molecular species of diglycerides, diglyceride acetates, or diglyceride heptadecanoates from natural sources; the subfractions obtained are still complex mixtures. Complementary separation is needed for further analysis. Since this complementary fractionation should preferably be based on differences in molecular size, the partition methods are tailor-made for this purpose. We have subjected various diglyceride acetates both to GLC (88) and to reversed-phase partition TLC (86). Our TLC system is of the same type as used by Privett and Blank (82) for oxidized lecithins. It consists of silica gel G plates (0.25 mm) impregnated with silicone (Dow Corning 200 fluid, 10 cs) and developed with acetic acid–water mixtures. This type of system appears to separate homologous diglyceride acetates of different unsaturation when the molecular size differs by two methylene units. Quite recently van Golde and van Deenen (88a) studied the relatively simple diglycerides derived from phosphatidyl ethanolamines of E. coli. They could separate the diglycerides on silver nitrate plates both according to number of double bonds and according to molecular size.

Carter et al. (89), Sweeley and Moscatelli (90), and Karlson (91) have provided evidence that the sphingosine bases form a heterogeneous group of substances, and it is well known that several different fatty acids are present in the sphingolipids. These lipids can be separated into molecular species quite advantageously when they are first converted into ceramide diacetates (84). However, the variation in unsaturation is far less in the sphingolipids than in the glycerolipids; the bases appear to contain only up to two double bonds, and the component acids are saturates or monoenes. Therefore the "silver nitrate-method" is of less importance with ceramide acetates than with glycerolipids. On the other hand, the sphingolipids reveal a great variation in the chain length and also in the number of hydroxyl groups; therefore, the partition and adsorption methods may prove to be exceptionally effective in the fractionation of molecular species of ceramide acetates.

The advantages of using nonpolar derivatives of polar lipids for the subfractionations appear clear from the examples given above. All these derivatives were rendered nonpolar by cleaving the polar parts away from the phosphatides. Another possibility that we have recently studied is to "mask" the polar parts, not to cleave them away. Promising methods for masking the phosphatides appear, for instance, in the papers of Collins (*91a*). TLC of phosphatidic acid dimethyl esters is illustrative, because they are the simplest possible masked phosphatides. In this very nonpolar form the plasmalogenic and ordinary phosphatidic acids are separated on silica gel G quite as are the corresponding diglyceride acetates; argentation chromatography too works admirably well with these lipids (*31a*). Partition methods are also available in a very recent paper on these lipids by Wurster and Copenhaver (*91c*).

V. Quantitative TLC of Polar Lipids

After TLC separation of the polar lipids various complementary methods including quantification procedures are easily applied. This task can be achieved either by removing the lipids from the plate or by applying proper procedures directly to the intact adsorbent layer on the plate.

A. Quantitative Methods Based on Removal of Lipids from the TLC Plates

This group of methods can be further divided into two slightly different types of approach. The first includes extraction and recovery of intact lipids before the quantitative analysis, whereas in the second type the mixture of lipid and adsorbent is subjected to analysis.

The methods belonging to the first group are the most cumbersome, but since intact lipids are recovered in pure form, several different methods of complementary analysis can be applied. As a matter of fact, this group of methods can be called as preparative thin-layer chromatography. The problems in this type of work consist of (1) localization of the spots, (2) complete recovery of the intact lipids from the adsorbent, and (3) avoidance of contamination, secondary reactions, and subsequent lipid degradation.

The localization of the lipids must be carried out without structural changes. The commonly used staining reagents are Bromthymol Blue, Rhodamine 6G and 2',7'-dichlorfluorescein, all of which can be eliminated from the lipids afterward. In some cases, especially when the aim of the work is preparative separation rather than quantitative estimation, it is advantageous to spray the plates with distilled water (*92*). When the water-moistened plate is held against a dark background, it is observed that most of the plate is translucent; however, the bands containing the separated substances are not translucent but white. This spray eliminates the removal of the dye from the lipids, but it is rather insensitive. Clear zones are sometimes obtained by first saturating the plate with water and then letting it dry until the zones are visible. The process may be repeated if necessary. This method of localization has been used for lipids, for example, by the Svennerholms (*56*). Another widely used spray in quantitative work is iodine, but as the polyenoic fatty chains (*93*) and plasmalogens (*94*) react with this agent, it should not be used for the detection whenever the fatty chains or their combinations are subsequently analyzed. In many analyses, however, the iodine does not interfere and its use may mean distinct advantages over the dyes or water. In such cases the excess of iodine is simply sublimed away before the quantitative assay.

Hydrolysis and oxidation are probably the most common among the secondary reactions that may cause trouble in this type of thin-layer work. It seems likely that the lipids are oxidized mostly during the drying period after the development. Thus it is good practice to shield the plates from light, and if possible from air, and never let them get completely dry. Use of antioxidants, for instance BHT, in the TLC solvents also appears to be a highly recommendable practice (*25a,25b*). BHT can be removed from the lipids afterward very easily when necessary; it moves on TLC faster than the methyl esters of fatty acids.

The extraction of the separated lipids can be carried out batchwise, but we have preferred a column chromatographic technique using very small tubes or injection syringes. The eluent used must be sufficiently polar; a good idea of the polarity required is obtained from the R_f of the component under study.

A good description of quantitative TLC of polar lipids by this "extraction method" is in a paper by Skipski et al. (*95*), who analyzed animal phosphatides. Their method is also applicable to glycolipids or to radioactive materials. These authors used thicker layers of silica gel than is general in qualitative analysis, since it allowed the chromatography of rather large samples and the analysis of minor components. The details of the procedure were as follows: 40 g of silica gel without calcium sulfate binder (Camag, Muttenz, Switzerland) was slurried with 90 ml of 1 mM Na$_2$CO$_3$ solution, and 0.5-mm-thick plates were prepared with the adjustable Desaga applicator. The plates were allowed to dry at room temperature for 1–2 hr, and just before the experiment they were activated at 110°C for 1 hr. The lipids were applied using 50 μl or less of solvent, and the chromatograms were developed with chloroform–methanol–acetic acid–water 25:15:4:2. The plates were air-dried at room temperature for 20 min, and the lipids were detected with iodine vapor and encircled with a fine needle.

The separation obtained with a 2-mg sample of total lipids of rat liver is shown in Fig. 20. Most of the iodine was allowed to evaporate before removal of the spots; the small amounts of iodine remaining did not interfere with the phosphorus determinations. For the removal of the spots the areas of silica gel under the origin and above the solvent front were first removed with a razor blade. Beginning with the origin of any one running lane a drop of water was placed on the silica gel area to be removed, and the adsorbent was transferred to a centrifuge tube with a spatula. Water makes the silica gel cohere, so that almost the whole spot can be transferred at the same time. To ensure complete transfer of the material, the remaining periphery of the spot was scraped off and the plate placed vertically and tapped to allow the scrapings to fall onto glazed paper; this powder was combined with the previously removed material. The areas in which there was no lipid were also removed for analysis, and thus on completion all of the silica gel was removed from the running lane. For extraction of the phosphatides the powder was suspended in 3 ml of the eluting solvent, chloroform–methanol–acetic acid–water 25:15:4:2, by gently tapping the tube. After centrifugation, the solvent was removed with a capillary pipette and the extraction was repeated once with 2 ml of

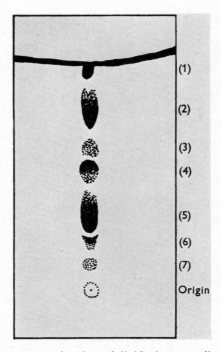

FIG. 20. TLC of a 2-mg sample of total lipids from rat liver on "basic" silica gel that does not contain binder (Camag). Developing solvent, chloroform–methanol–acetic acid–water 25:15:4:2. Detection by charring. 1, Neutral lipids and, possibly, cardiolipin and phosphatidic acid; 2, phosphatidyl ethanolamine; 3, phosphatidyl serine; 4, phosphatidyl inositol; 5, phosphatidyl choline; 6, sphingomyelin; 7, lysophosphatidyl choline. [From V. P. Skipski et al., *Biochem. J.*, **90**, 374 (1964).]

the eluting solvent, once with 2 ml of methanol, and once with 2 ml of methanol–acetic acid–water 94:1:5. All extracts were combined and assayed for phosphorus.

In Table 3 are listed the results obtained by Skipski et al. with the rat liver lipids that are shown in Fig. 20. Abramson and Blecher (*42*) have described a similar quantitative procedure for phosphatides in connection with their two-dimensional technique, and the Svenner-holms have used an essentially analogous procedure for the glycolipids of human serum (*56*).

It is clear that more rapid analyses can be carried out if it is not necessary to isolate the lipids. Numerous reports show that both phosphorus and the carbohydrate moieties of polar lipids can be assayed in the presence of silica gel, and this is also the case in the radioassay by liquid scintillation counting. A good example of glycolipid analysis of this type is described in a report of Suzuki (*96*), who analyzed ganglioside patterns in human brain. He used samples containing about 40 μg of sialic acid and developed the plates with the descending propanol–water 7:3 system of Korey and Gonatas (*65*).

TABLE 3

QUANTITATIVE ANALYSIS OF RAT LIVER PHOSPHATIDES[a]

Spot	Compounds	Average recovery of P, %
1	Phospholipids at the front (cardiolipin etc.)	5.12 ± 0.15
2	Phosphatidyl ethanolamine	25.32 ± 1.40
3	Phosphatidyl serine	3.02 ± 0.26
4	Phosphatidyl inositol	8.81 ± 0.36
5	Phosphatidyl choline	54.96 ± 1.40
6	Sphingomyelin	1.83 ± 0.07
7	Lysophosphatidyl choline	0.87 ± 0.13
8	Nonlipid phosphorus (at origin)	0.23 ± 0.08
	Total recovery	100.84 ± 3.3

[a] Details are given in the text. A total of 50.68 μg of P was applied. Results are given as means ± S.D. of four replicate experiments on lipids from five pooled livers.

The gangliosides were located with iodine vapor, and after complete sublimation of the iodine the areas of the individual gangliosides were scraped off into small centrifuge tubes. The quantitative analysis was carried out by sialic acid assay applying the Svennerholm resorcinol method (*97*) to the mixture of silica gel and gangliosides. For this, 0.5 ml of water was added to the scraped-off silica gel and the tubes were allowed to stand for a few minutes. Then 0.5 ml of resorcinol reagent was added and each tube was vigorously shaken with a vortex mixer for 1 min. The shaking is necessary to ensure maximum color development. After heating for 15 min in a boiling-water bath, the color was extracted into 1.5 ml of butyl acetate–butanol 85:15 as described by Miettinen and Takki-Luukkainen (*98*). The silica gel remained in the aqueous phase, leaving the organic phase perfectly clear even without centrifugation. The optical density was then read at 580 mμ. With a light path of 10 mm the optical density obtained with 10-μg sialic acid was about 0.200. The optical density of the reagent blank was 0.001–0.003 and that of the silical gel blank was

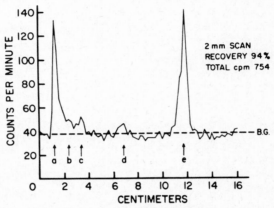

FIG. 21. Carbon-14 distribution on a TLC plate of bone marrow lipids after oral administration of palmitic acid-1-^{14}C to a rat. Peak identification: (a) phosphatides, (b) monoglycerides, (c) diglycerides, (d) free fatty acids, (e) triglycerides. The solvent front is at 15 cm. BG represents the background. [From F. Snyder, *Anal. Biochem.*, 9, 183 (1964).]

0.003–0.006. The recovery of sialic acid from the plates was checked by determination of the sialic acid content of the ganglioside mixtures used for the chromatography; the total yields were 96–102% when seven ganglioside components were assayed. The reproducibility of the procedure appears promising.

Phosphatide assay without prior elimination of the adsorbent has been reported, e.g., by Doizaki and Zieve (*99*). Radioassay of thin-layer chromatograms of lipids by direct counting in liquid scintillation solutions is an approximately analogous procedure; the lipid and the adsorbent are removed from the plate, but the lipid is not actually isolated before the assay. Snyder (*100*) used a thixotropic gel to give a suspension system for scintillation counting that can be used for both ^{14}C and ^{3}H. He showed that iodine vapor, Rhodamine 6G, and dichlorofluorescein can be used for the detection of lipids on the TLC plates that are to be subsequently radioassayed; no quenching is effected by these agents. His polar scintillation solvent consisting of dioxane, water, and naphthalene, deactivates silica gel and prevents, or at least decreases, adsorption of the polar lipids, and this limits the losses of count rate caused by self-adsorption. Mixed ^{14}C-liver lipids and mixed ^{14}C-bone marrow lipids showed in Snyder's system almost unchanged count rates (98–99%) when 200 mg of silica gel was added to the counting vial. Snyder also described a zonal scraper that rapidly and quantitatively transfers small zones of adsorbent from narrow glass plates into counting vials for assay (*100*). A TLC pattern obtained with ^{14}C-bone lipids by using this apparatus is shown in Fig. 21.

B. Quantitative Analysis on Intact TLC Plates

To maintain the leading principles of simplicity and rapidity of analysis, current TLC separations seek faster methods of quantitative evaluation than those described above. Therefore, increasing attention has lately been paid to the quantitative analysis performed on intact plates. Although quite promising results have been obtained with neutral lipids, the use of this type of analysis has been rather limited with the polar lipids.

The different techniques of analysis on intact plates include measurement of spot size (*101*), but the most widely used procedures appear to be those based on photodensitometric measurement of transmitted or reflected light (*102,103*).

In the densitometry of transmitted light some problems are caused by refracted light from the TLC plates. These can be eliminated by suitable slit adjustments in the commercial densitometers designed for paper chromatography (*104*) or by making the charred plates translucent by spraying them with, e.g., an etheral solution of mineral oil (*53*) or with methanolic glycerol (*104a*). For good analyses the carbon content of the sample and the peak areas (i.e., integrated optical density values) obtained by passage of the spot over the slit should give a linear relationship passing through the origin.

For densitometric analysis the lipids are made visible by charring them under standard conditions. The ideal charring technique would give quantitative conversion of the lipids to carbon. However, this is not easily achieved in practice. Privett and Blank (*82,105*) have shown that, when the charring is carried out at high temperature (250°C) and with a weak oxidizing agent (50% H_2SO_4), the amount of carbon formed depends on two competing processes, oxidation and evaporation. Since unsaturated compounds are oxidized much faster than their saturated counterparts, they give considerably higher yields of carbon. To limit the losses caused by evaporation, additional oxidizing agents and relatively low charring temperatures (under 200°C) are used, but these conditions again easily convert the sample partially to carbon dioxide. Blank et al. (*104*) used a saturated solution of $K_2Cr_2O_7$ in 70%, v/v, H_2SO_4 as the charring spray and heated the plates in an oven for 25 min at 180°C. In a later paper these authors working together with Rouser's group on the polar lipids reported a slight modification in which 1.2 g of $K_2Cr_2O_7$ was dissolved in 200 ml of 55% reagent grade H_2SO_4 (*106*).

Besides the degree of unsaturation, other factors influence the yields of carbon obtained by the charring reaction of different lipids. Thus it has been observed that many lipid classes give equivalent carbon yields, but this is not invariably the case; for instance, phosphatidyl ethanolamine and sphingomyelin appear to give higher yields than phosphatidyl choline (*104*). The R_f value of the lipid also affects the carbon yield obtained. As a spot migrates higher on the plate, it becomes larger; although the increase in size is compensated by decrease in density under suitable conditions ($0.3 < R_f < 0.7$) (*104*), this may not be always true. Some observations, in fact, could suggest that, with increasing distance traveled by a spot, *increasing* yields of carbon are obtained in the charring reaction (*20*).

To control the variable effects of the polar group of unsaturation, and of R_f on the carbon yield, it is necessary to run proper standards **on the same plates** as the samples under study. This is the difficulty not yet satisfactorily resolved as regards the polar lipids; the numerous reference standards needed are simply not available in sufficient purity and stability. Comparisons have been made, however, between column and TLC procedures using natural mixtures of polar lipids; the results may justify cautious optimism.

Besides charring, other detection methods can be used for the quantitative evaluation of the TLC plates. The sphingolipids, for instance, have been analyzed by densitometry after staining with the chlorox–benzidine reagent, which is specific for the amide groupings (*107*). Direct radioscanning of TLC plates has also become possible

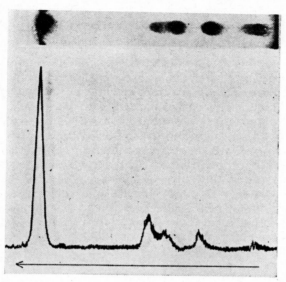

FIG. 22. TLC separation and direct radioactivity scan of lipids from *Anacystis nidulans* after incubation with (1-^{14}C) oleic acid. The plate was developed with chloroform–methanol–acetic acid–water 85:15:10:3, and visible components represent, from the origin: unknown, sulfolipid, digalactosyl diglyceride, phosphatidyl glycerol, mixed neutral lipids. [From B. W. Nichols, L. J. Morris, and A. T. James, *Brit. Med. Bull.*, **22**, 137 (1966).]

through the use of proportional gas-flow counter tubes (*108*). Nichols and co-workers (*109*) have described a scanner with a counting efficiency of about 30%; this instrument detects 100 $\mu\mu$c of activity, has a linear response, and gives integrated records of the separated components as shown in Fig. 22.

The natural mixtures of polar lipids are often too complex to be quantitatively analyzed by using TLC methods only. In such cases preliminary separations must be carried out by column chromatography, hydrolytic techniques, or otherwise; the DEAE-columns (*110*) in particular offer a valuable method to be used in conjunction with silica gel TLC in separation of lipid classes.

VI. Conclusion

The future development of polar lipid TLC will very likely go along many different lines, but we wish to draw attention to some

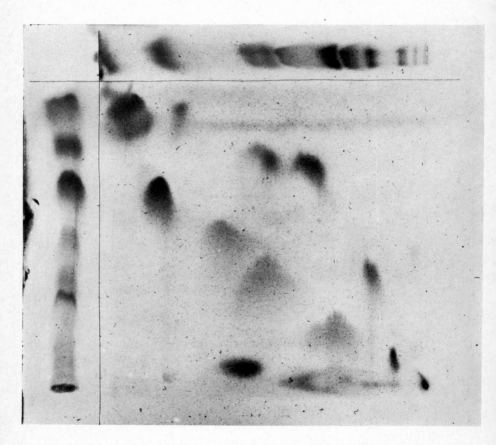

FIG. 23. The chromatographic separation of minor glycolipid and phosphatide components isolated from rat lungs. Adsorbent, silica gel H (Merck), 0.25 mm thick. Solvents: first direction, chloroform–methanol–water 65:25:4, v/v; second direction, tetrahydrofuran–methylal (dimethoxymethane)–methanol–water 50:30: 20:5 v/v. Spray, 50% sulfuric acid. For the identification of the spots see Fig. 24. With the second-direction solvent the chromatographic properties of compounds containing free hydroxyl groups may vary with different batches of silica gel H. However, separations are easily controlled by altering the ratio of methylal to methanol, and the separations shown in Fig. 23 have been consistently obtained with methylal/methanol ratios within the range 30:20 to 40:10. It is important to remove completely the first solvent from the plate (e.g., 20 min in a vacuum dessicator) before running in the second solvent.

areas of possible development where a beginning already has been
made. The problem of oxidation of the lipids during TLC has occupied
many lipid people, but not much direct experimental work has been
done. Most concerned with this problem are people who use multiple-
development, two-dimensional, or preparative TLC or analyze the
lipids for fatty chains after TLC. Such questions as how to use the

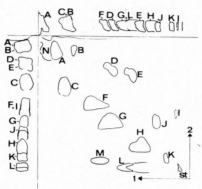

FIG. 24. A, cardiolipin; B, ceramide monohexoside; C, unknown phosphatide
(possibly phosphatidyl glycerol phosphate); D, ceramide dihexoside; E, ceramide
trihexoside; F, lyso-bis-phosphatidic acid; G, phosphatidyl glycerol; H, phosphatidyl inositide; I, unknown; J, ceramide hexosaminyl trihexoside (aminoglyco-
lipid); K, ganglioside; L, phosphatidyl serine; M, standard marker, phosphatidyl
ethanolamine; N, organic contaminants from silica gel at solvent front; St,
starting point; 1, first solvent system; 2, second solvent system.

antioxidants effectively, how to deliver unsaturated phosphatides onto
TLC plates with minimal oxidative damage, how to handle the plates
after development, whether the TLC tanks should be filled with inert
gas, and whether the trace metal ions of silica gel affect the oxidation
of lipids, are still unanswered. Satisfactory criteria for the oxidation
are badly needed; one possible solution might be obtained from the
decrease of polyunsaturated fatty chains.

Another area in which some development might be expected is direct
TLC of intact tissue slices or cell fractions. Curri et al. (*111*) have
provided a very interesting contribution by showing that normal TLC
pictures are obtained by simple development with chloroform–methanol–water 70:30:5 when histologic tissue slices are mounted on the

TLC plate and the adsorbent layer is subsequently applied and the plates are activated in vacuum. The quantitative analysis of the spots obtained from intact mitochondria agreed well with results of corresponding extracts. It is easy to see that such combination of morphologic and chemical analyses as has become possible through this type of technique will have quite interesting implications, especially when the tissue samples required become small enough. Their size again depends only on the sensitivity of the detection methods available. Neutron activation analysis, for instance, might have something to contribute in this area.

Acknowledgment

Dr. Maurice Gray, of the Lister Institute of Preventive Medicine, London, England, has kindly given permission to include in this chapter his newest work on the two-dimensional TLC separation of glycolipids and phosphatides, Figs. 23 and 24. This solvent pair shows promise of separating glycolipids and aminoglycolipids in the presence of phosphatides.

REFERENCES

1. J. Folch, M. Lees, and G. H. Sloane-Stanley, *J. Biol. Chem.*, **226**, 497 (1957).
2. E. G. Bligh and W. J. Dyer, *Can. J. Biochem. Physiol.*, **37**, 911 (1959).
3. G. J. Van Beers, H. De Iongh, and J. Boldingh, in *Essential Fatty Acids* (H. Sinclair, ed.), Butterworth, London, 1958, p. 43.
4. H. A. I. Newman, C.-T. Liu, and D. B. Zilversmit, *J. Lipid Res.*, **2**, 403 (1961).
5. D. S. Galanos and V. M. Kapoulas, *J. Lipid Res.*, **3**, 134 (1962).
6. H. E. Carter, K. Ohno, S. Nojima, C. L. Tipton, and N. Z. Stanacev, *J. Lipid Res.*, **2**, 215 (1961).
7. B. W. Nichols, in *New Biochemical Separations* (A. T. James and L. J. Morris, eds.), Van Nostrand, Princeton, N.J., 1964, p. 321.
8. H. E. Carter and E. J. Weber, *Lipids*, **1**, 16 (1966).
9. H. S. Hendrickson and C. E. Ballou, *J. Biol. Chem.*, **239**, 1369 (1964).
10. L. Svennerholm, *Acta Chem. Scand.*, **17**, 239 (1963).
11. M. A. Wells and J. C. Dittmer, *Biochemistry*, **2**, 1259 (1963).
12. A. N. Siakotos and G. Rouser, *J. Am. Oil Chemists' Soc.*, **42**, 913 (1965).
13. M. L. Vorbeck and G. V. Marinetti, *J. Lipid Res.*, **6**, 3 (1965).
14. N. S. Radin, F. B. Lavin, and J. R. Brown, *J. Biol. Chem.*, **217**, 789 (1955).
15. N. S. Radin, J. R. Brown, and F. B. Lavin, *J. Biol. Chem.*, **219**, 977 (1956)
16. G. Rouser, A. J. Bauman, G. Kritchevsky, D. Heller, and J. O'Brien, *J. Am. Oil Chemists' Soc.*, **38**, 544 (1961).
17. J. S. O'Brien and A. A. Benson, *J. Lipid Res.*, **5**, 432 (1964).
18. G. L. Feldman, L. S. Feldman, and G. Rouser, *Lipids*, **1**, 21 (1966).

19. B. W. Nichols, *Biochim. Biophys. Acta,* **70,** 417 (1963).

20. G. Rouser, G. Kritchevsky, C. Galli, and D. Heller, *J. Am. Oil Chemists' Soc.,* **42,** 215 (1965).

21. L. J. Morris, in *New Biochemical Separations* (A. T. James and L. J. Morris, eds.), Van Nostrand, Princeton, N.J., 1964, p. 295.

22. T. C. M. Schogt, in *Essential Fatty Acids* (H. Sinclair, ed.), Butterworth, London, 1958, p. 47.

23. W. Schlemmer, *Boll. Soc. Ital. Biol. Sper.,* **37,** 134 (1961).

24. H. Wagner, L. Hörhammer, and P. Wolff, *Biochem. Z.,* **334,** 175 (1961).

25. C. H. Lea, D. N. Rhodes, and R. D. Stoll, *Biochem. J.,* **60,** 353 (1955).

25a. J. J. Wren and A. D. Szczepanowska, *J. Chromatog.,* **14,** 405 (1964).

25b. T. S. Neudoerffer and C. H. Lea, *J. Chromatog.,* **21,** 138 (1966).

26. E. Borowski and T. Ziminski, *J. Chromatog.,* **23,** 480 (1966).

27. J. C. Dittmer and R. L. Lester, *J. Lipid Res.,* **5,** 126 (1964).

28. M. D. Bishel and J. H. Austin, *Biochim. Biophys. Acta,* **70,** 598 (1963).

28a. L. J. Morris, *J. Lipid Res.,* **7,** 717 (1966).

29. G. J. M. Hooghwinkel, P. Borri, and J. C. Riemersma, *Rec. Trav. Chim.,* **83,** 576 (1964).

30. O. Renkonen, *Acta Chem. Scand.,* **17,** 634 (1963).

31. O. Renkonen, S. Liusvaara, and A. Miettinen, *Ann. Med. Exptl. Fenniae (Helsinki),* **43,** 200 (1965).

31a. O. Renkonen, *Scand. J. Clin. Lab. Investig.,* in press.

31b. L. A. Horrocks and G. B. Ansell, *Biochim. Biophys. Acta,* **137,** 90 (1967).

32. P. D. S. Wood and S. Holton, *Proc. Soc. Exptl. Biol. Med.,* **115,** 990 (1964).

33. O. Renkonen, unpublished observations, 1966.

34. V. P. Skipski, R. F. Peterson, and M. Barclay, *J. Lipid Res.,* **3,** 467 (1962).

35. E. L. Gottfried and M. M. Rapport, *J. Lipid Res.,* **4,** 57 (1963).

35a. G. H. de Haas and L. L. M. van Deenen, *Biochim. Biophys. Acta,* **106,** 315 (1965).

36. L. A. Horrocks, *J. Am. Oil Chemists' Soc.,* **40,** 235 (1963).

37. W. D. Skidmore and C. Entenman, *J. Lipid Res.,* **3,** 471 (1962).

38. H. G. Müldner, J. R. Wherret, and J. N. Cumings, *J. Neurochem.,* **9,** 607 (1962).

38a. C. M. Redman and R. W. Keenan, *J. Chromatog.,* **15,** 180 (1964).

39. G. Rouser, J. O'Brien, and D. Heller, *J. Am. Oil Chemists' Soc.,* **38,** 14 (1961).

40. V. P. Skipski, R. F. Peterson, J. Sanders, and M. Barclay, *J. Lipid Res.,* **4,** 227 (1963).

40a. V. P. Skipski, M. Barclay, E. S. Reichman, and J. S. Good, *Biochim. Biophys. Acta,* **137,** 80 (1967).

41. A. F. Hofmann, in *Biochemical Problems of Lipids* (A. C. Frazer, ed.), Elsevier, Amsterdam, 1963, p. 1.

42. D. Abramson and M. Blecher, *J. Lipid Res.,* **5,** 628 (1964).

43. E. Mårtenson, *Acta Chem. Scand.,* **17,** 1174 (1963).

44. H. E. Carter, R. A. Hendry, S. Nojima, N. Z. Stanacev, and K. Ohno, *J. Biol. Chem.,* **236,** 1912 (1961).

45. P. S. Sastry and M. Kates, *Biochim. Biophys. Acta,* **84,** 231 (1964).

46. H. Wagner and W. Zofcsic, *Biochem. Z.,* **344,** 314 (1966).

47. H. E. Carter, P. Johnson, and E. J. Weber, *Ann. Rev. Biochem.* **34,** 109 (1965).

48. R. Ledeen, *J. Am. Oil Chemists' Soc.,* **43,** 57 (1966).

49. J. S. O'Brien, D. L. Fillerup, and J. F. Mead, *J. Lipid Res.,* **5,** 109 (1964).

50. E. Mehl and H. Jatzkewitz, *Z. Physiol. Chem.,* **339,** 260 (1964).

51. G. V. Marinetti, T. Ford, and E. Stotz, *J. Lipid Res.,* **1,** 203 (1960).

52. O. M. Young and J. N. Kanfer, *J. Chromatog.,* **19,** 611 (1965).

53. A. E. Thomas, J. E. Scharoun, and H. Ralston, *J. Am. Oil Chemists' Soc.,* **42,** 789, (1965).

54. C. F. Allen, P. Good, H. F. Davis, P. Chisum, and S. D. Fowler, *J. Am. Oil Chemists' Soc.,* **43,** 221 (1966).

55. M. M. Rapport, L. Graf, V. P. Skipski, and N. F. Alonzo, *Cancer,* **12,** 438 (1959).

56. E. Svennerholm and L. Svennerholm, *Biochim. Biophys. Acta,* **70,** 432 (1963).

57. G. M. Gray, *Nature,* **207,** 505 (1965).

57a. D. Shapiro and E. S. Rachaman, *Nature,* **201,** 878 (1964).

57b. D. Shapiro, E. S. Rachaman, Y. Rabinson, and A. Diver-Haber, *Chem. Phys. Lipids,* **1,** 54 (1966).

57c. H. M. Flowers, *Carbohydrate Res.,* **2,** 188 (1966).

57d. H. M. Flowers, *Carbohydrate Res.,* **2,** 371 (1966).

58. M. M. Rapport, *J. Lipid Res.,* **2,** 25 (1961).

59. R. Kuhn and H. Wiegandt, *Ber.,* **96,** 866 (1963).

60. R. J. Penick and R. H. McCluer, *Biochim. Biophys. Acta,* **106,** 435 (1965).

61. J. R. Wherret and J. N. Cumings, *Biochem. J.,* **86,** 378 (1963).

62. H. Weicker, J. Dain, G. Schmidt, and S. J. Thannhauser, *Federation Proc.,* **19,** 219 (1960).

63. E. Klenk and W. Gielen, *Z. Physiol. Chem.,* **323,** 126 (1961).

64. J. Eichberg, Jr., V. P. Whittaker, and R. M. C. Dawson, *Biochem. J.,* **92,** 91 (1964).

65. S. R. Korey and J. Gonatas, *Life Sci.,* **1,** 296 (1963).

66. R. Kuhn and H. Wiegandt, *Z. Naturforsch.,* **19b,** 80 (1964).

67. L. Svennerholm, *J. Neurochem.,* **10,** 613 (1963).

68. R. J. Penick, M. H. Meisler, and R. H. McCluer, *Biochim. Biophys. Acta,* **116,** 279 (1966).

69. G. V. Marinetti, J. Erbland, and J. Kochen, *Federation Proc.,* **16,** 837 (1957).

70. M. Lepage, *J. Chromatog.,* **13,** 99 (1964).

71. F. Haverkate and L. L. M. van Deenen, *Biochim. Biophys. Acta,* **106,** 78 (1965).

72. M. Faure and M.-J. Coulon-Morelec, *Ann. Inst. Pasteur,* **104,** 246 (1963).

73. M.-J. Coulon-Morelec, M. Faure, and J. Marechal, *Bull. Soc. Chim. Biol.,* **44,** 171 (1962).

74. A. A. Benson, S. Freer, and S. F. Yang, in *IXth International Conference of the Biochemistry of Lipids,* Noordwijk, Netherlands, 1965.

75. G. V. Marinetti, J. Erbland, and E. Stotz, *Biochim. Biophys. Acta,* **30,** 41 (1958).
76. L. Rathbone, *Biochem. J.,* **85,** 461 (1962).
77. N. A. Baumann, P.-O. Hagen, and H. Goldfine, *J. Biol. Chem.,* **240,** 1559 (1965).
78. Y. Inouye and M. Noda, *Arch. Biochem. Biophys.,* **76,** 271 (1958).
78a. M. L. Blank, L. J. Nutter, and O. S. Privett, *Lipids,* **1,** 132 (1966).
79. H. P. Kaufmann, H. Wessels, and C. Bondopadhyaya, *Fette Seifen Anstrichsmittel,* **65,** 543 (1963).
80. N. Pelick, T. L. Wilson, M. E. Miller, F. M. Angeloni, and J. M. Steim, *J. Am. Oil Chemists' Soc.,* **42,** 393 (1965).
81. G. A. E. Arvidson, *J. Lipid Res.,* **6,** 574 (1965).
82. O. S. Privett and M. L. Blank, *J. Am. Oil. Chemists' Soc.,* **40,** 70 (1963).
82a. C. F. Wurster and J. H. Copenhaver, Jr., *Biochim. Biophys. Acta,* **98,** 351 (1965).
83. O. Renkonen, *Acta Chem. Scand.,* **18,** 271 (1964).
84. O. Renkonen, *J. Am. Oil Chemists' Soc.,* **42,** 298 (1965).
85. O. Renkonen, *Biochim. Biophys. Acta,* **125,** 288 (1966).
85a. J. Elovson, *Biochim. Biophys. Acta,* **106,** 480 (1965).
86. O. Renkonen, unpublished observations, 1966.
86a. O. Renkonen and L. Rikkinen, unpublished observations, 1966.
87. L. M. G. van Golde, R. F. A. Zwaal, and L. L. M. van Deenen, *Koninkl. Ned. Akad. Wetenschap., Proc.,* **B68,** 255 (1965).
87a. L. M. G. van Golde and L. L. M. van Deenen, *Biochim. Biophys. Acta,* **125,** 496 (1966).
87b. W. E. M. Lands and P. Hart, *J. Am. Oil Chemists' Soc.,* **43,** 290 (1966).
87c. E. V. Dyatlovitskaya, V. E. Volkova, and L. D. Bergelson, *Bull. Acad. Sci. USSR, Div. Chem. Sci., English Transl.,* **1966,** 946.
88. O. Renkonen, *Ann. Med. Exptl. Fenniae (Helsinki),* **44,** 356 (1966).
88a. L. M. G. van Golde and L. L. M. van Deenen, *Chem. Phys. Lipids,* **1,** 157 (1967).
89. H. E. Carter, W. D. Colmer, W. E. M. Lands, K. L. Mueller, and H. H. Tomizava, *J. Biol. Chem.,* **206,** 613 (1954).
90. C. C. Sweeley and E. A. Moscatelli, *J. Lipid Res.,* **1,** 40 (1959).
91. K. Karlson, *Acta Chem. Scand.,* **19,** 2423 (1965).
91a. C. F. Wurster, Jr., and J. H. Copenhaver, Jr., *Lipids,* **1,** 422 (1966).
91b. O. Renkonen, *Acta Chem. Scand.,* in press.
92. R. J. Gritter and R. J. Albers, *J. Chromatog.,* **9,** 392 (1962).
93. M. Z. Nichaman, C. C. Sweeley, N. M. Oldham, and R. E. Olson, *J. Lipid Res.,* **4,** 484 (1963).
94. E. L. Gottfried and M. M. Rapport, *J. Biol. Chem.,* **237,** 329 (1962).
95. V. P. Skipski, R. F. Peterson, and M. Barclay, *Biochem. J.,* **90,** 374 (1964).
96. K. Suzuki, *Life Sci.,* **3,** 1227 (1964).
97. L. Svennerholm, *Biochim. Biophys. Acta,* **24,** 604 (1957).
98. T. A. Miettinen and I.-T. Takki-Luukkainen, *Acta Chem. Scand.,* **13,** 856 (1959).

99. W. M. Doizaki and L. Zieve, *Proc. Soc. Exptl. Biol. Med.,* **113,** 91 (1963).

100. F. Snyder, *Anal. Biochem.,* **9,** 183 (1964).

101. G. Schlierf and P. Wood, *J. Lipid Res.,* **6,** 317 (1965).

102. O. S. Privett, M. L. Blank, and W. O. Lundberg, *J. Am. Oil Chemists' Soc.,* **38,** 312 (1961).

103. C. B. Barrett, M. S. J. Dallas, and F. B. Padley, *J. Am. Oil Chemists' Soc.,* **40,** 580 (1963).

104. M. L. Blank, J. A. Schmit, and O. S. Privett, *J. Am. Oil Chemists' Soc.,* **41,** 371 (1964).

104a. O. S. Privett, M. L. Blank, D. W. Codding, and E. C. Nickell, *J. Am. Oil Chemists' Soc.,* **42,** 381 (1965).

105. O. S. Privett and M. L. Blank, *J. Am. Oil Chemists' Soc.,* **39,** 520 (1962).

106. G. Rouser, C. Galli, E. Lieber, M. L. Blank, and O. S. Privett, *J. Am. Oil Chemists' Soc.,* **41,** 836 (1964).

107. J. H. Austin, *J. Neurochem.,* **10,** 921 (1963).

108 P. F. Wilde, *Lab. Pract.,* **13,** 741 (1964).

109. B. W. Nichols, L. J. Morris, and A. T. James, *Brit. Med. Bull.,* **22,** 137 (1966).

110. G. Rouser, G. Kritchevsky, D. Heller, and E. Lieber, *J. Am. Oil Chemists' Soc.,* **40,** 425 (1963).

111. S. B. Curri, C. R. Rossi, and L. Sartorelli, in *Thin Layer Chromatography* (G. Marini-Bettolo, ed.), Elsevier, Amsterdam, 1964, p. 174.

RECENT REVIEWS

B. W. Nichols, in *New Biochemical Separations* (A. T. James and L. J. Morris, eds.), Van Nostrand, Princeton, N.J., 1964, p. 321.

O. S. Privett, M. L. Blank, D. W. Codding, and E. C. Nickell, *J. Am. Oil Chemists' Soc.,* **42,** 381 (1965).

B. W. Nichols, L. J. Morris, and A. T. James, *Brit. Med. Bull.,* **22,** 137 (1966).

L. J. Morris, *J. Lipid Res.,* **7,** 717 (1966).

D. C. Malins, in *Progress in the Chemistry of Fats and Other Lipids,* Vol. VIII, P. 3 (R. T. Holman, ed.), Pergamon Press, Oxford, 1966.

R. Ledeen, *J. Am. Oil Chemists' Soc.,* **43,** 57 (1966).

3

COLUMN CHROMATOGRAPHIC AND ASSOCIATED PROCEDURES FOR SEPARATION AND DETERMINATION OF PHOSPHATIDES AND GLYCOLIPIDS

George Rouser, Gene Kritchevsky, and Akira Yamamoto

DEPARTMENT OF BIOCHEMISTRY
CITY OF HOPE MEDICAL CENTER
DUARTE, CALIFORNIA

I. Introduction

A. General Comments

The purpose of this chapter is to present in detail the procedures for column chromatographic isolation and quantitative analysis of the polar lipids (phosphatides and glycolipids). Coverage has been limited in three major ways. First, no attempt was made to present a review of the literature. In general, only references pertaining directly to the procedures described are included. Second, only procedures used extensively in our laboratory are included, because assessments of advantages and limitations of procedures are derived only from routine application of procedures to many problems. Third, precise quantitative procedures are stressed. Procedures for isolation have been limited in general to those also suitable for quantitative analysis. With experience, the weight recovery of the sample applied should be $100 \pm 1\%$ or better and individual column fractions should vary by no more than $\pm 2\%$ of the mean value.

Descriptions of column chromatographic procedures are accompanied by descriptions of important steps prior to and after the column run. Without these associated procedures, the column chromatographic methods can not be applied successfully.

It is not always possible to describe a detailed procedure that can be followed without modification for lipids of animals, plants, and microorganisms. It is frequently necessary to present variations of a procedure for different applications. At times the general approach to be followed for modification of a procedure can be specified. This is the case when different batches of adsorbent, particularly from dif-

ferent suppliers, have significantly different chromatographic properties. In such cases, only a general procedure can be presented with a
discussion of the nature of the variability and its effects.

The reader is cautioned to avoid some common mistakes. The first
is to change a procedure in an apparently minor way only to find that
the results are not satisfactory. Procedures for a specified lipid mixture
should be followed as closely as possible. Second, there is a great
tendency to overload columns, not infrequently by a factor of ten,
with loss of resolution. Complete analytical separations of complex
mixtures of lipids can be obtained only with relatively low loads in
most cases. Nonquantitative preparative work must be distinguished
from analytical chromatography. Third, it is hazardous to assume
that a procedure described for one lipid mixture is applicable to other
lipid mixtures without modification. Procedures should be considered
only as guides where actual application to a specific problem has not
been reported. Modifications may be essential for optimum results.

Procedures presently available for quantitative analysis of polar
lipid composition have some rather well defined limitations. There are
no completely satisfactory procedures available for precise determination of all the individual types of gangliosides found in animal organs,
although major types can be determined in some cases. Methods for
the polyphosphoinositides (found in particular in brain) are not completely satisfactory. These lipids undergo rapid breakdown after death,
freezing in situ apparently being essential for their preservation.

Lipid extracts may contain polar pigments that are commonly excluded from analysis. Steroid glucuronides (a type of glycolipid) and
steroid sulfates occur in urine and may be present in extracts from
other sources. Satisfactory methods for the separation and determination of the intact conjugates are not available, although procedures
have been described for quantitative analysis of steroids by gas phase
chromatography after cleavage of the conjugates.

The individual molecular species of all lipid classes can not be
determined at present. Molecular species differ in length, degree of
unsaturation, and mode of attachment (ester, vinyl ether, ether) of
hydrocarbon chains. With the procedures described below, all molecular species of one lipid class are recovered together, although partial
separation on the basis of degree of unsaturation and length of hydrocarbon chains is sometimes obtained. The separations are not suitable
for quantitative analysis, and pure fractions of one molecular species

are not obtained. Determination of fatty acid and fatty aldehyde composition of a lipid class is presently the only measure of the types of molecular species. Because different portions of a peak from a column usually differ somewhat in composition, it is essential to collect all of a peak, including any leading and trailing portions, for accurate analysis of fatty acid composition.

Skill and experience are required for precise quantitative analysis of polar lipids. Initial applications of even well-standardized procedures may not always be successful, particularly if the details of a procedure are not followed carefully.

B. Column Dimensions and Sample Sizes

Stepwise elution of columns with increasing amounts of a more polar solvent in a less polar solvent may give rise to peaks with long trailing portions resulting in overlap of fractions. For isolations without quantitative recovery, the leading and tailing portions of peaks may be discarded to give pure fractions. Obviously, in quantitative analysis this is not permissible. Since tailing of peaks is very marked on Florisil and is also prominent with silicic acid, these columns are most useful in quantitative analysis when eluted in such a manner that widely different groups of lipids are recovered as separate fractions. Some closely related lipid classes are retained as a mixture for analysis by other means, even though they are partially or almost completely separable on the silicic acid column. As an example, brain lipids are separable on silicic acid into cholesterol, glycolipid, and phosphatide fractions by elution first with chloroform, then acetone, and finally methanol (Section IV). Although cerebrosides are almost completely separated from sulfatides by elution with chloroform–acetone 1:1 v/v, prior to acetone, the two lipids are eluted together with acetone for precise quantitative analysis by another procedure.

Columns of larger internal diameter are used primarily when larger sample loads are to be applied, rather than for improved resolution without change in sample size. When the proven load for one diameter is known, the load for another diameter can be determined from the ratios of the squares of the internal diameters. Thus, for example, if a 1-g sample is satisfactory for a column 2 cm i.d. and a 4-cm-i.d. column is available, from the squares of the diameters (4 and 16) it is determined that a 4-g sample can be applied to the larger column.

Elution volumes and the flow rate should also be increased by the same factor (4 in the example above).

Increase in column height without increase in sample load is used primarily to improve resolution, i.e., to reduce fraction overlap. When column length is increased without increase in diameter, the flow rate is not changed and elution volumes are increased in direct proportion to increase in length; e.g., for a column 20 cm long elution volumes used for 10-cm columns are doubled. It is usually wise to limit column length to 20–30 cm. Rather than attempt to complete resolution with extremely long columns, it is generally preferable to elute two or more lipid classes together from a short column for separation on a different type of column. Thus phosphatidyl ethanolamine and phosphatidyl serine are eluted together from silicic acid columns and the two lipids are then separated on a silicic acid–silicate column, where complete separation is readily accomplished with a short column (Section VII.A). Choice of column length is dictated in part by the type of separation desired. Thus, when elution of pure phosphatidyl ethanolamine from DEAE cellulose is desired, a column 20 cm high should be used. If elution of phosphatidyl ethanolamine along with phosphatidyl choline and sphingomyelin is acceptable, a 5- or 10-cm column may be satisfactory (see Section V.G).

Choice of sample load is influenced by the separations desired. The proper load for a given column is determined empirically by applying different loads until the desired results are obtained. Better resolution, particularly in quantitative analysis, may be obtained by decrease of sample load (or increase of column diameter) rather than by increase of column length. Thus at the proper load of brain lipid, ceramide is easily separated from cerebroside by eluting a 10-cm-high Florisil column with 5% methanol in chloroform (Section VIII). When the load is doubled, however, separation is incomplete even when column height is increased to 20 cm.

The column dimensions and sample loads recommended below for general use are more conservative than may be required for some applications. Generally, a somewhat smaller load and larger column dimensions should be used for samples not previously investigated, rather than run the risk of fraction overlap that may result with larger loads and smaller columns. On the other hand, much time can be saved when many runs of the same sample type are to be carried out by carefully determining the maximum load with the smallest size column and elution volumes.

C. Solvent and Reagent Specifications

All solvents should be reagent grade and redistilled in order to remove nonvolatile solids. Chloroform is stabilized by distillation into enough methanol to give a final concentration of 0.25% (by volume).

Concentrated (28% by weight) aqueous ammonia is prepared by bubbling gaseous ammonia (from a small cylinder) into ice-cold, freshly distilled water in a plastic bottle until the proper weight of ammonia is introduced. Ammonia should be prepared just before use; otherwise, nonvolatile solids may reappear. (Commercial concentrated aqueous ammonia contains a large amount of nonvolatile solids.)

Ammonium acetate used in ion-exchange cellulose column chromatography is preferably prepared, just prior to use, by mixing into the chromatographic solvents the calculated amounts of freshly prepared aqueous ammonia and redistilled glacial acetic acid.

Air should be removed from solvents just prior to use for column chromatography by placing under reduced pressure and returning to atmospheric pressure with pure nitrogen (at least 99.998%). Solvent volumes are specified as volume/volume ratios.

D. Terminology and Nomenclature Used in Column Chromatography of Lipids

The column diameter is always specified as the internal diameter of the chromatographic tube. A bed volume is the total volume of the chromatographic tube occupied by both adsorbent and solvent in the bed, the term commonly being used in directions for washing of adsorbents. A column volume is the volume of solvent in the chromatographic bed and is determined by noting the elution volume required for the first appearance of a substance not retained by the column. The value may be approximated by subtracting the volume occupied by adsorbent (calculated by assuming that each gram occupies 1 cc) from the total volume of the bed.

The term "phosphatidyl" is still most generally used in the literature to refer to all glycerol phosphatides with two hydrocarbon chains without regard to the nature of the linkages of the two chains to glycerol, and this usage is conformed to here. Thus, phosphatidyl ethanolamine refers to all derivatives of glyceryl phosphoryl ethanolamine with two hydrocarbon chains and includes diacyl, plasmalogen (vinyl or α,β-unsaturated ether linked), and alkoxy (ether-linked) forms. The prefix "lyso" is applied to glycerol phosphatides that have only one hydrocarbon chain. Thus, lysophosphatidyl ethanolamine refers to all derivatives of glyceryl phosphoryl ethanolamine that have one hydrocarbon chain regardless of the position (first or second carbon of glycerol) or mode of attachment (ester, vinyl ether, ether). Phosphatidyl ethanolamine is written for ease of reading as two words.

Reference to specific types of linkages of the hydrocarbon chains is made by the addition of appropriate prefixes. Thus, diacylphosphatidyl ethanolamine refers to the glyceryl phosphoryl ethanolamine derivative with both hydrocarbon chains in ester linkage. Plasmalogen phosphatidyl ethanolamine, phosphatidyl ethanolamine plasmalogen, or simply ethanolamine plasmalogen refers to the molecule with a vinyl (α,β-unsaturated) ether-linked hydrocarbon chain at carbon 1 and an ester-linked hydrocarbon chain at carbon 2 of glycerol. Alternatively, plasmalogen forms may be designated as "phosphatidal" derivatives (i.e., phosphatidal ethanolamine refers to ethanolamine plasmalogen), or the prefixes vinyl ether or α,β-unsaturated ether can be used before phosphatidyl.

Sphingolipids are referred to in some cases by well-established trivial names (gangliosides, cerebrosides, cerebroside sulfate or sulfatide, sphingomyelin), and in other cases by systematic derivatives of trivial names (ceramide dihexosides, ceramide polyhexoside, ceramide aminoethylphosphonate). This approach should be less confusing than substitution of more systematic names for the very commonly used trivial names.

II. Solvent Evaporation, Weighing, and Avoidance of Contaminants in Column Chromatography

A. Solvent Evaporation

Column fractions generally contain a small weight of lipid in a large volume of solvent. To avoid decomposition of lipids, the mildest conditions possible should be used for solvent evaporation. Efficient evaporation of large volumes of solvent is accomplished under reduced pressure rather than by applying heat. For such evaporations, two cold traps (methyl cellosolve–dry ice) in series are used with an efficient vacuum pump and a rotary-type evaporator. Almost all of the solvent collects in the first cold trap. The Buchler Model FE-2 flash evaporator (Buchler Instruments Inc., 514 West 147th Street, New York) is useful and is an example of the horizontal type of rotary evaporator. The Calab Model C Evaporator (California Laboratory Equipment Co., 1165-67th Street, Oakland, California 94608) is an example of an evaporator that stands at an angle. The latter apparatus facilitates evaporation of solvent to a small volume without deposition of lipid over a large glass surface, particularly when a pear-shaped flask is used. Glass joints should not be greased. If a tighter fit is required, it can be obtained readily by using glass-grinding compound (Section II.C).

Evaporations are carried out by first reducing pressure to remove air from the system followed by return to atmospheric pressure with pure nitrogen. The pressure is then reduced for vacuum evaporation and the sample flask is rotated (in air or water). Rapid evaporation is achieved with proper cold traps and a good vacuum pump (e.g., Welch Model 1400 B) without application of heat to the sample flask. Evaporation of chloroform, etc., may be so rapid that the temperature may fall below 0°C and ice may form on the rotating sample flask. The ice is removed by pouring warm water over the cake or by immersion in a warm water bath, but heat should be applied only to the point where the ice melts. Freezing of water or glacial acetic acid during evaporation is prevented by application of just enough heat from a water bath to maintain the temperature slightly above the freezing point. When boiling under reduced pressure becomes so rapid that the sample solution may pass into the cold trap, the pressure is raised by introduction of nitrogen through a three-way stopcock that in one position connects the sample flask with the vacuum system and in the other with nitrogen.

Many lipids are particularly susceptible to autoxidation when spread as a thin film over a glass surface. Decomposition with formation of artifacts is reduced or prevented by avoiding evaporation to complete dryness, working as much as possible under pure nitrogen, and adding an antioxidant. BHT (butylated hydroxytoluene; 2,6-di-*t*-butyl-*p*-cresol, Eastman Organic Chemicals, Rochester, New York) added to the level of about 0.1% of the weight of the lipid is effective. The antioxidant has relatively nonpolar chromatographic characteristics and is thus easily distinguished and separated from polar lipids.

Solvent replacement rather than complete evaporation to dryness is recommended for the more labile lipids in particular. Methanol, water, and acetic acid are replaced by repeated addition of chloroform followed by evaporation to a small volume. The solvent trap must be washed free of acetic acid when removal of the last trace of this solvent from the sample is to be accomplished by solvent replacement. Pure toluene is particularly useful as a replacement solvent for acetic acid. When the solvent in the flask is reduced to a small volume of pure solvent (chloroform, methanol, etc.), the lipid is washed out of the flask with the desired pure solvent or solvent mixture and diluted to a known volume in a graduated glass-stoppered container. The weight of the sample is determined by weighing a small aliquot that is discarded (Section II.B).

B. Weighing Procedures

Procedures involving evaporation of an entire sample to dryness followed by drying to constant weight in a vacuum desiccator are frequently satisfactory for large samples but should be avoided with small samples, since small amounts of lipid spread over a dry surface are easily altered by exposure to air. Phosphatidyl ethanolamine and phosphatidyl serine are particularly labile in the dry state. These changes may be minor with relatively large amounts of lipid (10 g or more) spread over the surface of a 2-liter flask. With smaller amounts of lipid the undesirable changes are avoided by weighing the lipid from a small aliquot of a bulk solution. The aliquot is discarded, and the weight of the total lipid in solution is calculated. Weighing is easily accomplished with a modern microbalance (Cahn Electrobalance, Cahn Instrument Co., Inc., 15505 Minnesota Avenue, Paramount, California). The Cahn microbalance is very simple to use and can be placed in almost any available space, unlike the older type of microbalance.

A weighing procedure suitable for routine use is as follows. Small aluminum pans weighing about 6 mg are either prepared from sheet aluminum or purchased. The tare weight of the empty pan is obtained by heating it on a hot plate or melting-point block, usually at 60–80°C, for 2–5 min, transferring with forceps to a desiccator containing potassium hydroxide pellets, cooling for 2–3 min, and then weighing. Rapid cooling and convenient transfer of the pan into and out of the desiccator is accomplished by placing the pan on a piece of aluminum foil supported on a beaker to bring the support up to the top of the desiccator. An appropriate volume (usually 50–200 μl) of the sample solution is transferred carefully to the pan from a Hamilton microsyringe of 50- to 100-μl capacity (Hamilton Co., P.O. Box 307, Whittier, California). The steps should be repeated to ensure that the correct weight has been obtained. The pan and sample may then be discarded. The procedure is not suitable for the more volatile lipids such as short-chain fatty acid methyl esters and some of the shorter-chain fatty acids, although these may not be lost if present in a mixture of lipids. Accuracy can be checked by weighing aliquots from a standard solution of cholesterol (weighed on an ordinary analytical balance).

C. Avoidance of Contaminants

Many lipid or lipid-soluble contaminants exist in every laboratory (*18*). One of the most important types is grease used for lubrication and sealing of joints, etc. Grease can be eliminated almost completely from the lipid laboratory. All stopcocks on chromatographic tubes, separatory funnels, etc., should be Teflon. All glass-stoppered vessels, glass joints (as on flash evaporators), and chamber tops for thin-layer chromatography should be used without grease. These can be ground to fit very tightly by using aluminum oxide grinding compound (800 grit obtainable from Optical Equipment Co., 6910 Santa Fe Avenue, Huntington Park, California). Desiccators are sealed with desiccator rings made of Neoprene, silicone rubber, or Viton (Des-O-Ring from La Pine Scientific Co., 6001 South Knox Avenue, Chicago, Illinois) rather than grease. Solvents are redistilled, adsorbents are washed carefully, and care is taken in general laboratory operations. Cork and rubber stoppers should not be used in the lipid laboratory (*18*).

III. Extraction of Lipids and Removal of Nonlipid Contaminants by Column Chromatography

A. General Comments

Partition column chromatography is useful routinely as a part of the procedure for preparation of lipid extracts (*1,2*). Sephadex, a cross-linked dextran gel, is preferable to cellulose because its use permits more complete and reproducible separations and larger sample loads. Cellulose columns, however, provide a relatively inexpensive means for removal of nonlipid contaminants. For quantitative analysis, column chromatographic procedures are superior to the aqueous salt wash procedures (*3*) that formerly provided the only means for removal of water-soluble nonlipid contaminants from lipid extracts. With the wash procedure, gangliosides present in extracts from different organs do not always partition completely into the upper phase. The wash procedure seldom, if ever, yields the precise quantitative results obtainable by column chromatography. Some loss of lipid into the upper phase and retention of nonlipid in the lower phase may be observed with the aqueous salt wash procedure (*4*). Furthermore, the wash procedure is difficult to apply to large volumes of extracts, whereas the column procedure is equally applicable to small- and large-scale extractions.

B. Preparation of Lipids for Column Chromatography

Specimens for quantitative analysis are preferably removed and extracted as quickly as possible to avoid enzymatic breakdown. If storage or shipment is necessary prior to analysis, fresh specimens should be frozen quickly by immersion in dry ice or liquid nitrogen and stored at $-20°C$ or below. Shipment should be by air with specimens at dry-ice temperature. This is easily achieved by using commercially available thermos bottles (small samples) or foam plastic shipping containers (large samples) containing dry ice. Specimens of human and many animal organs can be stored in plastic bags (large organs) or screw-capped plastic bottles (smaller specimens) for many years at $-20°C$ or below without apparent changes in lipid or water content.

Preliminary workup prior to column chromatography is sometimes simple. With gall bladder bile, evaporation of water followed directly by Sephadex column chromatography is all that is required (Section X.B.2). Most specimens, however, must be extracted. Organs that are difficult to grind (skin, heart, and peripheral nerve) can be treated in one of two ways. Small samples may be frozen and pulverized or large samples may be passed through an electric meat grinder with uniformity of sample being achieved by passage of the paste back through the grinder two or three times. The average composition of an organ, like brain, that varies in composition from region to region is obtainable by extraction of the entire organ (small) or by passage through a hand meat grinder several times to achieve uniformity prior to removal of an aliquot (large organ).

After preparation of a uniform homogenate of the sample, three small portions (50–500 mg each) are removed for moisture determination, e.g., by drying to constant weight over potassium hydroxide pellets, to make possible expression of quantitative results on both a wet- and dry-weight basis. Another portion of the sample is extracted in a blender (large samples) or glass homogenizer with a motor-driven pestle (small samples). An exhaustive extraction sequence with solvent volumes in parentheses referring to milliliters per gram of fresh tissue is (1) chloroform–methanol 2:1 twice (20 and then 10 ml); (2) chloroform–methanol 1:2 (10 ml); (3) chloroform–methanol 7:1 saturated with 28% (by weight) aqueous ammonia (10 ml); (4) chloroform–methanol–glacial acetic acid–water 4:2:1:0.5 (10 ml); and (5) chloroform–methanol–concentrated hydrochloric acid 200:100:1 (10 ml). Even after such exhaustive extraction some fatty acid, presumably from oxidized lipid, may be released by alkaline hydrolysis of residues. In bacteria, peptides or glycoprotein (peptide) that contain fatty acids may not be extracted by the above procedure, and the study of these substances is by procedural definition a separate problem from general lipid analysis.

The use of the entire sequence of extracting solvents is not required for most samples. The first three mixtures can be used routinely for animal organ extractions. The exact details of chloroform–methanol extraction can be varied to advantage. Thus for initial extraction of erythrocytes with chloroform–methanol 2:1 the proper amount of methanol for the final mixture is added first, the cells are homogenized, the proper amount of chloroform to give a final chloroform–methanol ratio of 2:1 is added, and the mixture is again homogenized. This procedure avoids the formation of the coarse rubbery protein mass obtained when chloroform–methanol 2:1 is used directly.

The solvent is evaporated at low temperature with care (Section II.A), acidic solvents being evaporated separately to avoid hydrolysis of labile lipids.

C. Cellulose Column Chromatography

1. GENERAL COMMENTS

Cellulose columns can be used for the separation of lipids from water-soluble nonlipid contaminants (*2*). Gangliosides are eluted with nonlipid; attempts to separate them quantitatively as separate fractions have failed. Cellulose columns have a lower capacity for lipids and separations are not as complete as on Sephadex, but cellulose is readily available and less expensive.

2. PREPARATION OF COLUMNS

A generally useful column size is 2.5 cm i.d. $\times$ 20 cm prepared in a tube 30 cm long with a Teflon stopcock, a glass wool plug for retention of adsorbent, and a 500- to 1000-ml solvent reservoir. Whatman standard grade ashless cellulose powder (20 g) is suspended in methanol–water 1:1 and passed into the chromatography tube. One column volume is about 70 ml. The packed bed is washed at a flow rate of 3 ml/min with the following solvents (number of column volumes in parentheses): methanol–water 1:1 (7), chloroform–methanol 1:1 (3), and chloroform–methanol 9:1 saturated with water (4).

3. SAMPLE APPLICATION, ELUTION, AND SOLVENT EVAPORATION

The sample (100–500 mg) is transferred to the column in 5–10 ml of chloroform–methanol 9:1 saturated with water, quantitative transfer being ensured by carefully washing the sample container and transfer pipette with the same solvent. Two fractions are eluted at a flow rate of 3 ml/min.

1. Chloroform–methanol 9:1 saturated with water; 20 column volumes, lipids.
2. Methanol–water 9:1; 12 column volumes; water-soluble nonlipid contaminants, gangliosides, and traces of some phosphatides.

Solvent is evaporated in the cold under reduced pressure (Section II.A). After removal of nonlipid contaminants, small quantities of lipid are more readily decomposed when spread over solid surfaces such as glass, and care should be taken not to evaporate completely to dryness (Section II.A). Methanol and water are removed by repeated addition of chloroform and evaporation to a small volume. The lipid is then dissolved in the desired volume of the appropriate solvent, and the amount present is determined by weighing an aliquot using the microbalance technique (Section II.B).

D. Sephadex Column Chromatography

1. GENERAL COMMENTS

Sephadex column chromatography is the preferred quantitative procedure for separation of water-soluble nonlipid contaminants from lipids (*1*). Gangliosides are eluted as a separate fraction and bile acid conjugates are separable from other lipids—making Sephadex columns useful for bile separations (Section X.B.2).

2. PREPARATION OF COLUMNS AND ELUTING SOLVENTS

Sephadex (G-25, coarse, beaded, Pharmacia Fine Chemicals, 800 Centennial Avenue, Piscataway, New Market, New Jersey 08854) is placed in methanol–water 1:1, the smallest particles of adsorbent (fines) are removed by decantation, and additional solvent is added. Dissolved gases are removed from the gel suspension under gentle suction from a vacuum pump with swirling (to prevent bumping) for 30–60 sec, and the gel is equilibrated for several hours at room temperature. Equilibration is more rapidly attained when dissolved gases are first removed. The equilibrated gel is degassed again and then poured into the chromatography tube to the desired height. A 2.5- to 5-cm layer of clean sand (J. T. Baker, purified, washed, and ignited) is gently poured on top of the gel. Sand is used to prevent floating of the gel in solvent mixtures that contain chloroform. The bed is then washed with two complete sequences of the solvent mixtures and solvent volumes to be used for column chromatography. Columns are left standing in mixture 4 and washed with mixture 1 just prior to sample application. Washing removes impurities and fines. A column can be reused repeatedly if not allowed to run dry. If even the smallest portion of the gel bed becomes dry, the gel should be extruded and the column repacked and washed as before. Eluting and wash solvents are:

1. Chloroform–methanol 19:1 saturated with water (about 5 ml/liter)

2. Chloroform–methanol 19:1, 850 ml; glacial acetic acid , 170 ml; water, about 25 ml (added to point of saturation)
3. Chloroform–methanol 9:1, 850 ml; glacial acetic acid, 170 ml; water, about 42 ml (added to point of saturation)
4. Methanol–water 1:1

Mixtures 1, 2, and 3 are shaken vigorously several times in a separatory funnel and the clear lower phase is used. Mixture 4 must be mixed and allowed to stand for an hour or more before use to ensure temperature equilibration. When mixing solvents 1, 2, and 3, it is best to add water slowly in order to avoid formation of a large upper layer. The solvents may be slightly undersaturated with water without appreciable change in elution characteristics. Slightly undersaturated mixtures are recommended to prevent separation of the mixture into two phases as a result of temperature fluctuations.

Columns of various diameters and heights are useful for various quantities of lipid and for different purposes. Columns 1 cm in diameter and 10–30 cm in height are useful for 250 mg or less of sample, and columns 2.5 cm i.d. and 10–30 cm in height are useful for larger samples. Samples weighing several grams can be applied to 2.5 cm i.d. × 30 cm columns. Directions are given for the generally useful 2.5 × 30 cm and 1 × 10 cm columns. Solvent volumes and sample loads for other column sizes are estimated as described in Section I.B. For 2.5 × 30 cm columns, a chromatographic tube 2.5 i.d. and approximately 40 cm long fitted with a Teflon stopcock and a 1-liter reservoir for solvent is used with a glass-wool plug (moistened with methanol–water 1:1) to retain the gel. For a 1 × 10 cm column, a tube 1 × 15–20 cm with a Teflon stopcock, a 100-ml solvent reservoir, and a glass-wool plug is used.

3. SAMPLE APPLICATION AND ELUTION

The sample is transferred to the column in the first eluting solvent. Some insoluble materials are conveniently transferred as a suspension by using a wide-mouth pipette. After transfer of insoluble particles to the bed, a plug of solvent-washed glass wool should be placed over the bed to prevent the insoluble matter from rising into the reservoir. When insoluble solids cling to the wall of the sample flask and cannot be transferred as a suspension, the flask contents are treated with each of the eluting solvents and soluble and/or particulate matter is transferred to the column just before application of the particular solvent for elution. This procedure is necessary, for example, for beef bile (Section X.B.2), since salts of taurine-conjugated bile acids are not soluble in the first two elution mixtures and do not suspend but remain spread over the surface of the glass container. After the sample is applied, it is washed in by addition of several small volumes of eluting solvent, allowing each small volume to pass just to the top of the column before the next addition.

When the sample has been transferred quantitatively to the column, the proper volume of the first eluting solvent is added. Bulk fractions are collected. For 2.5 cm i.d. × 30 cm columns 500, 1000, 500, and 1000 ml, respectively, are used for eluting solvents 1–4 at a flow rate of 3 ml/min. Sample sizes for lipid extracted from animal organs are preferably 5 g or less. Larger samples (12 g of solids) have been applied and good separations obtained. With large samples the initial flow rate may be slow until one or two column volumes of solvent have passed through the column. It is best to double the elution volume of the first solvent with such very large sample loads.

The scale-down factor for 1 cm i.d. × 10 cm columns as compared to 2.5 × 30 cm columns is almost exactly 20, and thus one-twentieth of the sample load and eluting solvent volumes are used, although flow rates are reduced only to one-sixth of the flow rate for the 2.5-cm column (i.e., from 3 to 0.5 ml/min). The 1 × 10 cm columns are particularly useful for removal of salts (e.g., ammonium acetate) and adsorbent (Florisil, silicic acid, etc.) from fractions obtained by column chromatography on other adsorbents. For this purpose columns are eluted with mixture 1 and then with mixture 4.

4. COMPONENTS OF COLUMN FRACTIONS

The following have been shown to appear in the Sephadex column fractions:

Fraction 1. Hydrocarbons; mono-, di-, and triglycerides; sterols; sterol esters; waxes; all phosphatides including lysophosphatides; cerebrosides; sulfatides; ceramide polyhexosides; mono- and diglycosyl diglycerides; plant sulfolipid; free fatty acids; unconjugated bile acids (cholic, deoxycholic, etc.) when applied in either salt or free-acid forms; conjugated (glycine, taurine) bile acids when applied in the free-acid form; uncharacterized organic-solvent-soluble nonlipids, particularly in fecal lipid extracts.

Fraction 2. Gangliosides; glycine-conjugated bile acids when applied as salts; traces of altered forms (nature of changes unknown) of some acidic phosphatides (diphosphatidyl glycerol and phosphatidyl serine in particular); urea and other organic-solvent-soluble substances (uncharacterized).

Fraction 3. Traces of gangliosides; taurine-conjugated dihydroxycholanic (deoxy and chenodeoxy) acids when applied as salts; some of the less polar amino acids (phenylalanine, tyrosine, etc.); and some uncharacterized organic substances.

Fraction 4. Taurocholate and most water-soluble nonlipids (salts, amino acids, sugars, etc.).

IV. Silicic Acid Column Chromatography

A. General Comments

The separation on silicic acid of the less polar or neutral lipids (eluted with chloroform) from the polar lipids (phosphatides and glycolipids eluted with methanol) was the first generally useful column procedure for lipid separations (5). This procedure is still of great value.

The separation of individual polar lipids by stepwise or gradient elution with increasing amounts of methanol in chloroform, following the introduction of this principle by Lea et al. (6), is still perhaps the most commonly used separation procedure for polar lipids. The procedure does not in general give complete separation of the individual polar lipid classes. Silicic acid column chromatography must thus be combined with paper, thin-layer, or another type of column chromatography for complete separation of individual lipid classes. Elution of silicic acid columns with increasing amounts of methanol in chloroform is useful for some separations, particularly for preparative chromatography, but is less useful than other procedures for quantitative analysis (see Section X).

Elution of polar lipids from silicic acid is complicated by variations in the properties of the adsorbent. The most troublesome variation arises from the changes in elution properties of acidic lipids with respect to the other polar lipids. The acidic lipids are elutable in a constant order (diphosphatidyl glycerol and phosphatidic acid followed by cerebroside sulfate and sulfolipid, phosphatidyl serine, and finally phosphatidyl inositol). The other (nonacidic) lipids are also eluted in a constant order (ceramide followed by cerebroside (also glycosyl diglyceride), phosphatidyl ethanolamine, phosphatidyl choline, sphingomyelin). The relative elution order for lipids of the two groups is, however, different with different silicic acid preparations. Thus on "acidic" silicic acid (acid-washed) phosphatidyl serine is eluted in part before phosphatidyl ethanolamine, whereas on preparations not acid-washed the two lipids may be eluted with almost complete overlap or phosphatidyl serine may be eluted largely after phosphatidyl ethanolamine. Complete separation of these two lipids is not obtained, however, with any of the preparations. The elution properties of the other acidic lipids follow those of phosphatidyl serine. Thus cerebroside sulfate may be eluted completely before phosphatidyl ethanolamine with some silicic acid preparations, whereas with other adsorbent preparations the two lipids may not be completely separated.

The variability of preparations makes it impossible to specify one procedure for elution of silicic acid columns with increasing percentages of methanol in chloroform. Rather, the investigator is advised to determine performance for each lot of adsorbent. Air should be removed from both adsorbent and solvents (*7*).

Smith and Freeman (*8*) introduced the use of acetone for the elution of cerebroside from silicic acid columns, and Vorbeck and Marinetti (*9*) demonstrated that mono- and diglycosyl diglycerides are separable from each other and other lipid classes by elution with chloroform–acetone 1:1 followed by acetone. Rouser et al. (*10*) then demonstrated the elution of cerebroside sulfate, plant sulfolipid, and ceramide polyhexosides with acetone. These observations provide the basis for a useful separation procedure employing silicic acid that is essentially the original procedure of Borgström coupled with the acetone procedure of Smith and Freeman and followed by further chromatographic separation by thin-layer or another type of column chromatography (Sections X.E, X.F).

With lipid mixtures from some sources the elution of a silicic acid column with the sequence chloroform, acetone, then methanol provides an essentially quantitative separation into three groups: less polar (so-called neutral) lipids, glycolipids, and phosphatides. One phosphatide, diphosphatidyl glycerol (cardiolipin), is eluted in part with acetone, and thus acetone elution is not as useful with lipid mixtures containing very little glycolipid and relatively large amounts of diphosphatidyl glycerol (heart, liver, etc.). The procedure is useful for brain and spinach leaf lipids, in which glycolipid is high and diphosphatidyl glycerol is a very minor component (*10*). Some of the pigments in lipid extracts of erythrocytes are eluted with acetone, as is glucose if not previously removed by Sephadex column chromatography.

B. Preparation of Column and Application of Sample

Unisil (100–200 mesh, Clarkson Chemical Co., Inc., Williamsport, Pennsylvania) is a relatively pure grade of washed silicic acid with a rapid flow rate. Unisil is used without additional washing or other preliminary treatment except heat activation at 120°C if the adsorbent has been exposed to moisture. Other preparations may be used, but washing as well as heat activation may be required before use. A chromatography tube 2.5 cm i.d. and 10–20 cm long equipped with a 500 to 1000-ml solvent reservoir and a Teflon stopcock with a piece of glass wool to retain the adsorbent is generally useful. A bed 5 cm high is prepared by pouring a slurry (about 15 g of silicic acid) in chloroform into the tube, and the bed is washed with three column volumes of chloroform.

The solvent level is allowed to descend to the top of the bed and then 100–200 mg of lipid in 5–10 ml of chloroform is applied. Any suspended solid is transferred along with the soluble material, and quantitative transfer is ensured by thorough rinsing of all glassware with chloroform. If an insoluble film remains on the sample container after vigorous shaking with chloroform, the film is treated with each of the solvents to be used for elution of the column and the dissolved solids are applied to the column just prior to addition of the bulk of each eluting solvent.

C. Elution of Column

The following sequence and volumes of solvents are used for 2.5 × 5 cm columns:

1. Chloroform, 10 column volumes (about 175 ml)
2. Acetone, 40 column volumes (about 700 ml)
3. Methanol, 10 column volumes (about 175 ml)

Elution is accomplished at a flow rate of 3 ml/min, and bulk fractions of the indicated volumes are collected. The solvent is evaporated with care (Section II.A), and the weight of each fraction is determined (Section II.B).

D. Components of Fractions

Fraction 1. The chloroform eluate contains the less polar lipids (sterols, sterol esters, mono, di- and triglycerides, hydrocarbons, free fatty acids).

Fraction 2. Acetone elutes mono and diglycosyl diglycerides, sulfolipid, and small amounts of uncharacterized somewhat less polar lipids (plants). With bacterial lipids both neutral and acidic glycosyl glycerides are eluted, although these are not present in extracts from all microorganisms. With extracts of animal organs (particularly brain) cerebrosides, sulfatides, and ceramide polyhexosides are eluted with acetone. With fecal lipid extracts a large number of uncharacterized substances devoid of phosphorus are eluted with acetone. No more than traces of phosphorus are found in acetone eluates except with samples that contain a large amount of diphosphatidyl glycerol that is eluted in part with acetone.

Fraction 3. The methanol eluate from animal organ extracts, spinach leaves, and some bacteria contains phosphatide with, at most, minute traces of glycolipids. Fecal lipid extracts contain a variety of substances devoid of phosphorus that are eluted with methanol along with phosphatides.

E. Modifications

Chloroform–acetone 1:1 (5 column volumes, about 90 ml) may be used prior to acetone. This solvent mixture separates monoglycosyl diglyceride from diglycosyl diglyceride (eluted with acetone) as shown by Vorbeck and Marinetti (*9*). Other somewhat less polar lipids are eluted along with monoglycosyl diglycerides with spinach leaf lipids as sample. Chloroform–acetone is not as useful with animal organ extracts. Cerebrosides are almost completely separated from sulfatides by elution with chloroform–acetone 1:1. Complete separation of the two lipids is not obtained with increase of the column bed height to 10 or 20 cm.

Phosphatides can be separated into groups by using increasing amounts of methanol in chloroform. As noted above, this approach is usually less satisfactory for quantitative analysis. Optimum elution volumes will vary depending upon column height, sample load, components of the sample, and the exact properties of the particular lot of adsorbent. The most efficient elution scheme should be determined by collecting small fractions (e.g., 10 ml with a 2.5-cm-i.d. column) to define column performance, after which larger volumes can be collected on subsequent runs. Since with increase in column height large volumes of acetone must be used, elution of all phosphatides witih methanol followed by separation on some other type of column or by TLC is frequently the best approach.

F. Separation and Determination of Polar Lipid Components of Fractions

The mixture of diglycosyl diglycerides and sulfolipid (plants) or the corresponding mixture of cerebrosides, sulfatides, and ceramide polyhexosides (brain) may be separated into individual lipid classes by DEAE column chromatography (Section V). Determination of the components of these fractions is accomplished by quantitative one-dimensional TLC using the charring–transmission densitometry procedure (11). The phosphatides eluted with methanol can be determined by two-dimensional TLC and phosphorus analysis of spots (12) or separated by other types of column chromatography (Sections V–VII).

V. Diethylaminoethyl (DEAE) Cellulose Column Chromatography

A. General Comments

DEAE cellulose column chromatography is generally useful for separation of components of complex mixtures of lipids (2,13–15). Lipid classes are separated by DEAE column chromatography by ion exchange and by differences in polarity provided by nonionic groups, principally hydroxyl groups. Division of lipids into three groups (nonionic, nonacidic ionic, and acidic) facilitates discussion and understanding of the elution characteristics of DEAE. Nonionic lipids are eluted according to relative polarity, the least polar lipids (sterols and sterol esters, glycerides, hydrocarbons) are eluted with chloroform; cerebrosides and glycosyl diglycerides first appear in column effluents with 3–5% methanol in chloroform; 20–30% methanol in chloroform is required for elution of ceramide polyhexosides. All non-ionic and ionic nonacidic lipids are eluted with methanol, whereas acidic lipids are not. The ionic nonacidic lipid group begins to be eluted with 3–5% methanol in chloroform (phosphatidyl choline, lysophosphatidyl choline, sphingomyelin); phosphatidyl ethanolamine is retained to a greater extent and is first eluted at a concentration of about 10% methanol, while lysophosphatidyl ethanolamine is first eluted with about 50% methanol. Acidic lipids are not eluted with chloroform–methanol or methanol unless acid, base, or salt is added. This characteristic of DEAE is useful and makes this adsorbent quite different from silicic acid, where overlap of acidic and nonacidic lipid fractions is obtained. The order of elution of choline and ethanolamine phosphatides from DEAE is reversed from that obtained with silicic acid, and this difference is of advantage in some cases.

B. Selection of Proper Grade DEAE

Selectacel DEAE cellulose, regular grade (Brown Co., Berlin, New Hampshire) or a similar grade from other sources is preferred. DEAE preparations vary in coarseness of grade as well as in total ion-exchange capacity. The coarser preparations such as the regular grade Selectacel DEAE are the most generally useful for lipid separations. Finer grades (e.g., Types 20 and 40 Selectacels) are less satisfactory even though they are readily packed into a chromatography tube without the manual pressure required for the regular grade. With the finer grades, adsorbent may appear in the column effluent with certain solvents and columns pack more solidly upon repeated use until flow rates are seriously reduced and fraction overlap may be observed. It is desirable to select preparations that have the highest rated number of equivalents per gram, although the texture of the material is generally more important for successful chromatography than the relatively small differences in number of milliequivalents per gram of the various preparations.

C. Washing DEAE

DEAE preparations contain various impurities that must be removed prior to use for column chromatography. The adsorbent is washed with $1 N$ aqueous hydrochloric acid followed by water to neutral pH and then 0.1 N aqueous potassium hydroxide followed by washing with water. Three cycles of acid and base are used. One cycle of acid and base dissolved in methanol is preferable for some lots of adsorbent. Exposure to acid and base should be as brief as possible, and no more than 3 bed volumes of acid or base are required per cycle. Washing may be carried out on a Büchner funnel or a sintered-glass filter under gentle suction from a water aspirator and is most rapid when both filter paper and cheese cloth or gauze (4 layers) are used to prevent clogging of pores during filtration.

After washing, DEAE is converted to the acetate form by passage of 3 bed volumes of redistilled glacial acetic acid through the bed. Excess acid is removed by washing with 3 bed volumes of redistilled methanol. The adsorbent is removed from the filter, spread over a clean glass surface, air-dried in an area free of fumes and dust, and finally dried to constant weight in a vacuum desiccator over potassium hydroxide pellets. Pressure should be reduced and the desiccator disconnected from the pump to avoid contamination. It is important to weigh carefully the amount of DEAE used if reproducible results are to be obtained. Ion-exchange celluloses can hold surprisingly large amounts of water or other solvent, yet appear to be dry. Care must be exerted, therefore, to obtain a constant weight upon drying.

D. Packing Columns

The thoroughly dried DEAE preparation is weighed and then left overnight in glacial acetic acid. Acetic acid aids in breaking up aggregates and facilitates packing in a uniform manner into the chromatographic tube. Any visible aggregates remaining are very gently broken up with a glass rod prior to transfer of the slurry to the chromatographic tube. Vigorous procedures such as treatment in a blender are particularly to be avoided because, while packing may be facilitated, column performance is poor.

A tube of double-thickness Pyrex glass 30 cm in length fitted with a Teflon stopcock and a solvent reservoir is used. Columns of 1, 2.5, and 4.5 cm i.d. are most commonly used. Columns 2.5 cm i.d. with a 1-liter solvent reservoir are very convenient for general use. The amount of DEAE required for packing columns of various sizes is readily determined from the ratio of the squares of the internal diameters (Section I.B). For columns 1, 2.5, and 4.5 cm in diameter, 2.4,

15, and 50 g, respectively, of regular grade Selectacel DEAE are used to obtain a height of 20 ± 2 cm.

A clean glass-wool plug is placed in the chromatographic tube. The plug is held in place by a glass rod and a portion of the slurry of DEAE in glacial acetic acid is poured into the tube. After the first addition of DEAE, the rod is withdrawn and packing is continued. Approximately five equal portions can be packed to give a satisfactory column. After each addition of DEAE, the excess acid is forced out rapidly under nitrogen pressure, the DEAE bed is pressed lightly and uniformly with a large-bore glass rod, the uppermost portion of the bed after application of pressure is gently stirred to free the bed of very tightly packed adsorbent, and the next addition of slurry is made. The procedure is repeated until all of the slurry has been transferred quantitatively into the tube to give a bed height of 20 ± 2 cm. The bed should not be allowed to run dry at any stage of column preparation. Glacial acetic acid is allowed to fall just to the top of the column, and the bed is freed of acid by washing with 3–5 bed volumes of methanol. The bed is then washed with 3 bed volumes of chloroform–methanol 1:1 followed by 3–5 bed volumes of chloroform. The column thus prepared is ready for testing to determine uniformity of packing.

E. Testing of Columns

Each newly packed DEAE column should be tested to determine that it will give satisfactory performance. If the glass-wool plug is not firmly in place, some of the finer particles of adsorbent may appear in the column effluent. Persistence of adsorbent fines in the effluent is usually related to a faulty glass-wool plug or to the use of a poor grade of DEAE. If a column is packed in an uneven manner, overlap of fractions may result. Improper solvent changes (e.g., going directly from methanol to chloroform) or allowing a column to run dry may create channels (not visible to the eye) through which lipid may pass very rapidly, resulting in fraction overlap. These defects can be disclosed by a simple test with a lipid such as cholesterol. Cholesterol should be freshly recrystallized from ethanol to avoid the presence of oxidation products that may be eluted only slowly from the column.

Testing of a 2.5×20 cm DEAE column with cholesterol is accomplished as follows. A solution of 10 to 30 mg of cholesterol in 5 to 10 ml of chloroform is applied to column. Fractions of 10-ml volume are collected beginning at the point of application of the sample. Each fraction is tested for cholesterol either by evaporation of a 1-ml aliquot and observation of solids or by a color reaction. For the latter, 5 drops of acetic anhydride and 1 drop of concentrated sulfuric acid are added to a 1-ml aliquot of the column effluent. A green to blue color is a positive test. Columns are judged to be completely satisfactory when cholesterol first appears in tube 8. If the column has been packed more tightly, as for example to 18 rather than 20 cm, cholesterol may appear in tube 7, while if the packing has been somewhat looser, e.g., to a height of 22 cm, the first appearance of cholesterol in a fully satisfactory column may be in tube 9. If the appearance of cholesterol is earlier than tube 6, the column should be repacked. The cholesterol test should be carried out just prior to application of the sample. If the test shows the column to be satisfactory, cholesterol is eluted quantitatively and quickly with 3 bed volumes of chloroform, after which the sample can be applied. If the column must be repacked, the DEAE is extruded and repacked as a slurry in chloroform. Both testing of the column and elution of remaining cholesterol should be carried out at a flow rate of 2.5–3.0 ml/min (2.5-cm-i.d. column). The flow rate for 4.5-cm columns is 10 ml/min. One column volume is equal to 75 and 230 ml, respectively, for the 2.5- and 4.5-cm-i.d. columns. Azulene, a highly colored hydrocarbon, can be used to test columns and has the advantage that imperfections in the packed bed are readily visualized.

F. Sample Sizes for Columns

Optimum loads for 2.5 cm i.d. × 20 cm columns vary depending upon the sample. Thus with bovine brain lipid 300 mg may be used, whereas with human brain lipid 100 mg is preferable. Larger amounts can be applied when the sample contains a great deal of less polar lipid elutable with chloroform. The optimum load should be maintained in scaling up or down. The ratio of the squares of the column radii or internal diameters is used to determine the conversion factor for columns of various diameters. It is preferable to first free the lipid extract of protein and water-soluble nonlipid by Sephadex column chromatography (Section III.D), although e.g., crude bovine brain lipid extracts containing about 10% nonlipid have been separated directly by DEAE column chromatography with nonlipid contaminants (other than protein) being eluted almost quantitatively with methanol. The optimum load for each column is determined empirically for each mixture of lipids. Once the characteristics of a particular column for one type of sample have been determined, very reproducible results are obtained and it is necessary only to collect bulk fractions.

In DEAE column chromatography the volumes of most eluting solvents can be increased without fraction overlap, since elution is generally an all-or-none phenomenon; i.e., regardless of the volume of solvent used, the next fraction is not eluted. The exception is the elution of phosphatidyl ethanolamine (or ceramide aminoethylphosphonate) with chloroform–methanol 9:1. The point at which phosphatidyl ethanolamine appears in the column effluent with this solvent is determined by the load of total lipid and more particularly by the amount of phosphatidyl ethanolamine. The greater the amount of phosphatidyl ethanolamine, the more rapidly this lipid appears in the column effluent. At very high loads, phosphatidyl ethanolamine may not be separated completely from phosphatidyl choline, whereas at suitably low loads separation is wide and easily obtained. If phosphatidyl ethanolamine is found in the chloroform–methanol 9:1 fraction, the load of lipid should be reduced and the exact elution volume required for separation on a particular column determined first by collecting small fractions (e.g., 10 ml for a column 2.5 cm i.d.).

The appearance of phosphatidyl ethanolamine in a column effluent is detected by reaction with ninhydrin (7). For the test 0.1 ml of column effluent is mixed with 0.1 ml of 0.1% ninhydrin in 1-butanol and 0.1 ml of lutidine or pyridine in a 2- or 3-ml tapered-bottom test tube and heated to boiling in a sand bath. Loss of sample is prevented by rotating the tube during heating and periodically removing the tube and tapping with the finger to mix the contents of the tube. The appearance of a purple color is a positive ninhydrin test and demonstrates the presence of phosphatidyl ethanolamine in the eluate. The test is more sensitive when the color is viewed by looking down into the tube against a white background. After the lipid has been detected or when all of the preceding fraction has been eluted with chloroform–methanol 9:1, chloroform–methanol 7:3 is used for more efficient elution of phosphatidyl ethanolamine. Water-soluble nonlipid contaminants, if not previously removed, are eluted with methanol.

DEAE columns are most easily overloaded with respect to lipids having carboxyl groups as the only ionic groups (gangliosides, bile acids, free fatty acids). In general, at higher loads these lipids will be eluted quickly with neutral solvents rather than be retained for elution with acidic solvents. Samples containing large amounts of such lipids may be separated more conveniently on TEAE cellulose (Section VI), which has a much higher capacity for carboxylic acids.

Variability of elution and fraction overlap are avoided in DEAE column chromatography by careful adherence to the procedure described. Of great importance are selection of the proper grade of ion-exchange cellulose, proper washing, proper packing and sample load, and use of pure reagents. Impurities in commercially available ammonia and ammonium acetate can seriously impair column performance and cause variability of elution of acidic lipids. Extraneous substances that have a similar effect develop with time in ammonia prepared from gaseous ammonia. It is essential, therefore, that only carefully purified and fresh reagents be used.

G. Elution Sequences for DEAE Columns

Many elution sequences are possible with DEAE columns. Six useful sequences are shown in Table 1. In each case the essential details are provided in the table and legend. Columns 20 cm high are recommended for the more complex mixtures of lipids with sequence 3. Shorter columns can be used for the simpler mixtures (sequences 4–6) or when elution of phosphatidyl ethanolamine from other lipids in complex mixtures is not desired. In general, column height is most important for the separation of phosphatidyl ethanolamine or ceramide aminoethylphosphonate from other lipids, and thus shorter columns of larger diameter can be used if this separation is not desired (sequences 1 and 2). It is frequently more convenient to use 20-cm columns of smaller diameter rather than a 5- to 6-cm column of larger diameter to provide the same load equivalent, since 20-cm columns can be used for all types of separations.

Sequence 1 separates a lipid mixture into nonacidic and acidic fractions only. Sequence 2 separates mixtures into less polar or neutral lipids (sterols, sterol esters, glycerides, hydrocarbons, waxes), nonacidic lipids (polar ionic and nonionic), and acidic lipids. Sequence 3 is more extensive and is useful in conjunction with other types of column chromatography or TLC for quantitative analysis and complete separation of lipid mixtures into individual pure lipid classes (Section X.C). Sequences 4–6 are for special purposes. Sequence 4 is used for the separation of phosphatidyl ethanolamine from phosphatidyl serine after the two lipids are eluted together from a silicic acid column. Since sulfatide is eluted with the amino phosphatides with some lots of silicic acid, the separate elution of this glycolipid is shown as well. Sequence 5 is used for separation of cerebroside, ceramide dihexoside (and other ceramide polyhexosides), and sulfatide. This mixture can be obtained by silicic acid column chromatography by elution with acetone (Section IV). Sequence 6 is used for separation of mono- and diglycosyl diglycerides and plant sulfolipid obtained from a silicic acid column (Section IV) or from Florisil (Section VIII).

Columns may be reused repeatedly if not allowed to run dry. Before reuse, columns should be washed as directed in footnote 7 of Table 1. If impurities in solvents or samples cause column performance to vary, the column should be discarded or washed thoroughly. Salts in column fractions are removed by Sephadex column chromatography (Section III.D). Other eluting mixtures can be used to give additional separation of acidic lipids (*14*), but generally this is less economical of time and effort than use of the elution procedures shown in Table 1 combined with other fractionation procedures (see Section X.C).

H. DEAE Cellulose (Borate Form) Column Chromatography for Recovery of Ceramide Polyhexosides

The borate form of DEAE, in contrast to the acetate form, retains phosphatidyl ethanolamine when eluted with methanol or chloroform–methanol mixtures. Use of DEAE borate in combination with DEAE acetate thus provides a means for quantitative separation of phosphatidyl ethanolamine from ceramide polyhexosides (also obtained with TEAE, Section VI.B).

1. PRELIMINARY FRACTIONATION

Lipid extracted with chloroform–methanol is freed of nonlipid contaminants and gangliosides by Sephadex column chromatography (Section III.D) and then separated into fractions on DEAE (acetate form, Section V) using elution sequence 3. Ceramide polyhexosides when present are eluted partly with chloroform–methanol 7:3 and 1:1

TABLE 1

Elution of DEAE Columns (1)

Sequence	Col. ht., cm (1)	Col. diam., cm (1)	Flow rate ml/min (1)	Col. prepared in (2)	Sample applied in (2)	Elution solvents (2)	Elution volumes (3)	Lipids eluted (4)
1	6	4.5	10	C–M 2:1	C or C–M 2:1	[1] C–M 2:1 [2] M	4 } 10	All nonacidic lipids plus salt (5,6) (C–M 2:1 and M eluates pooled)
				Separation into acidic and nonacidic fractions (4)		[3] C–M–NH₃–salt [4] M	10 10	Acidic lipids and salts (6) Salts and possibly traces of lipids (7,8)
2	6	4.5	10	C	C	[1] C [2] M	8–10 10	Less polar (neutral) lipids Nonacidic polar lipids + salts (5)
				Separation into less polar, other nonacidic, and acidic fractions (4)		[3] C–M–NH₃–salt [4] M	10 10	Acidic lipids and salt (6) Salts and possibly traces of lipids (7,8)
3	20	2.5	3	C	C	[1] C [2] C–M 9:1	8–10 9	Neutral lipids Cer, DGDG, LPC, MGDG, PC, Sph
				General elution scheme		[3] C–M 7:3 [4] C–M 1:1 [5] M [6] C–HAc 3:1 [7] HAc [8] M	9 9 10 10 10 4	CAEP, CDH, CPH, PE CPH, LPE, OPE OPE, salts (5,6) FFA, GCBA, UCBA PS (9) None (10)

[9] C–M–NH₃–salt	10	DPG, PA, PG, PI, SL, Su, and salt (6)
[10] M	10	Salts and possibly traces of lipids (7,8)

4

6	4.5	10	C or C–M 7:3
20	2.5	3	C or C–M 7:3

Separation of phosphatidyl ethanolamine, phosphatidyl serine, and sulfatide (11)

[1] C–M 7:3	9	PE
[2] M	8	Salts (5)
[3] HAc	10	PS
[4] M	8	Removal of excess HAc
[5] C–M–NH₃–salt	10	Su and salt (6)
[6] M	10	Salts and possible traces of lipid (7,8)

5

6	4.5	10	C or C–M 9:1
20	2.5	3	C or C–M 9:1

Separation of cerebroside, ceramide polyhexosides, and sulfatide (12)

[1] C–M 9:1	9	Cer
[2] C–M 7:3	9	CPH
[3] M	10	Salts (5,6)
[4] C–M–NH₃–salt	10	Su, salts (6)
[5] M	10	Salt and possibly traces of lipid (7,8)

6

6	4.5	10	C
20	2.5	3	C

Separation of mono and diglycosyl diglycerides and sulfolipid (13)

[1] C–M 98:2	8	MGDG
[2] C–M 9:1	10	DGDG
[3] M	10	Salts (5)
[4] C–M–NH₃–salt	10	SL and salt (6)
[5] M	10	Salts and possibly traces of lipid (7,8)

1. The column dimensions and solvent volumes in Table 1 are for columns prepared from 15 g of DEAE cellulose. Although the solvent volumes are the same for both sizes, the flow rates are different (3 ml/min for columns 2.5 cm i.d. and 10 ml/min for those 4.5 cm i.d.). Maximum sample sizes vary from about 100 to 600 mg. See text for additional discussion. In addition, columns 1 and 15.2 cm i.d. have been used with flow rates of 0.5 and 110 ml/min, respectively. Column volumes for a height of 20 cm with these diameters are 12 and 2700 ml, respectively.

2. Abbreviations for solvents: C, chloroform; M, methanol; HAc, glacial acetic acid; C–M–NH₃-salt, chloroform–methanol 4:1 made 0.01 to 0.05 M with respect to ammonium acetate or potassium acetate to which is added 20 ml of freshly prepared 28%, w/w, aqueous ammonia per liter (Section I.C).

3. Elution volumes are specified in column volumes. Column volumes for the 2.5 × 20 cm and 4.5 × 6 cm columns are the same, since the same quantity of adsorbent is used for their preparation (see footnote 1).

4. Nonacidic lipid refers to all less polar (neutral) lipids plus nonionic glycolipids (cerebrosides, etc.) and phosphatides without a net negative charge. Less polar (neutral) lipid refers to nonionic lipids that are less polar than cerebrosides and glycosyl diglycerides and includes hydrocarbons, sterols, and sterol esters, glycerides, etc. Acidic lipid refers to substances having only negatively charged groups (fatty acids, cerebroside sulfate, etc.) or at least one more negatively charged than positively charged group (phosphatidyl serine etc.).

Abbreviations for lipid classes: CAEP, ceramide aminoethylphosphonate; CDH, ceramide dihexosides; Cer, cerebrosides; CPH, ceramide polyhexosides (three or more carbohydrate moieties); DGDG, diglycosyl diglycerides; DPG, diphosphatidyl glycerol (cardiolipin); FFA, free fatty acids; GCBA, glycine conjugated bile acids; LPC, lysophosphatidyl choline; LPE, lysophosphatidyl ethanolamine; MGDG, monoglycosyl diglyceride; OPE, oxidation products (uncharacterized) from phosphatidyl ethanolamine; PA, phosphatidic acid; PC, phosphatidyl choline; PE, phosphatidyl ethanolamine; PG, phosphatidyl glycerol; PI, phosphatidyl inositol; PS, phosphatidyl serine; SL, sulfolipid (plants); Sph, sphingomyelin; Su, sulfatide (cerebroside sulfate); UCBA, unconjugated bile acids.

5. Methanol elutes salts (acetates) formed by ion exchange of DEAE with acidic lipids and other salts present in the sample.

6. Salt in fractions removed by Sephadex column chromatography (Section III.D).

7. Before reuse the column should be washed with the following solvents and indicated column volumes: methanol, 3; acetic acid, 3; methanol, 3; chloroform–methanol 1:1, 3; chloroform, 3–4.

8. Methanol should be used to clear columns to ensure complete removal of lipid. Lipid may appear in the final methanol eluate if impurities in samples or solvents cause tighter binding of lipid.

9. Gangliosides when present are eluted with acetic acid but should be removed by column chromatography on Sephadex (Section III.D) prior to DEAE column chromatography.

10. Methanol is used to remove excess acetic acid, but little or no lipid is eluted.

11. The mixture is obtained by silicic acid column chromatography. With brain lipid and some preparations of adsorbent only the mixture PE plus PS is obtained, whereas with others Su is also mixed with PE and PS.

12. The mixture is derived from brain lipids by elution from silicic acid with acetone (Section IV).

13. The mixture is derived from plant lipids by elution from silicic acid with acetone (Section IV).

and partly with methanol. The fractions are pooled and the mixture is applied to a column of DEAE in the borate form.

2. PREPARATION OF COLUMN

A DEAE cellulose column is prepared as described in Section V.D. The acetate form of the packed column is converted to the borate form by washing with a saturated solution of sodium tetraborate in methanol (3 bed volumes). Excess borate is removed by washing with methanol (4 bed volumes), and the methanol is replaced by washing with chloroform–methanol 2:1 (4 bed volumes).

3. SAMPLE APPLICATION AND ELUTION

The mixture of phosphatidyl ethanolamine and ceramide polyhexosides obtained from the DEAE-acetate column (weight 70 mg or less for a 2.5 cm i.d. × 20 cm column) is applied and washed into the column with a small volume (10–15 ml) of chloroform–methanol 2:1. Columns 2.5 cm i.d. are eluted at a flow rate of 3 ml/min as follows (1 column volume is about 70 ml).

1. Chloroform–methanol 2:1 (10 column volumes, ceramide polyhexosides)
2. Methanol (6 column volumes, salts, miscellaneous impurities)
3. Chloroform–methanol 2:1 plus 1% by volume glacial acetic acid (8 column volumes, phosphatidyl ethanolamine)

Before reuse the column should be washed with 3 column volumes of methanol saturated with sodium borate, 4 column volumes of methanol, and 4 column volumes of chloroform–methanol 2:1. The ceramide polyhexoside fraction can be separated into its components by thin-layer chromatography (Section IX) with one or more of the solvent mixtures recommended for general use. Each spot can be eluted from the adsorbent for further study as described in Section IX.K.

VI. Triethylaminoethyl (TEAE) Cellulose Column Chromatography

A. General Comments

TEAE cellulose differs from DEAE cellulose in two very important respects. First, TEAE has a much higher capacity than DEAE for lipids having carboxyl groups as the only ionic group (fatty acids, bile acids, gangliosides), making it ideal for use with mixtures that contain a large proportion of such lipids. TEAE is thus useful for fecal lipids high in free fatty acid and bile lipids rich in bile acids. Second, phosphatidyl ethanolamine is not eluted from TEAE with chloroform–methanol mixtures or methanol (*14*) as it is from DEAE (acetate form). Salt, acid, or base must be added to the solvent before phosphatidyl ethanolamine is eluted from TEAE. This property makes TEAE useful for separation of phosphatidyl ethanolamine quantitatively from ceramide polyhexosides (*24*) that are not completely separated on DEAE in the acetate form. TEAE thus provides a useful alternative to DEAE in the borate form (Section V.H) for the latter separation.

TEAE is best used in the hydroxyl form. TEAE columns can be eluted as described for DEAE columns (Section V.G), but the elution sequences described in Sections VI.B and C are recommended. Procedure 1 is most useful when ceramide polyhexosides are of primary interest, and procedure 2 is useful when it becomes necessary to remove ceramide polyhexosides from a DEAE (acetate form) column fraction.

B. Procedure 1. Application of Total Lipid

1. PREPARATION OF COLUMN

TEAE (Selectacel regular grade, Brown Co., Berlin, New Hampshire) is washed and packed into the chromatographic tube to a height of 20 cm as described for DEAE (Section V). About 12 g of TEAE is required for a 2.5 cm i.d. × 20 cm column. The same quantity of TEAE (about 12 g) can be packed into a 4.5-cm-i.d. tube to a height of about 6–7 cm to give a column that has the same capacity but can be eluted at a more rapid flow rate. After packing, acetic acid is removed by washing with 3 column volumes of methanol and the adsorbent is converted to the hydroxyl form by washing with 4 column volumes of 0.1 N potassium hydroxide in methanol followed by 6–8 column volumes of methanol and 4 column volumes each of chloroform–methanol 1 : 1 and chloroform.

2. SAMPLE APPLICATION AND ELUTION

Nonlipid contaminants and gangliosides are preferably first removed by Sephadex column chromatography (Section III.D). From 100–300 mg of sample is recommended for lipids extracted from animal organs and applied to columns prepared from 12 g of TEAE (2.5 cm i.d. × 20 cm or 4.5 × 6 cm). Smaller loads can be used, and in some cases larger loads may be permissible. The sample is applied in 5–10 ml of chloroform and washed in quantitatively with the same solvent.

Elution of 2.5 cm i.d. × 20 cm columns is accomplished at a flow rate of 3 ml/min (1 column volume is approximately 75 ml). If the same amount of TEAE is packed into a larger tube, the same capacity is obtained and thus the same load can be applied and the column volume remains the same. The flow rate can be increased, however, in proportion to increase in the square of the internal diameter, thus decreasing the time required for elution. Eluting solvents and volumes are:

1. Chloroform; 5 column volumes; cholesterol and other less polar (neutral) lipids.

2. Chloroform–methanol 9:1; 8 column volumes; cerebrosides, glycosyl diglycerides, phosphatidyl choline, and sphingomyelin.

3. Chloroform–methanol 2:1; 8 column volumes; ceramide polyhexosides.

4. Methanol; 8 column volumes; inorganic substances formed by ion exchange with acidic lipids.

5. Chloroform–methanol 2:1 plus 1% glacial acetic acid; 6 column volumes; phosphatidyl ethanolamine, ceramide aminoethylphosphonate, free fatty acids, free and glycine-conjugated bile acids, *N*-monomethyl and *N*,*N*-dimethyl phosphatidyl ethanolamines.

6. Glacial acetic acid; 6 column volumes; phosphatidyl serine.

7. Methanol wash; 3 column volumes; no lipid, used for removal of excess acetic acid.

8. Chloroform–methanol 4:1 made 0.01–0.05 M in ammonium or potassium acetate to which 20 ml/liter of 28% aqueous ammonia is added (8 column volumes) followed by methanol (6 column volumes). After elution, the latter two solvents are pooled as the final acidic lipid fraction in which phosphatidic acid, diphosphatidyl glycerol, phosphatidyl glycerol, cerebroside sulfate, plant sulfolipid, and phosphatidyl inositol and uncharacterized acidic lipids appear.

C. Procedure 2. Separation of Phosphatidyl Ethanolamine from Ceramide Polyhexosides

1. PREPARATION OF COLUMN

This is accomplished as described for procedure 1 except that the column is placed in chloroform–methanol 2:1 before sample application. Columns 5, 10, or 20 cm high may be used. The description below is for elution of a 2.5 × 20 cm column that is generally useful. The conversion factor for a tube of different diameter is obtained from the ratio of the squares of the internal diameters of the columns.

2. SAMPLE APPLICATION AND ELUTION

A sample weighing about 60 mg is recommended for a 2.5 × 20 cm column, although the load will vary with different samples. The sample is applied and washed in with chloroform–methanol 2:1. One column volume is 70–75 ml. Elution at a flow rate of 3 ml/min is as follows:

1. Chloroform–methanol 2:1; 8 column volumes; ceramide polyhexosides
2. Methanol; 5 column volumes; salts
3. Chloroform–methanol 2:1 plus 1% by volume glacial acetic acid; 6 column volumes; phosphatidyl ethanolamine

The ceramide polyhexosides may be separated by thin-layer chromatography (Section IX). A shorter column of larger diameter packed with the same quantity of TEAE has the same column volume and load capacity, but the flow rate can be increased, thus shortening the time for elution.

VII. Silicic Acid–Silicate Column Chromatography

A. Separation of Phosphatidyl Ethanolamine from Phosphatidyl Serine

The complete and quantitative separation of a mixture of phosphatidyl ethanolamine from phosphatidyl serine eluted from a silicic acid column was first achieved with silicic acid columns treated with aqueous ammonia (7).

1. PREPARATION OF COLUMN

Silicic acid (Unisil, 100–200 mesh, Clarkson Chemical Co. Inc., Williamsport, Pennsylvania) without additional treatment is prepared as a slurry in oxygen-free (degassed) chloroform–methanol 4:1 and poured into a chromatographic tube to a height of 20 cm. The chromatographic tube should be about 25 cm long and equipped with solvent reservoir (1 liter for 2.5-cm-i.d. column) and a Teflon stopcock with a glass-wool plug for retention of adsorbent. The solvent is allowed to fall just to the top of the bed, and one column volume (70 ml for a 2.5-cm-i.d. tube) of chloroform–methanol 4:1 containing 1% by volume concentrated (28% by weight) aqueous ammonia (see Section I.C for preparation) is passed through the bed. The upper portion of the bed removes most of the aqueous ammonia and becomes white in appearance.

2. SAMPLE APPLICATION AND ELUTION

About 100 mg of a mixture of phosphatidyl ethanolamine and phosphatidyl serine is applied to a 2.5 cm i.d. × 20 cm column in a few ml of chloroform–methanol 4:1. Elution of a 2.5 cm i.d. × 20 cm column is accomplished at a flow rate of 3 ml/min as follows:

1. Chloroform–methanol 4:1; 8 column volumes (about 560 ml); pure phosphatidyl ethanolamine
2. Methanol; 3 column volumes (about 225 ml); pure phosphatidyl serine

Solvents should be free of air (Section I.C), and solvent evaporation should be conducted with great care (Section II.A), since both lipids undergo rapid autoxidation (7). Storage of both lipids without decomposition for prolonged periods is possible at −20°C or below in the absence of light when the lipids are dissolved in redistilled cyclohexane and sealed in a tube after flushing with pure nitrogen.

B. Separation of Phosphatidyl Choline (Lecithin) from Sphingomyelin

Silicic acid–silicate columns provide the only quantitative column procedure reported for complete separation of phosphatidyl choline from sphingomyelin (14). The columns are useful for larger-scale isolation of these two lipids. For quantitative lipid class and fatty acid analysis thin-layer chromatography (Section IX) is more rapid.

1. PREPARATION OF COLUMN

Silicic acid (Unisil, 100–200 mesh) is placed on a sintered-glass filter of coarse porosity and washed with 3 bed volumes of 6 N hydrochloric acid and then with 5 bed volumes of water. The adsorbent is heated at 120°C for 6 hr and cooled out of contact with air. For a 2.5-

cm-i.d. column 25 g of silicic acid is prepared as a free-flowing slurry in chloroform–methanol 1:1 and 4 ml of concentrated (28%) aqueous ammonia (see Section I.C for preparation) is added. The slurry is passed into a chromatographic tube to a height of 10 cm. The tube, about 20 cm long, should be fitted with a 1-liter solvent reservoir, a Teflon stopcock, and a glass-wool plug for retention of adsorbent. The bed is washed with 4 column volumes of chloroform for removal of methanol and water.

2. SAMPLE APPLICATION AND ELUTION

About 50 mg of a mixture of phosphatidyl choline and sphingo-myelin is recommended for a 2.5 cm i.d. × 10 cm column. If a DEAE column fraction containing cholesterol and cerebroside in addition to the two phosphatides is applied as sample, a load of about 150 mg is satisfactory. The sample is applied in 5–10 ml of chloroform. Elution volumes are variable with different preparations of adsorbent, and the volumes given below should be used as a guide only, the exact volumes for optimum results being determined by collection of small fractions (10 ml for a 2.5-cm-i.d. column) and TLC separation of components. Elution of a 2.5-cm-i.d. column at a flow rate of 3 ml/min is accomplished as follows:

1. Chloroform; 2.5 column volumes (about 50 ml); cholesterol and other less polar lipids
2. Chloroform–methanol 4:1 plus 0.5% water; 4 column volumes; cerebroside
3. Chloroform–methanol 4:1 plus 1.0% water; 8 column volumes; phosphatidyl choline
4. Chloroform–methanol 4:1 plus 1.5% water; 11 column volumes; sphingomyelin
5. Methanol plus 2.0% water; 5 column volumes to clear the column; lysolecithin and oxidation products when present

Traces of adsorbent may appear in fractions 3 and 4. The lipid can be freed of adsorbent by passage through a Sephadex column 10 cm high (Section III.D). Ceramide can be obtained as a separate fraction by elution with chloroform–methanol 19:1 prior to elution of cerebroside.

C. Separation of Acidic Lipids

Acidic lipids recovered from DEAE cellulose columns can be separated into useful groups by following the elution sequence recommended in Section B.2. The approach has not been used extensively for quantitative analysis and recommendations for sample load and exact proportions of eluting solvents and their volumes cannot be made with certainty.

VIII. Florisil Column Chromatography

A. General Comments

Florisil, a synthetic magnesium silicate, is useful for isolation of cerebroside, sulfatide, and ceramide (*13,14*). Chromatography on Florisil with even traces of water in the adsorbent or solvents results in elution of phosphatide along with glycolipid, more phosphatide being eluted when more water is present. Complete retention of phosphatides is possible with very dry adsorbent (heat activation) and dry solvents (obtained by addition of 2,2-dimethoxypropane). By elimination of water, however, the chromatographic system is converted from one partly partition in nature to one almost entirely of adsorption (direct attachment of lipid to adsorbent rather than through water molecules as in partition chromatography). As is characteristic for adsorption chromatography, the peaks are broad with a long trailing portion. Sulfatide with hydroxy fatty acid is not eluted quantitatively from such dry columns. The significance of this retention depends upon the load applied. At low loads a sizable percentage of the sulfatide is left, but at higher loads the percentage retained is lower. Since low loads are necessary for complete separation of ceramide from cerebroside, loading at two levels is essential for quantitative analysis. These difficulties are avoided by using the silicic acid column procedure (Section IV), with which glycolipids can be recovered quantitatively with relative ease.

B. Preparation of Adsorbent

One pound of Florisil (Floridin Co., Pennsylvania Glass Sand Corp., Two Gateway Center, Pittsburgh, Pennsylvania) is placed on a large sintered-glass filter of medium porosity and washed with 8 bed volumes of water. The adsorbent is then heat-activated at 100°C for 6 hr and cooled without exposure to air.

C. Packing of Columns

The desired amount of adsorbent is weighed quickly, water (1% by weight) is added, and adsorbent and water are allowed to equilibrate for at least an hour in a closed container. A slurry of adsorbent is prepared in chloroform containing 5% (by volume) 2,2-dimethoxypropane and poured into the chromatographic tube to a height of 10 cm. The chromatographic tube should be about 20 cm long and fitted with a solvent reservoir (1-liter size for a 2.5-cm-i.d. tube) and a Teflon stopcock with a glass-wool plug for retention of adsorbent.

D. Column Diameters and Loads

Various column diameters are useful for different purposes. We have used columns 0.8, 1.0, 2.5, 4.5, 10.2, and 15.2 cm i.d. The column 10.2 cm (4 in.) i.d. with the specified load is suitable for preparative isolation from bovine brain lipid of a "ceramide fraction" containing uncharacterized components but free of cerebroside and sulfatide and a fraction composed of the latter two glycolipids free of phosphatides.

E. Sample Application and Elution of 10.2-Cm-I.D. Columns

Crude chloroform–methanol 2:1 extract of brain (about 3 g) is applied in chloroform containing 5% by volume 2,2-dimethoxypropane and eluted as follows. (All solvents contain 5% by volume 2,2-dimethoxypropane.)

1. Chloroform; 10 column volumes; cholesterol and other less polar lipids when present
2. Chloroform–methanol 19:1; 10 column volumes; ceramide plus other uncharacterized lipid
3. Chloroform–methanol 70:30; 20 column volumes; cerebroside plus sulfatide

One column volume is about 480 ml. The flow rate should be maintained at about 50 ml/min. Phosphatides can be eluted from the column with 20 column volumes of chloroform–methanol 2:1 saturated with water. The latter fraction contains adsorbent that can be removed quantitatively by Sephadex column chromatography (Section III.D).

F. Recovery of Pure Lipids from Fractions

After evaporation of solvent (Section II.A), the lipids in each fraction can be recovered in pure form by additional chromatography. For the ceramide fraction, preparative TLC with 5–10% methanol in chloroform is suitable. Lipid is localized with a water spray and eluted with chloroform–methanol 1:1 saturated with water (Section IX.K). Cerebroside and sulfatide are separated on DEAE cellulose by using elution sequence 5 (Section V.G), and glycosyl diglycerides and plant sulfolipid are separated by using DEAE elution sequence 6.

IX. Thin-Layer Chromatography (TLC)

A. General Comments

TLC is widely used as an adjunct to column chromatography and is discussed by Renkonen and Varo in Chapter 2. It is a very versatile technique, since the composition and thickness of adsorbent layers can be varied readily and compact spots with good resolution of components are routinely obtainable. Samples are examined by TLC prior to column chromatography to determine the number and nature of components present, and TLC of fractions obtained by column chromatography aids in definition of the composition of the fractions. TLC is useful for isolation of small amounts of lipids and for quantitative analysis of column fractions. Determination of the extent of charring by transmission densitometry provides a general means for determination of all lipid classes after TLC separation (11,16), and phosphorus analysis of spots is applicable for phosphatide determinations (12).

Spot overlap by TLC is prevented by use of column chromatography prior to TLC. Fatty acid composition of lipids can be determined after TLC by micro techniques (17), and small amounts of pure lipids can be recovered after TLC, particularly when traces of adsorbent and other salts are removed by Sephadex column chromatography (18).

B. Adsorbents

Many adsorbents are available for TLC, and they may be modified in several ways. We prefer the mixture of Silica Gel plain (no binder) plus 10% by weight magnesium silicate (11) for polar lipids. Merck Silica Gel G containing calcium sulfate as a binder is useful, but greater spot spread for some acidic lipids is noted than with the magnesium silicate–silica gel adsorbent (11). The Adsorbosils (Applied Science Laboratories) have not been as useful for polar lipid separations, since spots tend to be more diffuse and some of the preparations tend to fall from the plate during chromatography with the polar solvent mixtures used for phosphatides and glycolipids. Merck Silica Gel plain (no binder) is a stronger adsorbent but is frequently useful, particularly when the air is very humid and the adsorbent layer is thus more rapidly deactivated. Commercially available precoated glass plates have been used successfully for some applications.

The following procedure is recommended for routine use. Adsorbent so prepared has a high capacity and excellent resolving power. Silica Gel plain (180 g; Merck, obtainable from Warner-Chilcott Laboratories, Instrument Division, 200 South Garrard Boulevard, Richmond, California) and magnesium silicate (20 g; Allegheny Industrial Chemical Company, P.O. Box 786, Butler, New Jersey) are placed in a wide-mouth jar of 2-liter capacity with a screw cap along with 80–90 porcelain grinding balls ½ in. in diameter (weight about 350 g) and heated at 120–150°C for 6 hr. Immediately upon removal from the oven, the bottle is tightly capped and cooled out of contact with air with an occasional brief opening of the cap to equalize pressure. The adsorbent is then placed on a ball mill and ground for 2 hr at a relatively slow rotation speed. The speed of rotation should be adjusted carefully for maximum efficiency of mixing and grinding of adsorbent. After approximately 2 hr the adsorbent is very finely powdered, thoroughly mixed, and devoid of aggregates that produce uneven adsorbent layers. If the adsorbent is not thoroughly dry, grinding is less successful. After the mixing and grinding operation, the adsorbent is passed through a coarse-mesh screen for removal of the porcelain grinding balls and stored in a tightly capped bottle.

C. Spreading of Adsorbent

A conventional Desaga spreading board with a fixed-distance (0.25 mm) Desaga spreader is generally useful. Glass plates 0.3 cm (⅛ in.) thick and 20 × 20 cm are cut from plate glass, and the corners and edges are ground smooth with sandpaper. Such plates are obtainable from many local glass companies. Since plates prepared from one piece of plate glass are of very uniform thickness, each set of plates cut from one piece of plate glass is marked with an identification number and spreading is accomplished using five plates from the same set.

Plates must be scrupulously clean. After chromatography, the plates are freed of adsorbent under running tap water with a brush (relatively stiff bristles of plastic rather than metal) and placed in a strong solution of detergent (Lakeseal detergent, Peck's Products Co., 610 East Clarence Street, St. Louis, Missouri, is satisfactory). After standing in the detergent solution for at least an hour, the plates are scrubbed with the brush, rinsed thoroughly with tap water and then distilled water, and placed on a drying rack. Just before use the dry plates are rinsed with chloroform sprayed on from a Teflon wash bottle, dried in air, and then placed on the spreading board. The plates are then washed with a small amount of chloroform that is wiped off with a paper towel to remove any remaining adherent lipophilic material that prevents the adsorbent from spreading uniformly over the glass plate. The plates on the spreading board are allowed to stand for 10–15 min before spreading of adsorbent.

A uniform slurry of adsorbent, 20 g in about 65 ml of water for five plates, is prepared by mixing for about 30 sec in a mortar with a pestle. The slurry is quickly transferred to the spreader, and the adsorbent is applied with an even motion to the glass plates (mixing and spreading should be accomplished in about 60 sec). Immediately after spreading and before the adsorbent has begun to set, the spreading board is vibrated manually to eliminate unevenness in the adsorbent layer. Plates are dried in an oven (e.g., at 120°C) or in air and stored out of contact with contaminating vapors.

Adsorbent layers of different thickness are produced with a fixed-distance (0.25-mm) spreader depending upon the thickness of the slurry and the speed with which the spreader is moved across the plates. Thin slurries and rapid movement of the spreader both yield relatively thin layers, whereas thicker layers are obtained with more concentrated slurries and a slower pass of the spreader over the plates. Different relative migrations are obtained with adsorbent layers of different thickness under otherwise similar conditions. The acidic lipids are most affected, and the relative migration of sulfatide is most markedly altered. With very thin layers of adsorbent and chloroform–methanol–water 65:25:4 as solvent, sulfatide migration is less than that of phosphatidyl choline (lecithin). Under the same conditions but with intermediate layer thicknesses, sulfatide migrates with phosphatidyl choline, and with thick layers sulfatide migration is greater than that of phosphatidyl choline. These changes in relative migration can be used to place lipids in positions convenient for isolation or quantitative analysis.

D. Control of Activity (Moisture Content) of Adsorbent Layers

Moisture content of the adsorbent must be controlled carefully for reproducible TLC. A small humidity-indicator meter that can be read whenever plates are being processed for chromatography is an invaluable aid for control of moisture content.

Immediately before use the plates are heat-activated at 120°C for 20 min and then cooled. Cooling in air yields variable results, since the moisture content of air is variable. On extremely dry days very active plates will be obtained, whereas on days when the humidity is high plates are less active. Such variations can be controlled by cooling the plates in air for about 5 min (or just to the point where they can be handled without a glove) and placing them in a chamber that has a vented cover and is equipped with a line for flushing with gas (air or nitrogen). Air from a compressed-air cylinder is convenient. The gas is bubbled through a column of water to introduce water vapor and then passed through the chamber containing the plates. The chamber is equipped with two glass or plastic tubes, one on each side of the chamber, containing a number of holes down their length to ensure uniform conditions throughout the chamber. With the particular apparatus and settings used in this laboratory, 10 min or less is required. The time and exact moisture content of the flushing gas are varied to ensure uniform activity of plates under different conditions. When humidity is very high, adsorbent composition can be altered to advantage. Thus 2.5 or 5% instead of 10% magnesium silicate can be used. At extremely high humidity levels Silica Gel without binder (a stronger adsorbent than when mixed with magnesium silicate) can be used to obtain migrations obtainable at lower humidity levels with the less active adsorbent. When the moisture content of the air is very low, plates can be spread with 0.01 M KOH rather than water. The potassium silicate thus formed holds more water than the plain silica gel, thus providing a means of holding water in the adsorbent under very dry conditions.

Control of moisture content changes during application of sample solutions is readily accomplished by construction of a small glass or plastic chamber from which only the end of the plate protrudes for spotting. A glass plate (placed just above but not touching the adsorbent layer) supported on three sides by other strips of glass is adequate.

Reproducibility of two-dimensional TLC is accomplished by carefully controlling the manner of removal of the solvent used in the first dimension. After development in the first dimension, the chromatogram can be placed in a chamber of the type used for cooling under constant-humidity conditions prior to spotting (see above); the system is flushed with dry nitrogen for about 10 min. The exact drying time should be adjusted carefully and will depend upon the flow rate of gas through the chamber and the solvents used for chromatography. Drying time should not be prolonged, since lipids decompose quite rapidly on dry adsorbent and extraneous spots may be obtained in the second dimension.

E. Spotting of Samples

Hamilton microsyringes (usually of 10 or 50 μl capacity, Hamilton Co., P.O. Box 307, Whittier, California) are preferable for sample application. For two-dimensional TLC in particular, the sample is applied as a row of slightly overlapping very small spots extending over a length of about 1 cm to give a long narrow band. For the charring-densitometry procedure after one-dimensional TLC (*11*) the sample should be applied as a single small spot. When relatively large volumes are to be applied, the initial applications are allowed to dry and spotting over the same area is repeated until the desired volume of solution has been applied. Solutions of polar lipids in chloroform, chloroform–methanol 9:1, or chloroform–methanol 2:1 are satisfactory for spotting. The spotting solution should not contain a large amount of water (no more than 2% is desirable). Some crude lipid extracts may contain a small amount of suspended solid unless 1–2% water is present, and it is preferable to add water to prevent clogging of the syringe needle.

F. Development of Chromatograms

Chambers $10\frac{3}{4} \times 2\frac{3}{4} \times 10\frac{1}{2}$ in. ($l \times w \times h$; Brinkmann Instruments Inc., 115 Cutter Mill Road, Great Neck, Long Island, New York) are useful and relatively inexpensive, since they are prepared from commercial glass building blocks. Chambers are lined on all sides with Whatman No. 3 or 3MM paper. Other chamber sizes may be used, but different results may be obtained. Just before spotting of a chromatogram is begun, 200 ml of the desired solvent mixture is added to the chamber and the chamber liner is saturated with solvent by tilting the chamber first to one side and then the other. The chamber top is then put into place and opened only for a few seconds when a plate is to be inserted. Chamber tops should fit well and be put into place dry, never with grease. The top can be held in place with a weight such as a large bottle of water. One plate only is placed in a chamber when plates are leaned against one side of the chamber for support. If a metal or glass support rack is placed in the center of the chamber, two plates may be developed in the same chamber when adsorbent layers face the chamber liner. A uniform solvent front is generally ensured with a chamber liner and a line marked across the top of the plate about $\frac{1}{2}$ in. from the edge. When the solvent has just reached the line indicating complete development, the chromatogram is removed, dried as described above for two-dimensional TLC, and placed in the second chamber with the desired solvent. Solvent should be discarded and paper liners dried or replaced after each run if most reproducible results are to be obtained.

G. Developing Solvents

Two different solvent pairs have been found useful for routine two-dimensional TLC of polar lipids. With system 1 the chromatogram is developed first with chloroform–methanol–water 65:25:4, dried, and developed in the second dimension with 1-butanol–glacial acetic acid–water 60:20:20 (1,11,15). With system 2 initial development is with chloroform–methanol–concentrated (28%) aqueous ammonia 65:35:5, followed by drying and development in the second dimension with chloroform–acetone–methanol–glacial acetic acid–water 5:2:1:1:0.5 (1,15). The individual solvent mixtures can be used for one-dimensional TLC. On occasion it may be advantageous to use the first developing solvent of system 1 with the second solvent of system 2 and the first solvent of system 2 with the second of system 1, i.e., to use the chloroform–methanol–water mixture with the chloroform–acetone–methanol–acetic acid–water mixture and the ammoniacal solvent mixture with the butanol–acetic acid–water mixture. An additional mixture, 1-propanol–water 70:30 is used for ganglioside separations. Chloroform–methanol–water 65:25:4 followed by 1-propanol–water 70:30 and chloroform–methanol–28% aqueous ammonia 65:35:5 followed by 1-propanol–water 70:30 are useful two-dimensional TLC solvent pairs for separation of gangliosides in some complex mixtures of lipids.

The true complexity of samples is best disclosed by two-dimensional TLC, and the technique is recommended for routine use, since resolution is much better than by one-dimensional TLC. Only neutral or basic solvents rather than acidic solvents are used for development in the first dimension in two-dimensional TLC, since decomposition of some lipids is observed during drying of chromatograms developed with acidic solvents.

Relative migrations of natural and synthetic lipid classes in the standard two-dimensional systems are illustrated schematically in Fig. 1. Two-dimensional TLC results with various types of natural lipid extracts are shown in Figs. 2 to 10. Solvent pairs (Figs. 2 to 10): (1) chloroform–methanol–water 65:25:4 followed by 1-butanol–acetic acid–water 60:20:20; (2) chloroform–methanol–28% aqueous ammonia 65:35:5 followed by chloroform–acetone–methanol–acetic acid–water 5:2:1:1:0.5; (3) chloroform–methanol–28% aqueous ammonia 65:35:5 followed by chloroform–acetone–methanol–acetic acid–water 5:1:0.5:0.5:0.2; (4) chloroform–methanol–28% aqueous ammonia 65:35:5 followed by 1-butanol–acetic acid–water 60:20:20. Spots were developed with (1) 55% sulfuric acid–0.6% potassium dichromate spray or (2) concentrated sulfuric acid–30% formaldehyde solution 97:3, followed by heating at 180°C. The following abbreviations are used in Figs. 2 to 10: CAEP, ceramide aminoethylphosphonate; CN, cerebroside with normal (nonhydroxy) fatty acids; CH, cerebroside with hydroxy fatty acids; CPH, ceramide polyhexosides; DGDG, diglycosyl diglyceride; FA, free fatty acid; G, gangliosides; MGDG, monoglycosyl diglyceride; NL, neutral lipid; P, protein; PA, phosphatidic acid; PC, phosphatidyl choline; PE, phosphatidyl ethanolamine; PG, phosphatidyl glycerol; PI, phosphatidyl inositol; PS, phosphatidyl

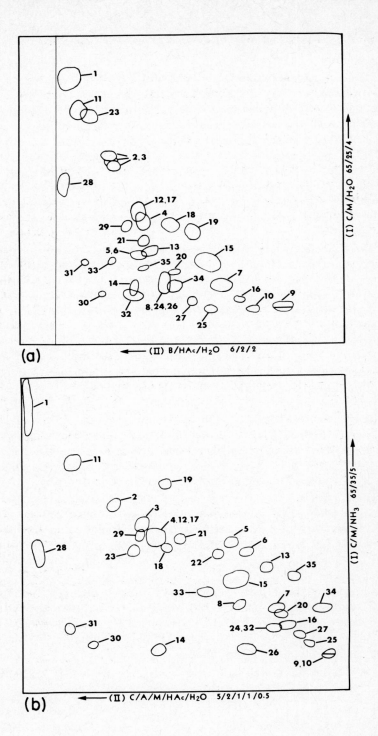

(a)

(I) C/M/H₂O 65/25/4 →

(II) B/HAc/H₂O 6/2/2 ←

(b)

(I) C/M/NH₃ 65/35/5 →

(II) C/A/M/HAc/H₂O 5/2/1/1/0.5 ←

serine; Pig, pigment; SH, sulfatide with hydroxy fatty acids; SN, sulfatide with normal (nonhydroxy) fatty acids; SL, sulfolipid; Sph, sphingomyelin; X, uncharacterized.

H. Detection Reagents

There are a number of useful general detection reagents (applicable to all lipids) and several more specific reagents that provide valuable additional information. The reagent most specific for lipids is alkaline Rhodamine 6G (*14*). Chromatograms are viewed wet under ultraviolet light. Exposure to iodine vapors evolved at room temperature in a closed container (or by heating to obtain intense vapors) is a useful general means of detection that is somewhat less specific for lipids. Iodine spots fade rapidly unless chromatograms are covered with a clean glass plate. For routine use, general detection reagents that are sensitive and that give relatively stable spots visible in daylight rather than the ultraviolet are preferable. Two reagents are recommended. A spray of 55% (by weight) sulfuric acid containing 0.6% potassium

---

FIG. 1. Maps of spots by two-dimensional TLC with chloroform–methanol–water 65:25:4 followed by 1-butanol–acetic acid–water 60:20:20 (a) and chloroform–methanol–28% aqueous ammonia 65:35:5 followed by chloroform–acetone–methanol–acetic acid–water 5:2:1:1:0.5 (b). Sample application is to the lower right side of the chromatogram. All lipids migrate behind the second solvent front with butanol–acetic acid–water. See text for details. The positions designated for the lipid classes are approximations only, since the exact migrations vary with unavoidable variations in chromatographic conditions. *Lipid class designations:* 1, neutral (less polar) lipids (sterols, sterol esters, mono-, di-, and triglycerides, etc.). *Sphingolipids:* 2, cerebrosides with normal (nonhydroxy) fatty acids; 3, cerebrosides with hydroxy fatty acids; 4, ceramide dihexoside; 5, sulfatide with normal (nonhydroxy) fatty acids; 6, sulfatide with hydroxy fatty acids; 7, sphingomyelin; 8, ceramide aminoethylphosphonate; 9, gangliosides (major types in brain); 10, Tay-Sachs ganglioside. *Glycerol lipids:* 11, monoglycosyl diglyceride; 12, diglycosyl diglyceride; 13, sulfolipid (plants); 14, phosphatidic acid; 15, phosphatidyl choline (lecithin); 16, lysophosphatidyl choline; 17, phosphatidyl ethanolamine; 18, *N*-monomethyl phosphatidyl ethanolamine; 19, *N,N*-dimethyl phosphatidyl ethanolamine; 20, lysophosphatidyl ethanolamine; 21, phosphatidyl glycerol; 22, phosphatidyl glycerol glycine ester; 23, diphosphatidyl glycerol (cardiolipin); 24, phosphatidyl inositol; 25, lysophosphatidyl inositol; 26, phosphatidyl serine; 27, lysophosphatidyl serine. *Other:* 28, free fatty acids; 29, cholesterol sulfate; 30, cholic acid; 31, deoxy and chenodeoxy cholic acids; 32, glycocholic acids; 33, glycodeoxy and glycochenodeoxy cholic acids; 34, taurocholic acid; 35, taurodeoxy and taurochenodeoxy cholic acids.

dichromate (*11*) is fairly specific for lipids (carbohydrates also react). A spray prepared by mixing 3 volumes of 30% formaldehyde solution and 97 volumes of concentrated sulfuric acid with spot development by heating at 180°C for 30 min is quite sensitive, but intense spots are

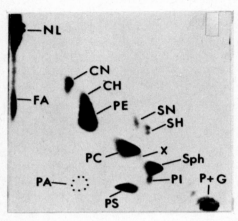

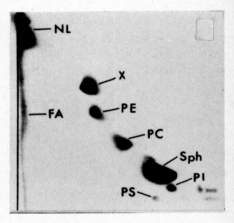

FIG. 2. Human brain lipid extract (800 μg) containing nonlipids including proteins. The patient suffered from a hereditary developmental defect apparently related to Niemann-Pick disease as judged from the approximately twofold elevation of Sph in organs other than brain. See the text. Chromatography with system (2) and spots located with spray (1). The brain lipid pattern showed developmental arrest with abnormally low levels of cerebrosides and sulfatides. The presence of abnormal amounts of CPH was not demonstrable by direct TLC but was disclosed by TLC of the fraction eluted from a silicic acid column with acetone (Section IV of text) as shown in Fig. 4.

FIG. 3. Extract (800 μg) of lipids from the liver in Niemann-Pick disease. See the text. Chromatography with system (2) and spray reagent (1). The extremely high sphingomyelin level (69% of the phosphatide) is readily apparent as is the presence of a high level (10% of the total phosphatide) of an uncharacterized phosphatide normally present as a minor component of liver. TLC is useful as an aid in the diagnosis of such hereditary metabolic defects.

obtained for organic compounds other than lipids. The latter spray is useful for detection of miscellaneous organic compounds found in lipid extracts and is particularly useful with fecal lipid extracts which contain many organic compounds with TLC behavior similar to that of lipids. Some substances give spots that vary in color during heating, and the color changes may prove to be useful for identification purposes. Nondestructive reagents are used when it is desired to locate lipids for isolation. Spraying with water is useful but not very sensitive. A spray of 0.001% Rhodamine 6G in water followed by viewing while wet under ultraviolet light is somewhat more sensitive.

More specific reagents are useful for division of polar lipids into groups. The phosphatide spray (*19*) is sensitive and specific. Phosphorus analysis of spots (*12*) is a useful alternative that gives a quantitative measure of phosphatides after TLC. The α-naphthol spray (*1*) is useful for glycolipids, but sterols and steroid conjugates

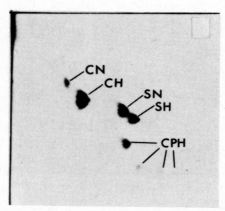

FIG. 4. Chromatogram illustrating the value of silicic acid column chromatography (Section IV of text) followed by TLC. The brain specimen, the total lipid of which is shown in Fig. 2, was applied to a silicic acid column and the glycolipid fraction obtained was examined by TLC (see the text; solvent pair (2), spray (1), 200 μg of fraction applied). Note the spots from CPH not visualized by TLC of the total lipid extract prior to column chromatography (Fig. 2).

also give colored spots and thus the reagent must be used with caution. Ninhydrin is useful for detection of lipids with free amino groups under proper conditions. Phosphatidyl serine may not be detected unless chromatograms are developed with a solvent mixture containing acetic acid and sprayed with reagent before all of the acid has evaporated from the plates. A suitable reagent is prepared from a stock solution of 0.1% ninhydrin in chloroform–methanol–water 4:4:1 that is mixed 3:1 with redistilled lutidine or pyridine just before use. Spots are developed by heating at 120°C for several minutes with care to avoid

overheating and decreased color yield. The orcinol spray for gangliosides (20) is useful, but the various reagents reported for the detection
of choline-containing lipids are not sufficiently sensitive or specific for
routine use.

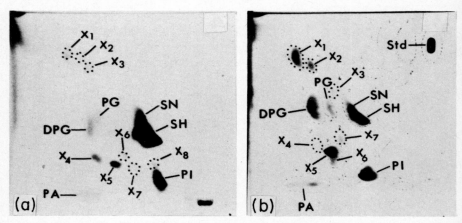

FIG. 5. Chromatograms illustrating the value of DEAE column chromatography
combined with TLC (Sections V and X.C of text). Chromatograms developed
with system (2) and sprayed with reagent (1) (see the text, 400 µg of each
sample applied). The samples were acidic lipids of brain eluted from DEAE
with chloroform–methanol–ammonia–ammonium acetate (Section V of text).
Figure 5(a) was from a child 8 years of age and has a high sulfatide level with
smaller amounts of phosphatides as compared to lipids of a child 5 months old
[Fig. 5(b)]. The uncharacterized components (X's) are not visualized by TLC
prior to column chromatography. Spot X₅ was shown to be an artifact, the
formation of which can be prevented using only freshly prepared ammonia and
ammonium acetate and solvent replacement rather than complete evaporation
to dryness (Section II.A of text). The spot marked Std in Fig. 5(b) was from
an aliquot of the sample applied after TLC as a standard for phosphatide
analysis.

A permanent record suitable for filing in compact form along with
other data is very desirable for TLC results. Such a record is provided
by photography with a Polaroid camera (21).

I. Standard Lipid Preparations for TLC

Mixtures of lipids are required for control of two-dimensional TLC
performance, since satisfactory performance is defined by migration

of different lipid classes relative to each other. Natural lipid mixtures are satisfactory and are less expensive than mixtures prepared from pure reference standards. Bovine brain lipid extracts provide a convenient reference mixture that contains cerebrosides and sulfatides in addition to phosphatides. Diphosphatidyl glycerol is conveniently placed on chromatograms with lipid extracted from whole beef heart

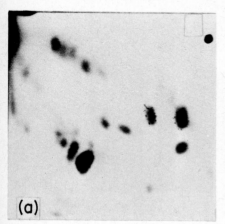 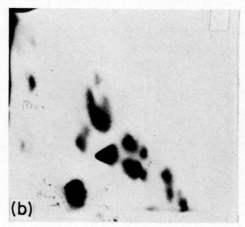

FIG. 6. TLC of substances in human feces eluted from a silicic acid column with acetone (a) and methanol (b) as described in Section IV of the text. Solvent pair (3) was used for TLC of the acetone eluate (a) and solvent pair (2) for the methanol eluate (b). Both chromatograms were sprayed with the formaldehyde–sulfuric acid reagent [spray (2)]. The numerous spots on both chromatograms are uncharacterized but illustrate the value of column chromatography combined with TLC for recognition of components of complex mixtures of organic compounds. The formaldehyde–sulfuric acid spray yields many spots of different colors that appear dark in the black and white reproductions. This reagent is particularly useful for detection of organic solvent-soluble (as opposed to water-soluble) nonlipids.

or beef heart mitochondria, in which it is a major component. A commercial soybean phosphatide mixture (Asolectin, obtainable from Associated Concentrates, Woodside, Long Island, New York) is useful for placement of phosphatidyl inositol (a major component of the mixture along with phosphatidyl ethanolamine and phosphatidyl choline). Spinach leaf extracts are useful for placement of mono- and diglycosyl diglycerides and sulfolipid. Pure reference standards of cerebroside, sulfatide, phosphatidyl choline (lecithin), phosphatidyl ethanolamine, phosphatidyl serine, sphingomyelin, and gangliosides can be isolated chromatographically from bovine brain. Preparations of many polar lipids that are satisfactory as reference standards are now available commercially (e.g., from Applied Science Laboratories, State College, Pennsylvania).

J. Fatty Acid Analysis after TLC

One-dimensional TLC after column chromatographic separation rather than two-dimensional TLC is generally most convenient for

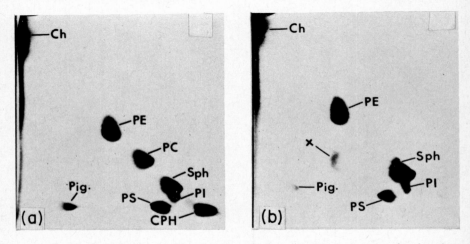

FIG. 7. Chromatograms developed with solvent pair (2) and spots developed with spray reagent (1). Figure 7(a) is from 600 μg lipid of pig erythrocytes and Fig. 7(b) from the same amount of sheep erythrocyte lipids. Both extracts were prepared from well-washed cells (to remove plasma) and passed through Sephadex (Section III.D of text) for removal of gangliosides and nonlipid contaminants. Note the large CPH spot in the pig cells that is absent from the sheep cells and the complete absence of PC from the sheep cells. Pigment in the extracts gives a spot.

isolation of lipid classes for fatty acid analysis (see Chapter 5). As soon as most of the developing solvent has evaporated, the chromatogram is sprayed with water or 0.001% aqueous Rhodamine 6G for spot localization. Larger amounts of lipid can be seen after spraying with water (giving white spots), but smaller amounts are best located under ultraviolet light after spraying with the fluorescent dye. The spots are scraped from the plate while the adsorbent is wet, transferred to test tubes, and dried over potassium hydroxide pellets in a vacuum desiccator flushed with pure nitrogen. Methyl esters are prepared without prior elution.

Esterification is accomplished as follows: 2 ml of 14% boron trifluoride in methanol (prepared by bubbling gaseous BF_3 into ice-cold methanol until the proper weight is obtained) is added to the dry adsorbed lipid obtained from the two plates. The tube is flushed with

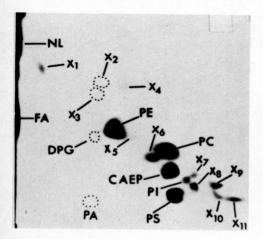

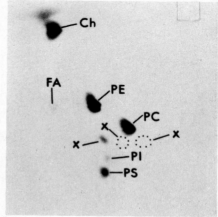

FIG. 8. Chromatogram of sea anemone lipids [solvent pair (2), spray reagent (1), 800 μg lipid]. Note that in addition to the common animal cell lipids several uncharacterized substances are present and CAEP is a major component.

FIG. 9. Chromatogram of octopus brain lipids [solvent pair (4), spray (1), 400 μg of sample applied]. Note the presence of several uncharacterized minor components.

nitrogen, sealed, and heated at 110°C for 90 min. The tube is cooled and opened; 1 ml of water is added; and the methyl esters are extracted with three 0.5-ml portions of *n*-hexane. The hexane solution is used for analysis by gas-liquid chromatography.

K. Isolation of Lipids after TLC

Elution of substances after separation by TLC is frequently convenient to provide small amounts of lipids to be used as standards or for identification, particularly by infrared spectrophotometry (18). Unaltered lipids can be eluted after TLC if the chromatograms are not allowed to stand after removal from the chamber (decomposition takes place upon exposure of lipid on dry adsorbent to air). The chromatogram is removed from the chamber and, immediately upon evaporation of most of the solvent (a few minutes only), sprayed with distilled water for localization of lipid components. The desired areas are scraped from the plate with a razor blade while the adsorbent is still moist. The areas from two one-dimensional TLC plates are scraped into a small glass-stoppered flask, and 10 to 15 ml of chloro-

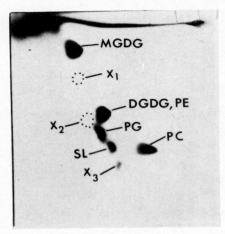

FIG. 10. Chromatogram of spinach leaf lipids [solvent pair (1), spray (1), 400 μg sample].

form–methanol 2:1 saturated with water or chloroform–methanol 1:1 containing about 10% water is added. The mixture is shaken vigorously and filtered through a sintered-glass filter of medium porosity. The extraction is repeated at least once and preferably twice for quantitative recovery. Water should not be allowed to evaporate completely from the plate prior to transfer to the extraction flask, since decomposition of some lipids may occur. Solvent is removed by evaporation in the cold under reduced pressure (Section II.A) taking care to minimize the time the lipid remains in the dry state.

The lipid dissolved in chloroform–methanol 19:1 saturated with water is freed of adsorbent and salts by passage through a Sephadex column 1.0 cm i.d. × 10 cm (Section III.D) using chloroform–methanol 19:1 saturated with water (50 ml) for elution of lipids and methanol–water 1:1 (50 ml) to clear the column of salts before reuse (flow rate 0.5 ml/min). The solvent is carefully removed by evaporation under reduced pressure in the cold (Section II.A); the lipid is dissolved in the desired solvent; and the amount of material is determined by weighing on a microbalance (Section II.B). Aliquots are then removed for TLC, hydrolysis, or infrared spectroscopic examination. Quantitative recovery of pure lipids is routinely obtainable with the procedure.

X. General Approaches to Quantitative Chromatographic Analysis of Complex Mixtures of Lipids

A. General Comments

Various combinations of column chromatography with thin-layer chromatography (TLC) provide the basis for chromatographic analysis of complex mixtures of lipids. TLC is used whenever possible as a substitute for column chromatography for quantitative analysis, since TLC is more rapid. Small quantities of unaltered lipids can be isolated after TLC (Section IX.K) for hydrolysis, infrared spectrophotometric examination, etc. Fatty acid analysis of spots after TLC is possible without prior elution from the adsorbent for methyl ester formation (Section IX.J). Gas phase chromatography in the nanogram range provides a means sufficiently sensitive for analysis of fatty acid composition of even a small spot after TLC (*17*).

Column chromatography prior to TLC possesses two major advantages. First, column fractions contain a smaller number of components. Separation by TLC is thus facilitated. Second, minor components can be concentrated in column fractions for determination by TLC. Some minor components are not visualized by TLC without prior column chromatography. Some components of column fractions can be determined by direct spectrophotometric assay without TLC.

Various combinations of chromatographic procedures useful for different purposes are presented below. The choice of procedure will depend upon the components of the particular sample to be analyzed and upon which components are to be isolated or determined.

B. Procedure 1. Sephadex Column Chromatography Followed by Thin-Layer Chromatography (TLC)

1. GENERAL COMMENTS

Total lipid is accurately determined by Sephadex column chromatography (Section III.D), since water-soluble nonlipid contaminants are quantitatively separated from lipids. TLC can be used for determination of the lipid classes in the first fraction from Sephadex, but no fully satisfactory TLC procedure is available for all types of gangliosides recovered in fraction 2.

Phosphatides are easily and accurately determined in Sephadex fraction 1 by two-dimensional TLC and phosphorus analysis of spots (*12*). Determination of glycolipids (cerebrosides, sulfatides, ceramide polyhexosides, glycosyl diglycerides, and sulfolipids) is less satisfactory without additional column chromatographic separation. The charring-transmission densitometry procedure provides accurate quantitative values only after one-dimensional TLC (*2*). Since spot overlap is usually obtained by one-dimensional TLC of complex mixtures of lipids, the charring procedure is not as generally applicable without prior column chromatography. Carbohydrate spectrophotometric assay procedures (with anthrone, etc.) when used with care can yield accurate results in some cases. Several problems are noted, however. Blank values may be high (and variable), and a decreased color yield may be observed in the presence of some phosphatides. Since the color yields from different sugars are different, the structures of the glycolipids must be known and suitable standards must be used if accurate analytical data are to be obtained. At present, glycolipids, particularly ceramide polyhexosides, are best determined and/or isolated by procedures employing column chromatography for separation of lipid classes prior to TLC (Sections X.C to X.F). The Sephadex column–TLC procedure is suitable for analysis of bile lipids (*22*) as described in Section X.B.2.

2. APPLICATION TO BILE

A Sephadex column 2.5 cm i.d. × 30 cm is prepared as described in Section III.D. Enough bile to yield about 0.5 g of solids is dehydrated (by evaporation in a rotary evaporator or by lyophilization). The sample should be spread evenly over the surface of a 2-liter flask to facilitate solution. The solids are first treated with chloroform–methanol 19:1 saturated with water, and the soluble material is transferred to the column. Some components of bile, particularly taurine conjugates of bile acids, are insoluble in the first solvent mixture. It is thus essential to treat the solids remaining in the sample flask with each of the different solvent mixtures used for elution of the column and to apply the material dissolved in each solvent just before elution with that solvent.

The column is eluted with the four standard solvent mixtures (Section III.D.3). The first fraction contains phosphatides and less polar (neutral) lipids along with traces of unconjugated bile acids and glycine-conjugated acids. The second fraction contains most of the glycine-conjugated acids. The third fraction contains the taurodihydroxycholanic acids, and the fourth fraction contains taurocholic acid and water-soluble nonlipids. The small amount of bile acid in the first fraction from Sephadex is removed by DEAE–Sephadex column chromatography. The DEAE–Sephadex column is prepared as described for ordinary Sephadex. The solids from Sephadex fraction 1 are applied to the DEAE–Sephadex column in chloroform–methanol 19:1 saturated with water, and the first two eluting mixtures and volumes used for Sephadex are used for elution of DEAE–Sephadex fractions: the first fraction contains lipids free of bile acids and the second contains free and glycine-conjugated bile acids. The second fraction from DEAE-Sephadex is pooled with the second fraction from Sephadex as the total free plus glycine-conjugated bile acid fraction.

The individual lipid classes of the four fractions are determined by quantitative TLC. Charring and transmission densitometry (*11,16,22*) after separation with *n*-hexane–diethyl ether–glacial acetic acid 70: 30:1 or 60:40:1 is used for neutral lipids. Phosphatides are determined after TLC by phosphorus assay of spots (*12*). Bile acids are first separated by TLC with chloroform–methanol–water 65:25:4 or chloroform–methanol–28% (by weight) aqueous ammonia 65:35:5 and determined by charring–transmission densitometry using pure bile acids as standards (*22*).

C. Procedure 2. Sephadex and DEAE Cellulose Column Chromatography with Thin-Layer Chromatography

The use of Sephadex (Section III.D) followed by DEAE column elution sequence 3 (Section V) and then TLC (Section IX) is the most versatile general analytical procedure with which accurate results are obtainable. The chief advantage of this approach compared with the Sephadex–TLC procedure is that the DEAE column fractions contain fewer components and accurate quantitative analysis by TLC is facilitated. The lipid mixture is freed of water-soluble nonlipids by Sephadex column chromatography, and gangliosides are recovered as a separate fraction. The less polar (neutral) lipids are separated from the polar lipids by elution from DEAE with chloroform.

The components of DEAE column fractions can often be determined by one-dimensional rather than two-dimensional TLC, and several alternatives are usually possible. With brain lipids the chloroform–methanol 9:1 fraction contains cerebrosides, phosphatidyl choline, and sphingomyelin. All three can be determined after one-dimensional TLC by the charring–transmission densitometry procedure (11). As an alternative, cerebroside can be determined in the fraction without TLC with anthrone or other carbohydrate assay reagents and the two phosphatides can be separated by one-dimensional TLC, the spots aspirated, and the amounts determined by phosphorus assay (12). For fatty acid analysis, the components are separated by one-dimensional TLC with chloroform–acetone–methanol–acetic acid–water 5:2:1:1:0.5 as developing solvent, the chromatogram is sprayed with water for spot localization, and fatty acid methyl esters for gas phase chromatographic analysis are prepared without prior elution from adsorbent (Section IX.J). The spots for cerebrosides with hydroxy and nonhydroxy fatty acids are completely separated. The hydroxy fatty acids are analyzed by gas phase chromatography as methyl esters after acetylation of the hydroxyl group (23) or formation of the trimethylsilyl derivative.

With brain lipids, the other fractions from DEAE are handled as for the first fraction, i.e., by carbohydrate assay, phosphorus analysis, or charring–transmission densitometry after TLC. Lipids of plants including glycosyl diglycerides and sulfolipid are determined in DEAE column fractions by spectrophotometric carbohydrate assay, charring–transmission densitometry after TLC, or phosphorus analysis of spots after TLC. Fatty acid analysis is accomplished after TLC as noted above. Lipids of microorganisms are determined in the same general manner.

D. Procedure 3. Sephadex and TEAE Cellulose Column Chromatography with Thin-Layer Chromatography

The procedure is similar to procedure 2 employing DEAE. TEAE (Section VI) is substituted for DEAE when a ceramide polyhexoside fraction is to be separated from other lipids or when the sample contains a large amount of free fatty acid, bile acid, or other lipids with free carboxyl groups as the only ionic groups, since DEAE is easily overloaded with respect to such lipids. TEAE is thus preferable for lipids extracted from lung and the lens of the eye, since both contain appreciable amounts of ceramide polyhexosides. TEAE is also useful for fecal lipids that contain much free fatty acid. Quantitative analysis of components of fractions from TEAE columns is accomplished by quantitative TLC as described for fractions from DEAE columns (Section X.C).

E. Procedure 4. Sephadex and Silicic Acid Column Chromatography with Thin-Layer Chromatography

The use of silicic acid (Section IV) rather than ion-exchange cellulose column chromatography (as in procedures 2 and 3 above) may be advantageous with samples that contain a great deal of glycolipid but very little diphosphatidyl glycerol (cardiolipin) and pigments. The lipid fraction from Sephadex is divided by silicic acid column chromatography into three major fractions: less polar (neutral) lipids (eluted with chloroform), glycolipids (cerebrosides, cerebroside sulfates, ceramide polyhexosides, glycosyl diglycerides, and sulfolipid, all of which are eluted with acetone), and phosphatides (eluted with methanol). The amounts of the individual components of the fractions from silicic acid are then determined by spectrophotometric assay of phosphorus (*12*) or carbohydrate after TLC or by the charring–transmission densitometry procedure (*11*).

The silicic acid column procedure is particularly useful for lipid extracts of brain, spinach leaves, and feces. The acetone fraction from silicic acid with normal brain lipids as sample contains cerebrosides and cerebroside sulfates with traces only of ceramide dihexoside, and brain lipids from some pathological states (e.g., Niemann–Pick and Tay–Sachs diseases) contain ceramide polyhexosides as major components. With lipid extracts of spinach leaves, elution with chloroform–acetone 1:1, followed by acetone separates mono- and diglycosyl diglycerides (*9,10*). Fecal lipids are separated into three distinctly different groups by elution with chloroform followed by acetone and then methanol. Most of the components of the acetone and methanol fractions from fecal lipid extracts are unidentified organic compounds, many of which are not lipids. These nonlipids are soluble in organic solvents (as contrasted with water-soluble nonlipids removed on Sephadex) and are best detected after TLC with the formaldehyde–sulfuric acid reagent (Section IX.H) that yields a wide variety of colors.

Since extracts of most animal organs other than brain contain more than traces of diphosphatidyl glycerol (eluted in part with acetone), TEAE column chromatography of the acetone eluate (procedure 5 below) is required for recovery of neutral glycolipids free of phosphatide.

F. Procedure 5. Sephadex, Silicic Acid, and TEAE Cellulose Column Chromatography Combined with Thin-Layer Chrimatography

The use of the sequence Sephadex (Section III.D), silicic acid (elution with chloroform, acetone, and then methanol, Section IV), and TEAE column chromatography (Section VI) prior to TLC is advantageous in several cases and should be considered for some of the more complex mixtures of lipids. Thus, with lipid extracts from most animal organs diphosphatidyl glycerol is a contaminant of the glycolipid fraction eluted from silicic acid with acetone. The phosphatide can be removed on TEAE. Elution with 10 column volumes each of (1) chloroform–methanol 9:1 (cerebroside), (2) chloroform–methanol 2:1 (ceramide polyhexosides), (3) methanol (inorganic compounds released by ion exchange), and (4) chloroform–methanol 4:1 made 0.01 M with ammonium or potassium acetate to which is added 20 ml of 28% (by weight) aqueous ammonia per liter (diphosphatidyl glycerol, sulfatide, sulfolipid) is adequate in most cases.

With fecal lipid extracts, each fraction from the silicic acid column is divided into acidic and nonacidic lipids by TEAE column chromatography. In this case the general elution scheme (10 column volumes of each solvent) is (1) chloroform, (2) chloroform–methanol 2:1, (3) methanol, (4) acetic acid, (5) methanol for removal of acetic acid (4 column volumes is adequate), (6) chloroform–methanol 4:1 made 0.01 M in ammonium or potassium acetate to which is added 20 ml of 28% (by weight) aqueous ammonia per liter. Most of the components of the fractions from feces are unidentified.

G. Procedure 6. Total Fractionation by Column Chromatography

Procedures employing only column chromatography for separation of pure lipids are useful for isolation of large amounts of major components of mixtures and for isolation of adequate amounts of minor and trace components for structural analysis. Larger amounts of the major components of lipid mixtures will be preferred when it is desired to separate and determine the individual molecular species of each lipid class.

Sephadex column chromatography (Section III.D) for removal of nonlipids should be used as the first step. The most useful sequence for separation of lipid classes is DEAE or TEAE cellulose followed by silicic acid–silicate and/or silicic acid column chromatography (stepwise elution with increasing amounts of methanol in chloroform). The details of column procedures are given in other sections.

Acknowledgments

This chapter is based upon research work supported in part by the following grants: NB-01847-04, -05, -06, -07, -08 and NB-06237-10 from the National Institute of Neurological Diseases and Blindness; CA-03134 (C4), (C5), -07, -08, -09 from the National Cancer Institute; a City of Hope Medical Center Public Health Service general research support grant; Grants DA-CML-18-108-G7, -G29, DA-AMC-18-035-71(A), and Contract CA-18-035-AMC-335(A) from the U.S. Army, Edgewood Arsenal. Mr. Richard Baldwin provided valuable laboratory assistance, and Mrs. Dorothy Heller assisted in preparation of the manuscript.

REFERENCES

1. A. N. Siakotos and G. Rouser, *J. Am. Oil Chemists' Soc.,* **42**, 913 (1965).

2. G. Rouser, C. Galli, and G. Kritchevsky, *J. Am. Oil Chemists' Soc.,* **42**, 404 (1965).

3. J. Folch, M. Lees, and G. H. Sloane-Stanley, *J. Biol. Chem.,* **226**, 497 (1957).

4. D. Nazir and G. Rouser, *Lipids,* **1**, 159 (1966).

5. B. Borgström, *Acta Physiol. Scand.,* **25**, 101 (1952).

6. C. H. Lea, D. N. Rhodes, and R. D. Stoll, *Biochem. J.,* **60**, 353 (1955).

7. G. Rouser, J. O'Brien, and D. Heller, *J. Am. Oil Chemists' Soc.,* **38**, 14 (1961).

8. L. M. Smith and N. K. Freeman, *J. Dairy Sci.,* **42**, 1450 (1959).

9. M. L. Vorbeck and G. V. Marinetti, *J. Lipid Res.,* **6**, 3 (1965).

10. G. Rouser, G. Kritchevsky, G. Simon, and G. J. Nelson, *Lipids,* **2**, 37 (1967).

11. G. Rouser, C. Galli, E. Lieber, M. L. Blank, and O. S. Privett, *J. Am. Oil Chemists' Soc.,* **41**, 836 (1964).

12. G. Rouser, A. N. Siakotos, and S. Fleischer, *Lipids,* **1**, 85 (1966).

13. G. Rouser, A. J. Bauman, G. Kritchevsky, D. Heller, and J. O'Brien, *J. Am. Oil Chemists' Soc.,* **38**, 544 (1961).

14. G. Rouser, G. Kritchevsky, D. Heller, and E. Lieber, *J. Am. Oil Chemists' Soc.,* **40**, 425 (1963).

15. G. Rouser, G. Kritchevsky, C. Galli, and D. Heller, *J. Am. Oil Chemists' Soc.,* **42**, 215 (1965).

16. M. L. Blank, J. A. Schmit, and O. S. Privett, *J. Am. Oil Chemists' Soc.,* **41**, 371 (1964).

17. G. L. Feldman and G. Rouser, *J. Am. Oil Chemists' Soc.,* **42**, 290 (1965).

18. G. Rouser, G. Kritchevsky, M. Whatley, and C. F. Baxter, *Lipids,* **1,** 107 (1966).

19. J. D. Dittmer and R. L. Lester, *J. Lipid Res.,* **5,** 126 (1964).

20. L. Svennerholm, *Nature,* **177,** 524 (1956).

21. G. Rouser, A. J. Bauman, N. Nicolaides, and D. Heller, *J. Am. Oil Chemists' Soc.,* **38,** 565 (1961).

22. A. Yamamoto and G. Rouser, submitted for publication.

23. J. O'Brien and G. Rouser, *Anal. Biochem.,* **7,** 288 (1964).

24. G. L. Feldman, L. S. Feldman, and G. Rouser, *Lipids,* **1,** 21 (1966).

4

ANALYSIS OF PHOSPHATIDES AND GLYCOLIPIDS BY CHROMATOGRAPHY OF THEIR PARTIAL HYDROLYSIS OR ALCOHOLYSIS PRODUCTS

R. M. C. Dawson

BIOCHEMISTRY DEPARTMENT
AGRICULTURAL RESEARCH COUNCIL
INSTITUTE OF ANIMAL PHYSIOLOGY
CAMBRIDGE, ENGLAND

I. Introduction

The examination of lipid mixtures by analysis of the products formed by total hydrolysis in strong acids and alkalis has been used since the earliest days of lipid research (1). Since it is often impossible to obtain any idea as to which lipid the components have been derived from, the limitations of this approach are obvious. In 1954 Dawson (2) introduced the principle of using a very mild alkaline alcoholysis to deacylate the phosphatides followed by a chromatographic examination of the water-soluble P products. This allowed a much more positive identification of the parent phosphatide to be obtained. Although originally the method was only intended as a means of measuring the specific radioactivity of individual diacylated phosphatides, it was later extended to give a complete quantitative picture of the complex phosphatide mixture present in tissues (3,4), and the same principle has been used for studying plant glycolipids e.g., (11) and for structural studies on phosphatides, e.g. (41,42).

From the outset the method has been widely used and, as is to be expected, many variations and modifications have been introduced by various workers. For convenience and clarity in presentation it is proposed to make the method that has been gradually evolved for the examination of phosphatide mixtures in this laboratory the main theme and backbone of this chapter while, at the same time, describing and acknowledging the various important modifications that have been introduced by other workers.

A. Principle of the Method of Partial Degradation

The phosphatide sample is subjected to a series of successive chemical hydrolyses or alcoholyses that selectively remove various hydrophobic groups from the molecule and consequently change its solubility distribution between an aqueous phase and an immiscible organic solvent phase. The water-soluble phosphorus-containing products are separated by chromatography, usually on paper but sometimes on an anion-exchange resin (5). The first treatment is a mild alkaline alcoholysis that removes fatty acids esterified to glycerol and consequently liberates water-soluble "glycerylphosphoryl" derivatives from simple diacyl and monoacylphosphoglycerides (see Table 1). For example, phosphatidyl ethanolamine gives glyceryl phosphoryl ethanolamine, lecithin, and lysolecithin form glyceryl phosphoryl choline. At the same time the plasmalogens and glyceryl ether phosphatides lose their fatty acyl ester groups but retain their solubility in lipid solvents by virtue of the ether-linked long-chain hydrocarbon groupings still attached to the molecule (e.g., plasmalogens are converted to lyso-plasmalogens).

The glyceryl phosphoryl derivatives produced are unstable to prolonged contact with alkali (*3*), but with the short periods of incubation used further decomposition is minimal. However, a small formation of cyclic glycerophosphate, particularly from glyceryl phosphoryl choline and glyceryl phosphoryl inositol, does occur (*3,6,7,8*), but fortunately a simple correction can be applied to compensate for this loss.

TABLE 1

STEPWISE ELUTION OF THE PRODUCTS FORMED BY ALKALINE ALCOHOLYSIS OF PHOSPHATIDES FROM DOWEX-1–FORMATE COLUMNS[a]

Eluting solvent contains		
Na tetraborate, mM	Ammonium formate, mM	Phosphate ester removed[b]
5		Glyceryl phosphoryl choline
5	10	Glyceryl phosphoryl ethanolamine
5	40	Glyceryl phosphoryl inositol
		Cyclic glycerophosphate
5	60	Glyceryl phosphoryl serine
5	90	Glycerophosphate
5	150	1,3-Bis(glyceryl phosphoryl)glycerol
5	300	Glyceryl phosphoryl inositol phosphate
		Glyceryl phosphoryl inositol diphosphate

[a] Data taken from References (*5*) and (*43*).
[b] For parent lipid see legend of Fig. 2.

The phosphatides retaining their solvent solubility after alkaline alcoholysis are then subjected to an acid hydrolysis at 37°C, the lysoplasmalogens largely decomposing to form water-soluble glyceryl phosphoryl derivatives. However, it has been found that a proportion of the lysoplasmalogen cyclizes and forms the cyclic acetal derivative, which is much more stable to acid hydrolysis (*9*). Fortunately, this latter reaction can be prevented by including $HgCl_2$ in the acid hydrolysis medium (*9*), and by this means the liberation of glyceryl phosphoryl derivatives from the lysoplasmalogens is made much more quantitative (*4*).

The phosphatides retaining their solvent solubility after mild alkaline alcoholysis and Hg^{2+}–acid hydrolysis are heated at 100°C with methanolic-HCl, which breaks down the fatty acid–amide bonds of sphingomyelin and liberates predominantly phosphoryl choline but also some sphingosyl phosphoryl choline (*10*). These are water-soluble, whereas the unhydrolyzed glyceryl ether phosphatides (in the lyso form) retain their solubility in organic solvents and can be estimated by direct phosphorus assay of the solution.

B. Advantages and Disadvantages of the Method

All the techniques at present available for examining the phosphatide distribution in a given sample possess their advantages and disadvantages, and consequently practically all lipid laboratories now use a variety of techniques either to supplement and complement one another or to choose from, depending on the information required.

The great advantage of the "partial hydrolysis" method is that it enables the phosphatide composition to be examined in far greater detail than with other chromatographic methods. Thus on thin-layer or paper-impregnated silicic acid chromatography or formaldehyde-treated paper chromatography the choline, ethanolamine, and serine phosphoglycerides are not resolved into their diacyl, plasmalogen, and glyceryl ether forms. Furthermore, the partial hydrolysis method gives a greatly superior resolution of the acidic phosphatides such as phosphatidic acid, phosphatidyl glycerol, cardiolipin, and phosphatidyl glycerophosphate, which other methods either poorly distinguish or totally fail to separate. Thin-layer chromatography on silicic acid also gives a poor resolution of phosphatidyl serine and phosphatidyl inositol, whereas both diphosphoinositide and triphosphoinositide almost invariably remain at the origin. As the method of partial hydrolysis relies on the chromatography of water-soluble substances, there is almost a complete absence of cochromatography as often occurs during the chromatography of the intact phosphatides. This means that with direct chromatography of the phosphatides a main peak on a chromatogram of the whole lipids is often contaminated by small amounts of other phosphatides, a phenomenon probably due to a tendency of the lipids to form mixed micelles in the solvent (inverted micelles in nonpolar solvents). Furthermore, contaminating substances (e.g., inorganic P) can cochromatograph with the intact phosphatides, and both of these phenomena can lead to serious errors in isotope work.

The extra resolution obtained by the method of partial hydrolysis is paid for by the extra labor involved compared with the direct chromatography of the phosphatides. An intrinsic disadvantage of the method is that it does not allow an examination of the fatty acid components of any individual phosphatide and there is no resolution of the diacyl and lyso forms of lecithin and phosphatidyl ethanolamine. Fortunately, lysocompounds usually exist in very low concentrations in normal tissues except in certain cases, e.g., in blood plasma, where about 10% of the lipid phosphorus is present as lysolecithin. Lyso compounds can, however, be readily separated by direct chromatography of the phosphatides on silicic acid, although it is essential to do this in alkaline solvents, else the plasmalogens decompose and give an elevated level for the lysophosphatides (*12*).

Because of the shortcomings and advantages of the various chromatographic methods for examining phosphatides, efforts have been made to wed the techniques and eliminate the disadvantages from the union. Thus Magee (*13*) initially fractionates the phosphatides on a short silicic acid column before carrying out alkaline or acid degradation.

II. Method of Phosphatide Analysis

A. Preparation of Lipid Samples

The tissue† is extracted by any of the conventional techniques used for lipid extraction. Homogenization at room temperature in $CHCl_3$– methanol 2:1, v/v, 20 volumes (*14*) in a high-speed blender is very satisfactory. After standing 10 min, the suspension is centrifuged and the residue is reextracted several times with more of the same solvent until no more lipid phosphorus is removed. Such an extraction leaves behind the polyphosphoinositides and traces of other phosphatides, especially those of an acidic nature (*15*). Preextraction of the tissue with acetone and ethanol allows some extraction of the polyphospho-inositides by chloroform–methanol, but some decomposition can occur and for quantitative assessment successive extractions of the residue with chloroform–methanol 2:1, v/v, acidified with conc. HCl (0.25% by volume) are necessary (*15,27*). Unfortunately, this solvent cannot be used to extract directly the tissue, else extensive acid hydrolysis of the plasmalogens occurs.

---

† Special methods of extraction are needed for microorganisms; see e.g., (*56*).

It is beneficial to wash the lipid extract in chloroform–methanol at least once using 0.2 volume 0.9% NaCl† to prevent the loss of phosphatides to the aqueous phase (*14*). This removes contaminating water-soluble phosphorus compounds from the lipid extract. The phosphorus content of the lower chloroform-rich phase is determined, and it is stored at −10°C for analysis. If the extract contains appreciable amounts of proteolipid, it is best to remove this by evaporating the extract to dryness in vacuo in the presence of water and redissolving the lipid in chloroform or light petroleum.

B. Alkaline Alcoholysis

1. METHOD 1

Alkaline alcoholysis is carried out using a limited amount of alkali to prevent the alkali-metal ions from interfering with the subsequent chromatography of the water-soluble P compounds produced. If the extract contains a large amount of neutral fat and low concentrations of phosphatide (<40%, w/w, e.g., extracts of adipose tissue or fatty livers) it is first necessary to separate the phosphatides with a simple silicic acid column or by other means such as acetone–$MgCl_2$ precipitation (*18,19*).‡

† Calcium chloride or magnesium chloride solutions should not be used for washing, as this can lead to low recoveries of acidic phosphatides (*16*). With some washed lipid mixtures, especially those from brain, in which the polyphosphoinositides have a great affinity for divalent cations, the acidic phosphatides can be precipitated on adding alkali to their methanolic solutions, leading to a low (80–90%) recovery of their alkali-labile P (*17*). Brockerhoff (*17*) circumvented this difficulty by converting them to their triethylammonium salts. In our laboratory, we wash the original lipid extract with 0.02 *M* sodium cyclohexane 1,2-diaminetetraacetate (pH 8) and then with theoretical upper phase (*14*) made from 0.9% NaCl.

‡ Theoretically, the NaOH is only acting as a catalyst if the reaction is entirely an alcoholysis with the formation of methyl fatty acid esters. Nevertheless, it is found that the deacylation reaction does not go to completion in the presence of large amounts of neutral fat containing fatty acyl ester bonds (e.g., triglyceride), and it is necessary to remove this from the lipid mixture before the reaction is carried out. If, however, the alkali is subsequently to be removed with a cation-exchange resin or the deacylation products are to be separated on an anion-exchange resin, more can be added to take the reaction to completion in the presence of neutral fat. Thus Prottey and Hawthorne (*54*) used 4–8 moles of hydroxide per mole of phosphatide when examining lipid extracts of pancreas, which are very rich in triglyceride. Care should be taken not to use a large excess of alkali, else breakdown of the phosphodiesters occurs with the formation of alcohol esters of glycerophosphate.

The lipid sample containing not more than 550 μg P is pipetted into a 50-ml round-bottomed flask with a ground-glass joint. The solvent is removed below 50°C in a rotary evaporator or in vacuo with a standard splash head and swirling to avoid splashing. The residue is redissolved in 0.8 ml AR carbon tetrachloride.† Ethanol (7.5 ml) is then added, followed by 0.65 ml water‡ and 0.25 ml aq. N NaOH.§

The mixture is incubated at 37°C for 20 min and pH-tested with indicator paper. It should still be distinctly alkaline (>pH 11); and if not, it is probable that too much neutral fat is present in the sample, and this should be removed (as above) before the alcoholysis. The

---

† Certain technical grades give pink spots on the subsequently developed chromatogram.

‡ Many workers have left out water from the alkaline medium, but this leads to esterfication of glycerophosphate (*52*).

§ Many variations of this incubation mixture have been used by various workers. Each variation can change the percentage of cyclic glycerophosphate formed by decomposition of the glyceryl phosphoryl derivatives, and consequently a correction factor different from the present one has to be used. Brockerhoff (*17*) showed that, with lecithin, less cyclic glycerophosphate is formed with LiOH than with KOH, which itself produces less than NaOH. However, we have found that NaOH produces more rapid deacylation, a factor that can be of importance when dealing with acyl ester bonds resistant to hydrolysis, such as in the plasmalogens (*25*). It has also been shown that the more highly polar the solvent used the less cyclic glycerophosphate is formed (*17,21*); e.g., in pure methanol 0.5% cyclic glycerophosphate is formed from lysolecithin, whereas in chloroform–methanol 3:2, v/v, this is increased to 19.3%. Consequently, an incubation mixture containing 0.8 ml 0.125 M methanolic LiOH·H_2O added to 0.2 ml of a chloroform solution of the phosphatides (20–30 μmoles lipid P) was used by Brockerhoff (*17*) to avoid further decomposition of the glycerol phosphoryl derivatives. Benson (*20*) dilutes the lipids (=25–200 mg wet plant tissue or 0.1–1.0 g fresh mammalian tissue) in solution in 0.1 ml ethanol–toluene 9:1, v/v, with 0.1 ml, 0.2 M ethanolic KOH.

For successful deacylation a strong base with a high dissociation constant is necessary. Thus Hübscher et al. (*22*) found that hydroxylamine was not effective as a deacylating agent unless traces of sodium methoxide were present. Although hydroxylamine was used by Hanahan (*29*) to prepare glyceryl phosphoryl choline by deacylation of lecithin and by Brockerhoff and Ballou (*41*) to study the structure of the brain polyphosphoinositide complexes, it is most likely that the reaction was brought about by the excess NaOH used in preparing the hydroxylamine solution. Bases such as NH_3, diethylamine (*3*), triethylamine, and triethanolamine (*17*) are also ineffective. On the other hand, quaternary ammonium compounds are reasonably effective (*17*) and degradation of lecithin with tetrabutylammonium hydroxide has been used to prepare glyceryl phosphoryl choline (*23*).

liberation of water-soluble phosphorus† is virtually complete in this time period,‡ although some of the plasmalogen may not have been deacylated. The reaction is stopped by adding ethyl formate (0.4 ml). This immediately neutralizes the excess alkali by the reaction ethyl formate + NaOH = sodium formate + ethanol and avoids local concentrations of acid that might hydrolyze part of the lysoplasmalogen formed during the deacylation.§ The hydrolysate is taken to dryness in a rotary evaporator below 60°C. A volume of water is equilibrated with 2 volumes of isobutanol–CHCl₃ mixture, 1:2 v/v,¶ by shaking for a few minutes and allowing to settle. Samples (1 ml) from the aqueous layer and lower solvent layer (2 ml) are added to the residue, and the flask is shaken and warmed slightly to ensure the quantitative solubilization of the alcoholysis products and the residual phosphatides. The emulsion, after transferring as quantitatively as possible to a centrifuge tube, is spun hard and the upper aqueous layer is withdrawn with a Pasteur pipette and stored for chromatography at 0°C. Usually, the separation obtained on centrifuging is good, with almost clear

† Although the solvent contains water, the reaction is largely an ethanolysis with the formation of ethyl esters of the fatty acids. The water prevents the formation of ethyl glycerophosphate (*52*).

‡ The rate of deacylation varies with the phosphatide concerned. At 37°C lecithin is completely broken down within a minute or so, and even at 0°C the rate of the reaction is appreciable. Fatty acid is removed from both α and β positions, although preferentially from the α position, and lysolecithin accumulates as a transient intermediary (*21,24*). At room temperature lecithin is deacylated 13 times more rapidly than phosphatidyl serine and phosphatidyl ethanolamine, but this difference in rate decreases as the temperature is reduced (*21*). The acyl ester bond of plasmalogens and glyceryl ether phosphatides is much more resistant to alcoholysis (*25,26,71*), and for complete deacylation it is necessary to prolong the time to 1 hr under the present conditions (*25*). The rate of deacylation increases as the polarity of the solvent used to carry out the reaction decreases (*21*). Thus we have found that deacylation of the plasmalogens is increased up the series methanol, ethanol, isopropanol, propanol (R. M. C. Dawson and N. Hemington, 1962, unpublished observations) for a given concentration of alkali.

§ Neutralization with HCl to give a pH of 1 has been used (*22*) but this would undoubtedly lead to a hydrolysis of lysoplasmalogens. Marinetti (*55*) has used Na methoxide in methanol to deacylate the phosphatides and has then chromatographed the products on paper without removal of the alkali.

¶ Isobutanol is included to facilitate the subsequent separation of the two phases containing the water-soluble products of the hydrolysis. It also prevents the loss of lysoplasmalogen to the aqueous phase provided the alkali has been removed or previously neutralized (*28*).

upper and lower phases and a small amount of interfacial material. Occasionally difficulty arises in obtaining a clean separation, but this can be eliminated by adding cetyltrimethylammonium bromide (0.25% final concentration) to the water. This does not affect the eventual recovery. Alternatively, the emulsion can be broken by freezing in solid carbon dioxide and centrifuging after thawing.

2. METHOD 2

This alternative method completely removes alkali by a cation-exchange carboxylic acid resin. Better resolution of the alcoholysis products can be obtained, especially if chromatography rather than ionophoresis is being used for their separation. The total removal of alkali allows more material to be loaded on the chromatogram and also allows greater amounts of alkali to be used if the lipid extract is rich in neutral fat. Unfortunately, a small portion of the residual phosphatides is adsorbed on the column and it is difficult to remove it by washing.

The alkaline alcoholysis is conducted in the same way as in method 1. However, alkali is removed by passing the alcoholysate cooled to 0°C through a short column† of Amberlite IRC 50 H⁺ (4 cm long × 0.8 cm diameter).‡ The alcoholysis products are washed through the column with two successive 5-ml portions of 80% ethanol. The total effluent of the column is taken to dryness in vacuo and then distributed between water and isobutanol as in method 1.

C. Hydrolysis of Plasmalogens

The alkaline alcoholysis carried out as described does not completely deacylate the plasmalogens whose fatty acyl ester bonds are much more resistant to alkali (25). Consequently, when the phosphatide

† A batch principle has been used (20) in which the reaction mixture is treated with dry Dowex 50 H⁺ (4 mg/0.1 ml 0.2 M NaOH) and light petroleum is added to solubilize the residual phosphatides. However, it is often necessary to elute the resin with phenol–water chromatographic solvent to completely recover the deacylation products.

‡ The contact with the resin should be as short as possible, else further decomposition of the alcoholysis products may occur. Zeo Karb 226 (H⁺) has been used (5), but loss of the water-soluble P products occurred unless the column was well washed with water.

mixture remaining is treated with acid–$HgCl_2$ to hydrolyze the vinyl ether bonds of the plasmalogens, a small part of their P is not rendered water-soluble.† This residual plasmalogen P appears in the next fraction and a correction must be applied.

A 1.6-ml sample of the almost clear lower solvent phase from the alkaline alcoholysis is withdrawn by a bulb-operated pipette. To ensure that this fraction is not contaminated with any of the aqueous layer, the point of the pipette is inserted below the interface and any entrapped water is driven out by slightly warming the pipette with the hand. The sample is transferred to a 10-ml stoppered centrifuge tube to which is added 0.8 ml of 10%, w/v, trichoroacetic acid containing $HgCl_2$ (5 mM).‡ The mixture is then vigorously shaken at 37°C for 30 min. Ether saturated with water (2 ml) is added, and after the mixture has been shaken and centrifuged the upper layer is removed. The lower aqueous layer is then extracted twice with 2 ml of water-saturated chloroform–ether–isobutanol 1:1:2, v/v, and finally with 2 ml of water-saturated ether. It is then roughly adjusted to pH 6–9 by holding a wick soaked in dilute ammonia in the air of the tube,

---

† If the alkaline hydrolysis is made more vigorous, then more cyclic glycerophosphate is formed by decomposition of the glyceryl phosphoryl derivatives. However, if sufficient lipid material is available, it is possible to carry out two alkaline alcoholyses on equal portions of the sample; the first a very mild one to examine the diacylated phosphoglycerides and the second a more vigorous one to investigate the residual phosphatides (plasmalogens, sphingomyelin, glycerol ether phosphatides). The first is carried out in a manner similar to that described above, but using methanol in place of ethanol, using LiOH in place of NaOH, and shortening the incubation at 37°C to 5 min. The aqueous layer (A) formed on distribution of the alcoholysate between water–isobutanol–chloroform contains virtually no cyclic glycerophosphate—the solvent layer is rejected. More vigorous alcoholysis is carried out on a second portion of the lipid extract using NaOH, substituting propanol for ethanol, and prolonging the alcoholysis to 60 min. In this case the A layer is rejected and the residual phosphatides are investigated in the lower chloroform-isobutanol layer. In a similar procedure used by Wells and Dittmer (28), the residual phosphatides are recovered from the initial alcoholysate and treated more vigorously with alkali to complete the deacylation of the plasmalogens.

‡ McMurray (69) uses a reagent to cleave the lysoplasmalogens which contains ethanol, acetic acid, and $HgCl_2$. However, this has not given satisfactory results in the hands of Horrocks and Ansell (70), who have also found that substitution of hydrochloric acid (0.06 N) for trichloroacetic acid leads to a less complete hydrolysis.

which is then stoppered and shaken. The solution is stored at 0°C for chromatography.

D. Methanolysis of Phosphatides Remaining after Mild Alkaline and Acid Treatment

The final treatment of the residual phosphatides is a HCl methanolysis that converts sphingomyelin into water-soluble phosphoryl choline and sphingosyl phosphoryl choline† *(30–32)*. At the same time the glycerol ether phosphatide phosphorus retains its solubility in organic solvents even though some base may be lost with the formation of batyl or chimyl phosphates *(33,34)*.

The combined solvent layers from the acid–HgCl$_2$ hydrolysis are evaporated to near dryness in vacuo at 100°C, and the residual lipid material, trichloroacetic acid, and HgCl$_2$‡ are heated in a sealed tube for 4 hr§ at 100°C with 1.25 ml $2 N$ anhydrous methanolic HCl.¶ Precautions should be taken to shield the tube in the event of a pressure explosion, but provided the tube is thick-walled, this rarely occurs. After cooling in ice the tube is opened and the contents are evaporated to dryness in vacuo. Water (ether-saturated) 0.5 ml and ether (water-saturated) 2 ml are added, and after shaking and centrifuging, the upper ether layer is withdrawn. The aqueous layer is rewashed with 1 ml ether and then stored for chromatography. The combined etherial layers are taken to dryness and the phosphorus content of the residue (glycerol ether phosphatides) determined as below.

E. Separation of Phosphate Esters

Many methods have now been used to separate the water-soluble phosphate esters produced by alkaline alcoholysis; these include paper

---

† With anhydrous methanolic HCl, phosphoryl choline is the predominant product, but if 20% H$_2$O is added to the reaction mixture, equal amounts of the two water-soluble phosphate esters are produced *(32)*.

‡ HgCl$_2$ is extracted from the aqueous hydrolysis medium by ether.

§ Pries *(21)* claims that complete methanolysis of sphingomyelin is obtained in $1 M$ methanolic HCl for 90 min at 65°C, but in our experience this is not so.

¶ Dry HCl gas prepared by heating NaCl with sulfuric acid is adsorbed into dry methanol using an anti-sucking-back device. After titration, the solution is diluted to $2 N$ with methanol.

chromatography† with various solvents, paper ionophoresis, thin-layer chromatography, and ion-exchange chromatography.

1. SEPARATION ON PAPER OF THE ALKALINE AND ACID DEGRADATION PRODUCTS

For maximum resolution of the alkaline alcoholysis products we use in this laboratory a two-dimensional separation on paper by (1) descending chromatography in a solvent composed of phenol‡ saturated with water–acetic acid–ethanol 50:5:6, v/v, and (2) ionophoresis at pH 3.6§ with volatile buffer (pyridine–acetic acid–water 1:10:89, v/v). It is found that ionophoresis is much better than solvent chromatography in the second direction because, besides being quicker, it gives increased resolution and effectively removes Na ions, which tend to interfere with the spray used for detecting phosphorus. For successful chromatography of the alkaline alcoholysis products (Section II.B) only 0.2 ml can be used because of the presence of sodium formate produced by the neutralization of the sodium hydroxide with ethyl formate. If the alkali has been removed with ion-exchange resin, this limitation does not apply. The 0.2-ml ali-

† The original two-dimensional separation on paper (1,51) using phenol–NH₃ and t-butanol–trichloroacetic acid has now largely been superseded. Miani (67) reported that the substitution of formic acid for trichloroacetic acid in the second solvent gave a better separation. Phenol–acetic acid–ethanol–H₂O followed by methanol–formic acid has been used (2), and the R_f values are reported in Table 2. Ferrari and Benson (11) employed chromatography on Whatman No. 4 paper in phenol–H₂O 5:2, v/v, in the first dimension and n-butanol–propionic acid–H₂O 151:75:100, v/v, in the second dimension (Table 2). Letters (36) used propan-2-ol–water–0.880 sp.gr. NH₃ 7:2:1, v/v, followed by phenol–acetic–ethanol–H₂O in the second (Table 2). Spencer and Dempster (37) used single-dimensional paper chromatography in the lower phase of water–liquid phenol (88% + 12% H₂O)–acetic acid–ethanol freshly prepared at 25°C. Resolution is good for simple phosphatide mixtures, but it is not suitable for more complex mixtures or when radioactive studies are contemplated.

‡ Avoid phenol stabilized with hypophosphorus acid stabilizer or, if this is present, distill in glass.

§ Ionophoresis at pH 10.9 (0.08 M bicarbonate buffer) has been used to separate the phosphoesters (72). The mobilities relative to inorganic phosphate are as follows: glyceryl phosphoryl choline, 0; glyceryl phosphoryl inositol, 0.33; glyceryl phosphoryl ethanolamine, 0.49; glyceryl phosphoryl serine, 0.61; glycerophosphoric acid, 0.78.

TABLE 2

R_f Values on Paper Chromatography of the Water-Soluble Products Formed by Alkaline Deacylation of Phosphatides

Compound for parent phosphatide; see Fig. 2	In phenol–H₂O 5:2, w/w (11)	In phenol saturated with H₂O-acetic acid-ethanol 50:5:6, v/v (4)	In n-butanol–propionic acid–H₂O 151:75:100, v/v (11)	In methanol–98% formic acid–H₂O 80:13:7, v/v (3)	In propan-2-ol–H₂O 0.880 sp.gr. NH₃ 7:2:1ᵃ	In t-butanol–H₂O 62:38, v/v + trichloroacetic acid 10%, w/v (51)
Glyceryl phosphoryl choline	0.83	0.88	0.22	0.67	0.45	0.41
Glyceryl phosphoryl dimethyl ethanolamine		0.79ᵃ			0.68	
Glyceryl phosphoryl monomethyl ethanolamine		0.84ᵃ			0.56	
Glyceryl phosphoryl ethanolamine	0.62	0.66	0.17	0.47	0.45	0.41
Glyceryl phosphoryl serine		0.30		0.50	0.15	0.41
Glyceryl phosphoryl inositol	0.12	0.20	0.06	0.45	0.22	0.26
Glyceryl phosphoryl glycerol	0.36	0.51	0.11	0.54		0.56
Cyclic glycerophosphate		0.49		0.66		
Glycerophosphate		0.32		0.74	0.16	0.61
Bis(glyceryl phosphoryl)glycerol		0.26		0.60	0.22	0.45

ᵃ Information kindly supplied by Dr. R. Letters.

quot is applied to Whatman No. 1 paper† as a spot at least 2 cm‡ in diameter, drying after each application with warm air from a hair dryer. Descending chromatography§ is carried out overnight using the phenol–acetic acid–ethanol solvent, which must be freshly prepared. The solvent is dried from the paper in an oven at 80°C, or alternatively the water is removed by air-drying and the phenol is extracted by two washes with ether in a large photographic tray. The dried paper supported on glass rods is wetted with the pyridine–acetic acid buffer delivered from a hand-operated bulb pipette. The buffer is finally allowed to flow by capillary action into the theoretical line of phosphate esters that have been separated by the phenol solvent so this is wetted from both sides. Ionophoresis is carried out under white spirit (high-flash-point kerosene, varsol) for 1 hr¶ at 40 volts/cm in an apparatus similar to that described by Ryle et al. (*35*, Fig. 1). The paper is dried in a current of warm air.

The water-soluble products from the $HgCl_2$–trichloroacetic acid hydrolysis and the methanolic HCl treatment of the "alkali-stable" residual phosphatides are conveniently separated on paper by using descending single-dimensional chromatography with the same phenol solvent and conditions as employed for the resolution of the alkaline alcoholysis products. Aliquots of 0.3 ml of the water-soluble products from the $HgCl_2$–trichloroacetic acid hydrolysis (Section II.C) and 0.2 ml of the aqueous phase from the HCl-catalyzed methanolysis (Section II.D) are convenient volumes but more can be used if necessary.

2. SEPARATION OF ALKALINE ALCOHOLYSIS PRODUCTS BY ANION EXCHANGE CHROMATOGRAPHY

The products formed by alkaline alcoholysis have also been separated by anion-exchange chromatography, a method introduced by Hübscher et al. (*5,38,39*). This enables larger amounts of the water-soluble phosphate esters to be resolved and is not so affected by the

† Washing with 2 N acetic acid and distilled water reduces the paper blank in the subsequent P determination.

‡ If the spot is made smaller, there is streaking on the chromatogram due to the salt present.

§ If sodium formate is present in the hydrolysate, it is essential for good resolution to chromatograph the alcoholysate initially in the phenol solvent before the ionophoresis.

¶ If the room temperature is low, this time should be extended.

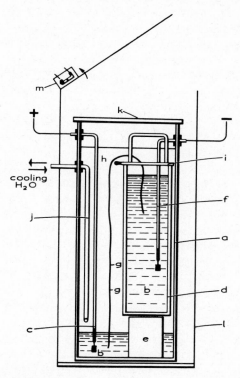

FIG. 1. Side view of high-voltage apparatus for paper ionophoresis. It consists of a rectangular glass tank $51 \times 50 \times 20$ cm, h $\times$ l $\times$ w, (a) containing about 8 cm deep pyridine–acetic acid–H_2O buffer pH 3.6 1:10:89, v/v, (b) connected to the anode of the power pack (3000 volts 100 ma) through a platinum foil–mercury in glass electrode (c). Contained in this is a second tank of glass or Tufnol (d) about $35 \times 45 \times 10$ cm, h $\times$ l $\times$ w, raised from the bottom of the large glass tank by two inverted beakers (e). This is filled with the pyridine–acetic acid buffer and is connected to the cathode of the power pack through a platinum foil–mercury in glass electrode (f) as before. The paper (g) is inserted between the two buffer electrode compartments under the white spirit (high-flash-point petroleum fraction) (h), which completely fills the tank. A rack (i) with glass rod and lifting handles at either end facilitates handling the paper. A serpentine glass coil (j) along one side of the tank cools the white spirit by running tap water. The tank is covered by a ground-glass lid (k). The whole apparatus is contained in a strong wire-mesh protective case (l) with lid and mercury safety switch that automatically cuts off the main supply to the power pack if it is accidentally opened without the high voltage being turned off. The apparatus requires little attention apart from occasional topping up of the buffers and white spirit.

presence of salt arising from the alkali. The method uses a stepwise elution from a Dowex-1–(formate) resin with sodium tetraborate solutions containing increasing amounts of ammonium formate. The alkaline alcoholysis products from 10 mg of lipid P† are taken up in 500 ml H_2O and 5 ml 0.1 M boric acid added to neutralize the excess alkali. 0.20 ml 0.1 M sodium tetraborate is then added, and the mixture is applied to a Dowex 1 × 2 column‡ (100–200 mesh, formate form 1 × 20 cm). The column is eluted as shown in Table 1, the fractions being collected in a fraction collector and their phosphorus content determined. The identity of the phosphate esters eluted by the various elution agents is shown in Table 1. While this method is suitable for fairly simple phosphatide mixtures, such as are obtained from liver, with more complex mixtures (e.g., from brain) it is necessary to extend the elution to recover the more acidic components. This can be done by following the Na tetraborate–ammonium formate mixtures with ammonium formate–formic acid mixtures and finally 0.3 M lithium chloride (40). The deacylation products of canine adrenal lipid have been fractionated in a similar manner, except that the phosphate esters from the more acidic lipids were eluted with higher concentrations of ammonium formate in the same tetraborate buffer (43) (Table 1). Lester (44) examined yeast lipid alcoholysates by eluting them from a Dowex 1 column with a continuously increasing gradient of ammonium formate in sodium tetraborate, and this principle has also been successfully applied to examine brain phosphatides (28), a somewhat better separation being obtained by elution at pH 8.5 than 9.5. The more acidic components obtained by deacylation of brain phosphoinositides have also been examined on a Dowex 1-Cl column eluted with a gradient of lithium chloride up to 0.4 M (41).

F. Detection and Identification of the Phosphorus-Containing Spots on Paper

A preliminary spray of the paper with 0.25% ninhydrin in acetone followed by heating at 100°C for a few minutes helps in the detection

---

† The phosphatide in 3 ml chloroform–methanol 2:1, v/v, is added to 2 ml 0.5 N NaOH in methanol. After standing 15 min at room temperature, 60 ml of water and 40 ml ether are added and the mixture is shaken and centrifuged. The ether layer and interfacial material are rejected.

‡ Nalcite SAR has also been used.

of glyceryl phosphoryl ethanolamine and its serine analog. Also, if a single ninhydrin-reacting compound not containing P is detected in the alkaline alcoholysate (Section II.B) obtained from the lipids of microorganisms, an amino acid ester of phosphatidyl glycerol should be suspected (*45*).† The phosphorus-containing spots are located by spraying with the acid–molybdate spray‡ of Hanes and Isherwood (*68*); the paper is then dried in air at room temperature§ and irradiated with a strong UV source¶ to develop the blue colors. However, for alcoholysates containing sodium formate (Section II.B.1) the spray reagent should be modified (see below).** Although this results in a slower development of the spots, it prevents the appearance on the chromatogram of a large green-blue spot due to the alkali metal. Very slow development of the spots is due to too much acid remaining on the paper from the acid–molybdate spraying, or, alternatively it may be due to phenol solvent left from the chromatographic run. ^{32}P spots can, of course, be easily located on the paper by radioautography. ^{31}P spots can also be made radioactive by neutron bombardment (*20*). Although this process is time-consuming, it allows detection and assay of 0.01–0.05 μg ^{31}P/cm^2 compared with 0.5–1 μg P by the spraying method. Location by the Wade and Morgan (*60*) procedure can also be used. In this the paper is dipped in a reagent prepared by dissolving 1.5 g FeCl$_3 \cdot 6$ H$_2$O in 970 ml acetone and 30 ml 0.3 N HCl. After drying, the chromatogram is immersed in a 1.25% solution of salicyl sulfonic acid in acetone.

The locations on the paper and identities of the water-soluble phosphate esters produced by the alkaline, HgCl$_2$–trichloroacetic acid, and methanolic HCl treatment of phosphatides are shown in Fig. 2 and Table 3. R_f values for paper chromatography in other solvent systems were given in Table 2.

† These lipoamino acids break down on alkaline saponification to give glyceryl phosphoryl glycerol and an amino acid. For confirmation the lipid extract should be rigorously freed from contaminating free amino acids by repeated washing or by the use of Sephadex (*46*).

‡ 5 ml 60% perchloric acid + 10 ml N HCl + 25 ml 4% w/v ammonium molybdate made to 100 ml with H$_2$O.

§ Drying at higher temperatures leads to an higher background color on the chromatogram.

¶ A Hanovia Chromatolite UV lamp with the glass filter removed is used in this laboratory.

** 10 ml 72% w/v perchloric acid + 20 ml 5 N HCl + 40 ml 5% w/v ammonium molybdate + 130 ml H$_2$O (*3*).

TABLE 3

WATER-SOLUBLE PHOSPHATE ESTERS DERIVED FROM VARIOUS PHOSPHATIDES[a]

Conditions of successive chemical treatments[a]	No. on Fig. 2	Compound	Parent phosphatide	Distribution of phosphatide in biological samples
A, 0.03 N NaOH in ethanol–water 4:1, v/v, at 37°C for 20 min	1	Glyceryl phosphoryl choline	Lecithin (lysolecithin)	Ubiquitous, apart from some bacteria; lysolecithin reported in blood plasma; chromaffin granules of adrenals; and traces in other tissues
	2	Glyceryl phosphoryl ethanolamine	Phosphatidyl ethanolamine	Widely distributed but low conc. in brain white matter; largely replaces lecithin in the domestic fly
	3	Glyceryl phosphoryl serine	Phosphatidyl serine	All animal tissues; high conc. in brain and erythrocytes
	4	Glyceryl phosphoryl inositol	Phosphatidyl inositol (monophosphoinositide)	All animal tissues; yeasts, higher plants
	5	Phosphoryl inositol		
	6	Unknown (cyclic derivative?)		
	7	Glycerophosphoric acid	Phosphatidic acid (traces can be formed from phosphoinositides)	Traces in animal tissues; presence in higher plants probably mainly due to phospholipase D action
	8	Bis(glyceryl phosphoryl)-glycerol[b]	Bis(phosphatidyl) glycerol (cardiolipin)	Animal tissues, photosynthetic plants, and bacteria; high conc. in heart; also concentrated in mitochondria
	9	Cyclic glycerophosphoric acid	Lecithin, phosphatidyl inositol	—

10	Inorganic phosphate	Traces often tenaciously accompany higher phosphoinositides of brain	—
11	Inositol triphosphate	Triphosphoinositide	Nervous tissue; electric organ; traces in other animal tissues
12	Glyceryl phosphoryl inositol diphosphate		
13	Inositol diphosphate	Diphosphoinositide	Nervous tissue; traces in other animal tissues
14	Glyceryl phosphoryl inositol monophosphate		
15	Bis(glyceryl) phosphate	Phophatidyl glycerol	Photosynthetic plants; bacteria; traces in animal tissues
16	Deacylation products of phosphatidyl inositotrimannoside (the slowest running is predominant)	Amino acid esters of phosphatidyl glycerol / Phosphatidyl inositotrimannoside[c]	Bacteria / Mycobacteria (bacteria)
17	Glyceryl phosphoryl dimethylethanolamine	Phosphatidyl dimethylethanolamine	Yeasts; *Neurospora crassa*; *Clostridium butyricum*; traces found in tissues and increased when animals given base (dimethyl ethanolamine, monomethylethanolamine)
18	Glyceryl phosphoryl monomethylethanolamine	Phosphatidyl monomethylethanolamine	
19	Glyceryl phosphoryl n-propanol	Phosphatidyl n-propanol	Formed by transfer of phosphatidyl unit to alcohol catalyzed by phospholipase D, e.g., when higher plant tissues are extracted with aqueous alcohol
20	Glyceryl phosphoryl methanol	Phosphatidyl methanol	
21	Glyceryl phosphoryl ethanol	Phosphatidyl ethanol	

continued

TABLE 3 (Continued)

Conditions of successive chemical treatments[a]	No. on Fig. 2	Compound	Parent phosphatide	Distribution of phosphatide in biological samples
B, 10%, w/v, trichloroacetic acid containing HgCl$_2$ (5 mM) at 37°C for 30 min	22	Glyceryl phosphoryl choline	Choline plasmalogen (phosphatidyl choline)	Many animal tissues, mammalian spermatozoa; highest conc. in heart; very low in brain and liver
	23	Glyceryl phosphoryl ethanolamine	Ethanolamine plasmalogen (phosphatidyl ethanolamine)	Many animal tissues, marine invertebrates; highest conc. in brain, especially myelin, low in liver
	24	Glyceryl phosphoryl serine	Serine plasmalogen (phosphatidyl serine)	Low conc. in brain and blood
	25	Glyceryl phosphoryl inositol	Inositol plasmalogen (phosphatidyl inositol)	Traces in brain
C, Methanolic 2 N HCl at 105°C for 4 hr	26	Phosphoryl choline + sphingosyl phosphoryl choline	Sphingomyelin	Widely distributed, but not in bacteria; largely replaces lecithin in ruminant erythrocytes
	27	Inorganic phosphate		
	28[d]	Glycerophosphate derivative (?)	Cyclic acetal plasmalogens mostly formed from natural plasmalogens by acid–HgCl$_2$ hydrolysis	Some may be present in marine invertebrates

[a] A, alkaline alcoholysates of phosphatides separated by chromatography and ionophoresis; B, acid–mercuric chloride hydrolysates of the phosphatides stable to mild alkaline alcoholysis separated by chromatography; C, methanolic-HCl-treated solutions of the phosphatides stable to mild alkaline alcoholysis and acid hydrolysis separated by chromatography. Experimental details are given in the text.

[b] Can be formed from bis(glyceryl) phosphate by drying under acid conditions or on prolonged contact with acidic ion-exchange resin (50).

[c] It is likely that the deacylation products of other phosphatidyl inositomannosides would occupy a similar position.

[d] Not the cyclic 1,2-phosphate or the methyl esters; sometimes free glycerophosphate (R_f 0.33) is formed as well.

G. Estimation of Phosphorus in the Chromatographic Spots and the Glycerol Ether Phosphatide Fraction

Each spot, plus a generous border, is cut from the paper, weighed on a torsion balance, and introduced into a Pyrex boiling tube. Perchloric acid (72%, w/w)† is added, the volume being roughly adjusted

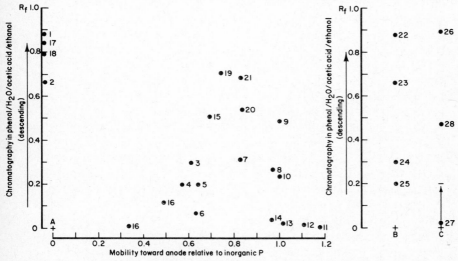

FIG. 2. Positions of the phosphate esters derived from phosphatides after separation on paper. A, B, and C were defined in Table 3, which also gives the identity of the spots and the parent phosphatide.

to correspond to the weight of paper being oxidized (1 ml/100 mg). The tube is heated in a suitably shielded‡ digestion rack. The oxidation is catalyzed by the molybdate present on the paper, and in 5–10 min the contents become colorless. At this stage the perchloric acid is reduced in volume, by strong heating, to 0.6–0.7 ml, as judged by eye. The subsequent P determination is independent of perchloric acid volumes between 0.5 and 0.9 ml. When cool, the contents are diluted with water to about 8 ml. If a slight precipitate is present at this stage,

† With 60% the oxidation takes longer.

‡ No explosion has occurred in many thousands of digestions. It is our opinion that, to avoid such mishaps, it is necessary to do the operation in a large enough tube and to add sufficient perchloric acid so that the mix becomes liquid and local overheating cannot occur.

it is removed by heating the tube for 10 min in a boiling-water bath. In any case the contents are allowed to cool to room temperature before proceeding with the estimation.†

To the solution are added 0.5 ml of 5% ammonium molybdate and, after mixing, 0.4 ml of 1 amino-2-naphthol-4-sulfonic acid reducing reagent [Fiske and Subbarow, (48)]. The mixture is made up to 10 ml and allowed to stand for 20 min, and the optical density is read at 660 mμ. Standards and paper blanks are included in each series, and the blank for each spot is calculated from its weight compared with the weight and color reading‡ of the blank. For greater sensitivity with spots containing less than 10 μg P the modification of Bartlett is used (49). The perchloric acid digest is diluted as before and then 0.75 ml 5% ammonium molybdate and, after mixing, 0.2 ml of the reducing agent are added. The mixture is heated in a boiling-water bath for 7 min and cooled for 20 min, and the optical density is determined at 830 mμ. This latter method gives a linear response between 1 and 20 μg P and has the same insensitivity to perchloric acid concentration as the original Fiske and Subbarow procedure.

The digestion of the glycerol ether phosphatide fraction is carried out after the thorough removal of the solvent ether present. The oxidation of the residue with perchloric acid tends to be more prolonged because of the fatty acids and aldehydes present, but it can be made more rapid by adding 0.05 ml of 1%, w/v, ammonium molybdate. The determination of the phosphorus in this digest is complicated by the presence of mercuric ions, that carry through from the lysoplasmalogen hydrolysis and are extracted by the etherial solvents used. This forms a faint cloudiness when the ammonium molybdate and reducing reagents are added. To remove this mercury prior to the phosphorus estimation, the perchloric acid digest is diluted to 6 ml with water in a stoppered graduated tube and 0.1 ml N HCl is added. The solution is then extracted twice with 6-ml volumes of ether and finally with 6 ml of light petroleum (b.p. 40–60°C). The mercury is removed in the ether, presumably as mercuric chloride, since it is essential to add the hydrochloric acid. The phosphorus in the solution is then estimated directly by the Fiske and Subbarow or Bartlett procedure as before.

---

† The color development is dependent on temperature (47).
‡ The color reading of the blank is usually low (O.D., 0.01–0.03).

H. Calculation of Phosphatide Concentrations

For the samples and dilutions used in the procedure described the percentage of phosphorus from a given phosphatide in the total phosphatide mixture is calculated as follows:

For alcoholysate A (ethanolic NaOH)

$$\% \text{ present in total phosphatide mixture} = \frac{P_X}{P_T} \frac{1}{0.2} 100$$

For hydrolysate B (HgCl$_2$–trichloroacetic acid)

$$\% \text{ present in total phosphatide mixture} = \frac{P_X}{P_T} \frac{0.8}{0.3} \frac{2}{1.6} 100$$

For methanolysate C (methanolic HCl)

$$\% \text{ present in total phosphatide mixture} = \frac{P_X}{P_T} \frac{0.5}{0.2} \frac{2}{1.6} 100$$

where P_T is the total phosphorus in the sample taken and P_X is the phosphorus in the spot. With a few of the phosphatides it is necessary to apply a small correction for the additional decomposition that occurs during the reactions. So long as the conditions are kept constant this breakdown is completely predictable and reproducible. Thus the glyceryl phosphoryl choline P is increased by 6% to allow for the formation of cyclic glycerophosphate. The sum of the glyceryl phosphoryl inositol P and phosphoryl inositol P is multiplied by 1.4 to give the phosphatidyl inositol P, and the glyceryl phosphoryl inositol diphosphate P plus the inositol triphosphate P is multiplied by 1.2 to give the triphosphoinositide P. These corrections are unnecessary if a milder alkaline alcoholytic procedure is employed (methanolic LiOH at room temperature), but of course in this case the plasmalogens are not deacylated to any appreciable extent. With the chromatograms of the methanolic HCl hydrolysate, if a minor spot appears at R_f 0.47, its phosphorus is added pro rata to the plasmalogens and any inorganic P (spot **27**, Fig. 2) to the sphingomyelin.

In our laboratory, the total recovery of phosphatide phosphorus in the various chromatographic spots is between **95** and **103%** of that taken for the analysis.

I. Micro Modification

With this method the composition of a phosphatide mixture can be measured with 100 μg phosphatide P obtained from 50–100 mg of tissue. The procedure is the same as described previously, except that 0.4 N NaOH is used in the alkaline alcoholysis in place of 1 N NaOH. For paper chromatography 0.5 ml of the mild alkaline alcoholysate (Section II.B) is used and all the other aqueous solutions, Sections II.C and II.D. Recoveries of the lipid P analyzed are approximately 80–90%, which is lower than in the macro method. Since the losses can be largely attributed to the nonquantitative transference of the water-soluble phosphate esters to the paper, the percentage composition of the phosphatides can be calculated with little error by expressing the recovery from the spots as a percentage of the total phosphorus recovered rather than of that analyzed as in the macro method.

III. Examination of Glycolipids by the Partial Hydrolysis Method

Some use of the partial hydrolysis method has been made for identifying and estimating certain glycolipids present in plants (*11,58,59, 61–65*) and animal tissues (*57,66*). Glycosyl diglycerides of various complexity are often the main neutral lipids present in plant tissue. They can be separated by chromatography on silicic-acid-impregnated paper and then deacylated by the same alkaline alcoholysis technique as described above for the phosphatides. On paper chromatography in solvents such as phenol–water and butanol–propionic acid a good separation of the glycosidic products can be obtained. Location of the spots on paper can be made by acetone–AgNO$_3$ and ethanolic NaOH treatment (*53*). Some confusion (*58*) [which has been partly resolved (*64*)] still exists as to the identity of some of the products, but the R_f values given in the literature are assembled in Table 4.

IV. Summary

It seems likely that the partial degradation method will continue to be of service in investigations on the structure, identity, and composition of complex lipid fractions, both as a primary method and as a supplement to others. At the moment, it is the method of choice when very detailed information on the phosphatide and probably the glyco-

TABLE 4

R_f Values of Some Glycosides Produced by the Deacylation of Glycolipids[a]

| Glycoside | In phenol–H₂O 5:2, w/w | In phenol saturated with H₂O, pH 5.4 | In n-butanol– | | In pyridine–n-butanol–H₂O 6:4:3, v/v | In sym-collidine saturated with H₂O |
			In n-butanol–propionic acid–H₂O, 151:75:100, v/v	In n-butanol–acetic acid–H₂O 5:3:1, v/v		
β-D-galactopyranosyl-1,1'-D-glycerol (G-Gal) (wheat flour, brain, algae, spinach leaves)	0.62 (*11*) 0.60 (*62*)	0.58 (*58*)	0.25 (*11*) 0.30 (*62*)	0.30 (*58*)		0.95 (*63*)
α-D-galactopyranosyl-1,6-β-D-galacto-pyranosyl-1,1'-D-glycerol (G-Gal-Gal) (wheat flour, spinach leaves, algae)	0.46 (*11*) 0.48 (*62*)	0.44 (*58,64*)	0.13 (*11*) 0.13 (*62*)	0.09 (*58,64*)		0.40 (*63*)
Next higher homolog (G-Gal-Gal-Gal) (*Chlorella*)	0.38 (*11*)		0.04 (*11*)			
6-sulfur-6-deoxyglycopyranosyl-1,1'-glycerol (*Chlorella*)	0.18 (*11*)		0.08 (*11*)			
Deacylated sulfolipid (spinach leaves)	0.1 (*62*)		0.1 (*62*)			
Deacylated sulfolipid (runner bean leaves)		0.19, 0.23 (*58*)		0.04 (*58*)		
Glycoside b (runner bean leaves)		0.45 (*58*)		0.09 (*58*)		
Glycoside c (spinach leaves)	0.63 (*62*)	0.55 (*58*)	0.31 (*62*)	0.28 (*58*)		
Glycoside containing 1 mole each glycerol and glucose (*St. faecalis*)					0.44 (*61*)	
Glycoside containing 1 mole each glycerol, glucose, and galactose (*St. faecalis*)					0.25 (*61*)	

[a] Italic numbers in parentheses indicate authors.

lipid composition of a complex mixture is required and is valuable in examining and confirming the group separation obtained by chromatography of the intact phosphatides and glycolipids on such media as silicic acid. Indeed, many laboratories are at present devising the best method of wedding the two techniques (i.e., chromatography of the intact lipids and their degradation products) to obtain the maximum information with the minimum labor.

REFERENCES

1. J. L. W. Thudichum, *A Treatise on the Chemical Constitution of the Brain,* Tindall and Cox, London, 1884.
2. R. M. C. Dawson, *Biochim. Biophys. Acta,* 14, 374 (1954).
3. R. M. C. Dawson, *Biochem. J.,* 75, 45 (1960).
4. R. M. C. Dawson, N. Hemington, and J. B. Davenport, *Biochem. J.,* 84, 497 (1962).
5. G. Hübscher, J. N. Hawthorne, and P. Kemp, *J. Lipid Res.,* 1, 433 (1960).
6. P. Fleury and L. LeDizet, *Bull. Soc. Chim. Biol.,* 36, 971 (1954).
7. T. Ukita, N. A. Bates, and H. E. Carter, *J. Biol. Chem.,* 216, 867 (1955).
8. B. Maruo and A. A. Benson, *J. Biol. Chem.,* 234, 254 (1959).
9. J. B. Davenport and R. M. C. Dawson, *Biochem. J.,* 84, 490 (1962).
10. F. Rennkamp, *Z. Physiol. Chem.,* 284, 215 (1949).
11. R. A. Ferrari and A. A. Benson, *Arch. Biochem. Biophys.,* 93, 185 (1961).
12. J. S. O'Brien, D. L. Fillerup, and J. F. Mead, *J. Lipid Res.,* 5, 329 (1964).
13. W. L. Magee, *Can. J. Biochem.,* in press (1967).
14. J. Folch, M. Lees, and G. H. Sloane-Stanley, *J. Biol. Chem.,* 226, 497 (1957).
15. R. M. C. Dawson and J. Eichberg, *Biochem. J.,* 96, 634 (1965).
16. A Sheltawy and R. M. C. Dawson, *Biochem. J.,* 100, 12 (1966).
17. H. Brockerhoff, *J. Lipid Res.,* 4, 96 (1963).
18. B. Borgström, *Acta Physiol. Scand.,* 25, 101 (1952).
19. E. J. Barron and D. J. Hanahan, *J. Biol. Chem.,* 231, 493 (1958).
20. A. A. Benson, *Methods Enzymol.,* 6, 881 (1960).
21. C. Pries, *Niet-enzymatische hydrolyse van fosfolipiden, Procfschrift, Rijksuniversitat,* Waltman, Leiden, Netherlands, 1965.
22. G. Hübscher, J. N. Hawthorne, and P. Kemp, *J. Lipid Res.,* 1, 433 (1960).
23. H. Brockerhoff and M. Yurkowski, *Can. J. Biochem. Physiol.,* 43, 1777 (1965).
24. G. V. Marinetti, *Biochemistry,* 1, 350 (1962).
25. G. B. Ansell and S. Spanner, *J. Neurochem.,* 10, 941 (1963).
26. D. J. Hanahan and R. Watts, *J. Biol. Chem.,* 236, PC59 (1961).
27. M. A. Wells and J. C. Dittmer, *Biochemistry,* 4, 2459 (1965).
28. M. A. Wells and J. C. Dittmer, in press (1966).
29. D. J. Hanahan, *Biochem. Prepn.,* 9, 55 (1962).
30. F. Rennkamp, *J. Physiol. Chem.,* 284, 215 (1949).
31. R. M. C. Dawson, *Biochem. J.,* 68, 357 (1958).
32. G. B. Ansell and S. Spanner, *Biochem. J.,* 79, 179 (1961).
33. L. Svennerholm and H. Thorin, *Biochim. Biophys. Acta,* 41, 371 (1960).

34. G. B. Ansell and S. Spanner, *Biochem. J.,* **88,** 56 (1963).
35. A. P. Ryle, F. Sanger, L. F. Smith, and R. Kitai, *Biochem. J.,* **60,** 541 (1955).
36. R. Letters, *Biochim. Biophys. Acta,* **116,** 489 (1966).
37. W. A. Spencer and G. Dempster, *Can. J. Biochem. Physiol.,* **40,** 1705 (1962).
38. J. N. Hawthorne and G. Hübscher, *Biochem. J.,* **71,** 195 (1959).
39. G. Hübscher and J. N. Hawthorne, *Biochem. J.,* **67,** 523 (1957).
40. R. B. Ellis, T. Galliard, and J. N. Hawthorne, *Biochem. J.,* **88,** 125 (1963).
41. H. Brockerhoff and C. E. Ballou, *J. Biol. Chem.,* **236,** 1907 (1961).
42. R. M. C. Dawson and J. C. Dittmer, *Biochem. J.,* **81,** 540 (1961).
43. T. L. Chang and C. C. Sweeley, *Biochemistry,* **2,** 592 (1963).
44. R. L. Lester, *Federation Proc.,* **22,** 415 (1963).
45. M. G. Macfarlane, *Nature,* **196,** 136 (1962).
46. R. Wuthier, *J. Lipid Res.,* **7,** 558 (1966).
47. R. J. L. Allen, *Biochem. J.,* **34,** 858 (1940).
48. C. H. Fiske and Y. Subbarow, *J. Biol. Chem.,* **66,** 375 (1925).
49. G. R. Bartlett, *J. Biol. Chem.,* **234,** 466 (1959).
50. D. E. Brundish, N. Shaw, and J. Baddiley, *Biochem. J.,* **97,** 37C (1965).
51. L. W. Wheeldon, *J. Lipid Res.,* **1,** 439 (1960).
52. R. Letters and E. Markham, *Biochim. Biophys. Acta,* **84,** 91 (1964).
53. W. E. Trevelyan, D. P. Procter, and J. S. Harrison, *Nature,* **166,** 444 (1950).
54. C. Prottey and J. N. Hawthorne, *Biochem. J.,* **101,** 191 (1966).
55. G. V. Marinetti, *J. Lipid Res.,* **3,** 1 (1962).
56. T. A. Pedersen, *Acta Chem. Scand.,* **16,** 374 (1962).
57. E. Mårtensson, *Biochim. Biophys. Acta,* **116,** 296 (1966).
58. M. Kates, *Biochim. Biophys. Acta,* **41,** 315 (1960).
59. M. Lepage, H. Daniel, and A. A. Benson, *J. Am. Chem. Soc.,* **83,** 157 (1961).
60. H. E. Wade and D. M. Morgan, *Nature,* **171,** 529 (1953).
61. M. L. Vorbeck and G. V. Marinetti, *Biochemistry,* **4,** 296 (1965).
62. L. P. Zill and E. A. Harman, *Biochim. Biophys. Acta,* **57,** 573 (1962).
63. H. E. Carter, R. H. McCluer, and E. D. Slifer, *J. Am. Chem. Soc.,* **78,** 3735 (1956).
64. P. S. Sastry and M. Kates, *Biochim. Biophys. Acta,* **70,** 214 (1963).
65. J. F. G. M. Wintermans, *Biochim. Biophys. Acta,* **44,** 49 (1960).
66. J. M. Steim and A. A. Benson, *Federation Proc.,* **22,** 299 (1963).
67. N. Miani, *Bull. Soc. Ital. Biol.,* **31,** 1008 (1955).
68. C. S. Hanes and F. A. Isherwood, *Nature,* **164,** 1107 (1949).
69. W. C. McMurray, *J. Neurochem.,* **11,** 287 (1964).
70. L. A. Horrocks and G. B. Ansell, *Biochim. Biophys. Acta,* **137,** 90 (1967).
71. O. Renkonen, *Acta Chem. Scand.,* **17,** 634 (1963).
72. A. A. Abdel-Latif and L. G. Abood, *J. Neurochem.,* **12,** 157 (1965).

5

THIN-LAYER CHROMATOGRAPHY
OF NEUTRAL GLYCERIDES

V. Mahadevan

UNIVERSITY OF MINNESOTA
THE HORMEL INSTITUTE
AUSTIN, MINNESOTA

I. Introduction

The distribution of fatty acids in individual triglycerides of natural fats has been a subject of fundamental interest to lipid chemists. Hilditch and co-workers separated glycerides by crystallization techniques and postulated the concept of "even distribution" as the generalized pattern of glyceride structure. However, recent studies have shown that the fatty acids of natural fats are not distributed among the triglycerides in accordance with any simple mathematical formula. The separation of natural fat into its component glycerides is essential to an understanding of the distribution of the fatty acids in the glyceride molecule. The application of thin-layer chromatography (TLC) to the separation of glycerides and fatty acids from complex lipids and their subfractionation into component glycerides and fatty acids according to chain length, degree of unsaturation, and the stereochemical configuration of the unsaturated fatty acids will be reviewed.

II. TLC of Triglycerides

A. Separation of Triglycerides from Total Lipids by TLC

Several reviews have provided detailed descriptions of apparatus and techniques used in TLC (1–3), which, therefore, will not be dealt with in this review. TLC permits rapid separation of triglycerides and free fatty acids from complex lipid mixtures as in tissue lipids. Malins and Mangold (4) effected this separation by adsorption chromatography on silica gel G using petroleum ether–ethyl ether–acetic acid 90:10:1 solvent system. Although no subfractionations within naturally occurring classes of these compounds occur, separations do occur within classes where great differences exist in chain length and degree of unsaturation of the individual constituents. Acetic acid is added to prevent the free fatty acids from streaking. Kaufmann and Viswanathan (5,6) used petroleum ether–benzene 4:7 to separate triglycerides from tissue lipids on silica gel G. Rao and Sreenivasan (7) found that a simpler solvent mixture of carbon tetrachloride and acetic acid 99:1 gives similar separations. This solvent system has the added advantage of being nonflammable and less volatile than the other systems.

B. Fractionation of Classes by Reversed-Phase Partition TLC

The fractionation of triglyceride classes into their constituents is accomplished by reversed-phase partition TLC. Malins and Mangold (4) prepared silicone-impregnated plates by slowly dipping activated silica gel G plates into a solution of 5% silicone (Dow Corning 200, viscosity 10 cs) in ethyl ether. After evaporation of the ether, the plate was ready for chromatography. Kaufmann and Makus (8) preferred undecane to silicone as the stationary phase for separation of fatty acids, since silicone interfered with the subsequent isolation of the compounds from the plate. They prepared undecane-impregnated plates by slowly dipping dry silica gel G plates in a 15% solution of undecane in petroleum ether and evaporating off the solvent from the plate at room temperature. After chromatography was carried out, in a suitable solvent, the plate was warmed at 110–120°C for 45 min, when the solvent and the undecane evaporated off. The lipid spots in the plate were revealed by spraying with dichlorofluorescein in the usual manner. This spray reagent was found to be unsatisfactory for locating lipid spots on siliconized chromatoplates because the entire plates fluoresced. The silicone-impregnated plates may be charred after spraying with aqueous or dichromate–H_2SO_4 reagents to reveal the lipid spots, whereas the undecane-impregnated plates may turn all black if the undecane is not completely removed. The α-cyclo-dextrin–iodine and phosphomolybdic acid indicators may be used on both plates for detection purposes. Unsaturated lipids may be detected on both plates with iodine vapors alone (4). Tetradecane was used as the stationary phase for the separation of triglycerides. After chromatography this was removed under vacuum at high temperatures. Liquid paraffin was recommended as the stationary phase for multi-development purposes.

The commercially available Kieselguhr G, which has no adsorptive capacity, is specially prepared for partition chromatography and should be highly useful, after impregnation with silicone, for the separation of triglycerides. It is important that the moving phase be 70–80% saturated with the stationary phase to prevent the latter being eluted during chromatography.

Kieselgel G plates silianized with dimethyl dichlorosilane (DMCS) or trimethyl chlorosilane (TMCS) were recently reported to have decided advantages over siliconized or paraffin-coated plates for reversed-phase partition chromatography (7a). Since silianization is accomplished by exposure of the plates to the vapors of DMCS or TMCS in a closed container, damage to the silica layer, which often occurs during impregnation by dipping, is avoided. Moreover, this procedure does not require saturation of the developing solvent with stationary phase and produces a permanent layer from which separated materials can be extracted free from the starting phase.

Kaufmann and co-workers undertook a systematic study of the separations of synthetic triglycerides. The saturated simple triglycerides† CyCyCy, CCC, LaLaLa, MMM, PPP, SSS were separated from one another on silica gel G plates impregnated with undecane using acetone–acetonitrile 7:3 solvent system (8). CCS, PLaLa,

---

† The component fatty acids combined with glyceryl radical are abbreviated as follows: Cy, caprylic; C, capric; La, lauric; M, myristic; P, palmitic; S, stearic; A, arachidic; B, behenic; O, oleic; L, linoleic; Le, linolenic; E, elaidic; St, saturated fatty acid.

MMM, PPLa, CSS, PPP, PSS, and SSS were separated on Kieselguhr G impregnated with a petroleum fraction (b.p. 240–250°C) using the solvent system acetone–acetonitrile 8:2. The R_f was inversely proportional to the total number of carbon atoms. Mixtures of LaLaS and MMM as well as LaSS and PPP possessing the same number of total carbon atoms were not resolved. The unsaturated glycerides COO, MOO, POO, SOO, and SSO were separated from one another. The separation of the critical pairs COO and MMM, MOO and CSS, OOO and PPO, POO, and PPP were also effected but not very distinctly. Using silicone-impregnated Kieselguhr G plates and methanol–acetonitrile 5:4 solvent system, the saturated triglycerides CyCyCy, CCC, LaLaLa, CCS, PLaLa, and SLaLa were separated from one another, but MMM, PPP, and SSS did not move from the origin. With the same plates but employing methanol–acetonitrile–propiononitrile 5:4:15 solvent system the unsaturated glycerides COO, MOO, OOO, POO, PPO, SOO, and SSO were separated from one another except OOO and POO, which formed critical pairs (9). However, using multiple development with the acetone–acetonitrile 8:2 solvent system, Kaufmann and Das (10) separated the critical pairs OOO and POO, OOO and PPO, OOO and PPP, as well as PPO and PPP.

The critical pairs LeLeLe and LaLaLa, LLL and MMM, OOL and MOO, and OOL and PPL were well separated even with single development, showing thereby that one double bond is not exactly equivalent to two methylene groups in the triglycerides as in fatty acids. Also separated were the *cis-trans* isomers, OOO and EEE, OOP and EEP, OOS and EES by three-stage development. Mixtures of triglycerides having the same total number of carbon atoms and total unsaturation were not separated. Kaufmann et al. (11,12) also effected a separation of critical pairs of triglycerides by two-dimensional reversed-phase partition chromatography in which the second stage included a bromation or hydrogenation reaction. Hydrogenation employing palladium as a catalyst on the plate and bromination by bromine dissolved in the mobile phase were carried out directly on the TLC plates. The bromo derivatives of the original unsaturated compounds also gave the same results.

C. Argentation TLC

Several excellent reviews of the method, including the principles involved and its application to the lipid field, have appeared recently

(13,14). The ability of the Ag⁺ ions to form coordination complexes with double bonds is the basis of the separation of lipids according to unsaturation. In argentation TLC silica gel G is the most frequently used adsorbent for the separation of triglycerides and fatty acids. The level of silver nitrate impregnation recommended has varied from 3 to 30%. However, for all practical purposes 5% impregnation is both economical and gives satisfactory separations. The use of iodine vapor for visualization is unsuitable in argentation TLC. Other reagents such as aqueous sulfuric acid, phosphomolybdic acid, and phosphoric acid followed by charring can be used for detection purposes and dichlorofluorescein and Rhodamine 6G for preparative purposes. Recently Wood and Snyder *(15)* recommended ammoniacal-silver-ion-instead of aqueous-silver-ion-impregnated silica gel G in argentation TLC. Improved resolutions, longer retention of resolving power, and less corrosion of conventional spreaders have been observed.

Two methods for impregnating silica gel G plates with silver nitrate have been described. An aqueous solution of silver nitrate instead of water is used to make a slurry of the silica gel in the usual manner for preparing TLC plates. Alternatively, spraying the dry silica gel plates with a 10–20% solution of silver nitrate in aqueous methanol is used. While the former technique needs proper care in handling of the silver nitrate solution because of its corrosive tendency on the applicator and staining property on a host of materials, the latter method is cumbersome and results in nonuniform impregnation. A simpler method of impregnation that we have used with success and that avoids the risk of staining and damage to the applicator consists in allowing a saturated solution of silver nitrate in 95% methanol to ascend the dry silica gel plates in the manner in which the development of a chromatogram is carried out. The plate is then removed, dried in the dark for 30 min, and activated at 110°C for suitable length of time.

Barrett et al. *(16,17)* resolved various glyceride mixtures using chloroform–glacial acetic acid 99.5:0.5 solvent system. This technique makes possible the resolution of isomeric 1- and 2-oleodistearin and of 1- and 2-linoleodistearin, as well as critical pairs. In contrast to the bromination and hydrogenation techniques used by Kaufmann and co-workers *(11,12)* for the separation of critical pairs by reversed-phase partition chromatography the triglycerides separated by AgNO₃–silica gel G adsorption chromatography can be recovered unaltered.

Morris (*13*) recommends the use of chloroform containing 1–2% methanol, ethanol, or isopropanol for better resolutions. A thin-layer chromatogram of glycerides and natural fats on a silica gel G–AgNO₃ adsorbent obtained by Barrett et al. is shown in Fig. 1.

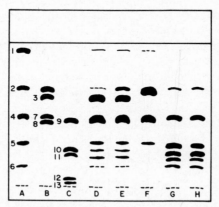

FIG. 1. Thin-layer chromatogram of glycerides and natural fats on a silica gel G–silver nitrate adsorbent (*17*). A,B,C, synthetic glyceride mixtures; D, lard; E, interesterified lard; F, cocoa butter; G, cottonseed oil; H, groundnut oil. 1, tristearin; 2, 2-oleodistearin; 3, 1-oleodistearin; 4, 1-stearodiolein; 5, triolein; 6, trilinolein; 7, 2-linoleodistearin; 8, 1-linoleodistearin; 9, 1,3-distearin; 10, 1,2-diolein; 11, 1,3-diolein 12, monostearin; 13, monoolein.

DeVries and Jurriens (*18*) separated triglycerides according to un-saturation and geometrical configuration. Mixtures of triglycerides SSS, PPO, SOO, and OOO and SSS, PPE, SEE, and EEE were separated from one another, as well as those of SSS, PPE, SEE, and EEE. PPO, PPE, SOO, SEE, OOO, and EEE were also well separated from one another, the *trans* compounds possessing higher R_f values than the *cis* isomers. Jurriens et al. (*19*) separated mixtures of triglycerides containing acyl groups with *cis* and/or *trans* double bonds into the groups StStSt, StStE, StEE + StStO, StOE + EEE, etc. A triglyceride with two *trans* double bonds came in the group as a triglyceride with one *cis* double bond.

Blank et al. (*28*) prepared two triglyceride mixtures A and B by interesterification of triacetin with an equimolar mixture of methyl palmitate, oleate, and linoleate and another mixture of palmitate, oleate, and linolenate. On subjecting these triglyceride mixtures to silver nitrate–silica gel G TLC by using a solvent system of 0.8% methanol in chloroform, nine separate bands were distinguished in the mixture A and eight bands in the mixture B. Fractionation of triglycerides did not occur strictly on the basis of total unsaturation. The type of fatty acids, as well as the total unsaturation, influenced the separation. An example of the type of overlapping of classes was that of triolein and monolinoleodipalmitin.

Schmid et al. (*38*) described a two-dimensional chromatography on silica gel, part of which was impregnated with $AgNO_3$ for scanning complex mixtures of lipids and for the separation of the triglycerides SSS, OSS, and OOO.

D. Preparative TLC

Triglycerides as a class can be separated from complex tissue lipids for further analysis by preparative TLC. Layers of silica gel G adsorbent 0.5–1 mm or even more thickness have been used for this purpose. Ordinary 0.25-mm layers have also been used (*6*). It is important that the adsorbent be washed with polar solvents to remove impurities or the plates be subjected to a preliminary development (washing) with ethyl ether to remove interfering contaminants.

In general, solutions of lipids are applied as a streak on the plate. Morgan (*20*) has described a special applicator for applying solutions on the plate in a series of closely spaced spots. Devices for the application of the solutions as a narrow even band on the chromatoplate have also been published (*20a,20b*).

Various spray reagents have been used to visualize the separated components. An alcoholic solution of dichlorofluorescein is usually sprayed, and the spots are viewed under UV light. Even though it is applied as a very dilute solution, the color is usually eluted with the glyceride and might interfere with subsequent analysis. Rhodamine 6G is recommended as a spray reagent for visualization, since it is not eluted by ethyl ether and should find extensive use for preparative TLC of neutral lipids. Spray reagents that cause chemical changes in the lipids or that are eluted should be used only along narrow strips of the adsorbent in the direction of development. The appropriate bands of adsorbent containing the lipids are marked and then scraped from the plate, and the products are eluted with a suitable solvent or solvents.

TLC affords a simple and quick means of purifying individual triglycerides in small quantities. It is possible to fractionate on one plate quantities of lipid mixtures of the magnitude encountered in radiochemical studies. Mangold et al. (*21*) found commercial tripalmitin-1-^{14}C and radioiodinated (^{131}I) triolein grossly impure. Purification of these samples was achieved by preparative TLC using silica gel G and petroleum ether–ethyl ether 95:5 solvent system. They also observed that ester linkages are hydrolyzed to a small extent during chromatography on freshly activated plates, whereas older and deactivated plates possessed no such disadvantages.

Compounds that are separable on silver nitrate-impregnated layers may be subjected to preparative TLC in the usual way and isolated in milligram amounts.

Goldrick and Hirsch (*22*) have described a technique for quantitative recovery of lipids from chromatoplates. A simple device for the quantitative recovery of silicic acid particles from the plate is also described by them.

E. Quantitative TLC

Various procedures have been used for the quantitative analysis of triglyceride spots on TLC plates. These procedures have been recently reviewed by Jurriens (*22a*). Many of these methods involve elution of the triglycerides from the adsorbent. Semiquantitative analysis is made by weighing the eluted material (*16,23*).

Komarek et al. (*23*) used alkaline silica gel G for the fractionation of triglycerides from total lipids. The lipid bands were visualized by spraying the plates with bromothymol blue. The triglycerides were eluted from the silica gel by diethyl ether. Erroneously high values were obtained owing to passage of dissolved $CaSO_4$ during filtration of the extract. A special filtering apparatus to remove $CaSO_4$ and silica gel from the lipid sample is described. Using 0.5-mm-thick layer silica gel G on a 20×24 cm plate, samples ranging from 80 to 110 mg were analyzed quantitatively by the gravimetric method with good precision.

Triglycerides of differing degrees of unsaturation were separated on silver-nitrate-impregnated silica gel, and each fraction was quantitatively determined by Jurriens et al. (*24*). A standard mixture of synthetic glycerides, PPP, PPE, PPO, EEE, SOO, and OOO was separated on 1-mm-thick layers of adsorbent. Each of the eluted fractions was saponified with alcoholic alkali, and after acidification, the fatty acids were extracted with ether. After removal of the alcohol in the aqueous layer, glycerol was determined by the periodic acid method. The mixture was analyzed four times to demonstrate the reproducibility of the method.

Quantification has been achieved in many other ways; by colorimetry after reaction with hydroxamic acid and ferric ions (*25*) or with chromatropic acid (*26*) and GLC of the component fatty acids after addition of a suitable internal standard (*27,28*). Quantitative GLC analysis of triglycerides separated by argentation TLC has also been described (*26*).

Direct quantitative determinations of triglyceride spots on TLC plates have been made by a number of methods including measurement of spot sizes or spot areas and photodensitometry of charred spots on photographs or X-ray prints. Purdy and Truter (*29*) and Seher (*30*) showed that the amount of material on a TLC plate can be determined from the size of the spot it forms. The measurement of spot size has been successfully employed for the quantitative analysis of lipids by TLC (*21*).

Photodensitometry after charring the glycerides on the TLC plate has been extensively applied by Privett et al. (*31,32*). After evaporation of the solvents the developed plate was sprayed with 50% aqueous H_2SO_4 and heated to char the glycerides and the charred spots were quantitated with a densitometer. The areas on the densitometer curves were found to be directly proportional to the amount of sample in a range of 5–30 mg of carbon. However, the areas given by compounds of one type of structure were not the same as those given by the same amount of compounds containing other structures even though the carbon densities were essentially the same. The sizes and densities of the spots for triolein bore a direct relationship to the amount of sample, but the area was much greater for triolein than for tripalmitin in spite of the fact that the compounds have essentially the same carbon densities. Privett and Blank (*33*) later found that saturated glycerides gave less intense spots because of their greater evaporation rate prior to their charring at 360°C. When the charring was carried out below 200°C after spraying with a more powerful oxidizing agent, chromic–sulfuric acid, both tripalmitin and triolein gave spots of equal intensity. Privett and Blank (*32*) described a method for the determination of component triglycerides in a mixture based on ozonization of double bonds and catalytic reduction of the ozonides followed by separation and quantitation of the glyceryl residues by TLC.

Barrett et al. (*17*) separated synthetic and natural glyceride mixtures according to their degree of unsaturation on silver-nitrate-impregnated silica gel G. Quantitation of the individual glycerides was accomplished after spraying the plate with phosphoric acid and charring the spots under carefully controlled conditions. Densities of the charred areas were measured with a densitometer. For any individual glyceride the area of the curve was proportional to its quantity, but the areas varied for various glycerides. Whereas Privett et al. (*31*) found that the densitometrically determined area for triolein was greater than that for tripalmitin, Barret et al. (*17*) found that the area for triolein was smaller than that for tristearin. The need for standardizing the conditions for charring is apparent from the above studies.

A direct and nondestructive method for the quantitative determination of triglycerides after their separation by TLC was reported by Krell and Hashim (*34*). The triglyceride fractions were dissolved in CS_2, and the IR spectrum of the sample was scanned from 4000–680 cm^{-1}. Absorbance values were measured at the 1742-cm^{-1} band associated with the ester group carbonyl stretching vibration. The IR spectra of solutions of synthetic triglycerides were obtained and baseline absorbance at 1742 cm^{-1} was plotted as a function of concentration. It was found that neither fatty acid chain length nor degree of unsaturation materially altered molar absorptivity of the carbonyl band. Concentrations of triglycerides were determined by reference to a working calibration curve.

F. Radiometric Techniques

TLC has been used to isolate and quantitate labeled triglycerides in metabolic experiments using labeled intermediates. It is also used for rapidly analyzing the radiopurity of commercially available triglycerides. For these purposes the triglyceride fraction is separated and visualized by aforementioned techniques. The specific area of silica gel G containing the triglyceride is scraped into a counting vial for liquid scintillation counting. Snyder and Stephens (*35*) suspended the scrapings in 15 ml of a 4%, w/w, Cab-O-Sil toluene scintillation solution and counted in a liquid scintillation spectrometer. Silica gel G was found to have no quenching properties (up to at least 100 mg). The quantitative radiopurity of a compound can be determined in less than 1 hr by this method. The technique has also been shown to be useful in demonstrating the incorporation of radioactivity into a given fraction when tissue homogenates are applied directly on the plate. The Cab-O-Sil system keeps the scrapings suspended, which makes it immaterial whether the labeled triglycerides are eluted or remain on the silica gel particles.

Brown and Johnston (*36*) have described in detail procedures for the separation and quantitation of ¹⁴C-labeled triglycerides on silica gel G. The silica gel containing the radioactive component was transferred to a counting vial containing 10 ml of a naphthalene–dioxane liquid scintillator. The samples were counted in a Tri-carb liquid scintillation counter. The activity recovered from the chromatograms was compared with that of a similar aliquot placed directly in the scintillation vial. The samples were checked for self-quenching by the internal standard method. No quenching was observed when the naphthalene–dioxane scintillation system was employed. When ¹⁴C-labeled tripalmin and palmitic acid were separated chromatographically as a mixture in 10 to 100-μg sample sizes, the recovery of radioactivity in the components was in excess of 90%. In many experiments with purified radioactive standards the recoveries ranged from 90 to 102%.

Goldrick and Hirsch (*22*) eluted the ¹⁴C-labeled triglycerides of rat adipose tissue from the TLC plates and determined the radioactivity after the addition of scintillation fluids to the dried aliquots of the eluate. Blank et al. (*37*) analyzed radioactive tripalmitin by densitometry of radiograms of chromatoplates developed from X-ray films. The length of time the plate was left in contact with the X-ray film depended on the activity of the spots. The peak areas of the spots on the X-ray film were proportional to the activities of the spots on the TLC plate.

G. Applications

Naturally occurring fats and oils have been fractionated into classes of compounds (*6,23,39–41*) by adsorption TLC. Glycerides of lipolyzed milk fat were fractionated on silica gel G by the same technique (42). The triglyceride content of livers, lungs, brain, and small intestine of rats aged 1 and 10 days postnatally and in adult animals was studied by Dobiasova et al. (43). The triglyceride content of the small intestine and the brain adipose tissue of day-old rats was negligible in comparison with older rats. Triglycerides of human serum were obtained by fractionation of serum lipids on 1–1.5 mm thick layers of aluminum oxide using petroleum ether–ethyl ether 95:5 solvent system (44). Total lipid extracts of human and animal blood sera, as well as various organs, have been fractionated by TLC to yield triglyceride fractions (45–49). Reversed-phase partition TLC has been used for characterizing the constituent glycerides of fats and oils (10–12,50). The triglycerides from livers and blood of animals, human urine, skin, and hair were fractionated by this technique (5,6). Fractionation of animal and vegetable fats has been accomplished by argentation TLC (17,26,51–53).

REFERENCES

1. H. K. Mangold, J. Am. Oil Chemists' Soc., 38, 708 (1961).
2. R. Maier and H. K. Mangold, Advan. Anal. Chem. Instr., 3, 369 (1964).
3. H. K. Mangold, H. H. O. Schmid, and E. Stahl, in Methods Biochem. Analy., 12, 393 (1964).
4. D. C. Malins and H. K. Mangold, J. Am. Oil Chemists' Soc., 37, 576 (1960).
5. H. P. Kaufmann and C. V. Viswanathan, Fette Seifen Anstrichmittel, 65, 607 (1963).
6. H. P. Kaufmann and C. V. Viswanathan, Fette Seifen Anstrichmittel, 65, 538 (1963).
7. P. B. Rao and B. Sreenivasan, Chem. Ind. (London), 1966, 1376.
7a. W. O. Ord and P. C. Bamford, Chem. Ind. (London), 1966, 1681.
8. H. P. Kaufmann and Z. Makus, Fette Seifen Anstrichmittel, 62, 1014 (1960).
9. H. P. Kaufmann, Z. Makus, and B. Das, Fette Seifen Anstrichmittel, 63, 807 (1961).
10. H. P. Kaufmann and B. Das, Fette Seifen Anstrichmittel, 64, 214 (1962).
11. H. P. Kaufmann, Z. Makus, and T. H. Khoe, Fette Seifen Anstrichmittel, 64, 1 (1962).
12. H. P. Kaufmann and T. H. Khoe, Fette Seifen Anstrichmittel, 64, 81 (1962).
13. L. J. Morris, in New Biochemical Separations, Van Nostrand, Princeton, N.J., 1964, p. 295.
14. L. J. Morris, J. Lipid Res., 7, 717 (1966).
15. R. Wood and F. Snyder, J. Am. Oil Chemists' Soc., 43, 53 (1966).
16. C. B. Barrett, M. S. J. Dallas, and F. B. Padley, Chem. Ind. (London), 1050 (1962).
17. C. B. Barrett, M. S. J. Dallas, and F. B. Padley, J. Am. Oil Chemists' Soc., 40, 580 (1963).
18. B. DeVries and G. Jurriens, Fette Seifen Anstrichmittel, 65, 725 (1963).

19. G. Jurriens, B. DeVries, and L. Schouten, *J. Lipid Res.,* **5**, 267 (1964).
20. M. E. Morgan, *J. Chromatog.,* **9**, 379 (1962).
20a. H. J. Monteiro, *J. Chromatog.,* **18**, 594 (1965).
20b. P. J. Curtis, *Chem. Ind. (London),* **1966**, 1680.
21. H. K. Mangold, R. Kammereck, and D. C. Malins, *Microchem. J. Symp.,* **2**, 1697 (1962).
22. B. Goldrick and J. Hirsch, *J. Lipid Res.,* **4**, 482 (1963).
22a. G. Jurriens, *Chem. Weekblad,* **61**, 257 (1965).
23. R. J. Komarek, R. G. Jensen, and B. W. Pickett, *J. Lipid Res.,* **5**, 268 (1964).
24. G. Jurriens, B. DeVries, and L. Schouten, *J. Lipid Res.,* **5**, 267 (1964).
25. E. Vioque and R. T. Holman, *J. Am. Oil Chemists' Soc.,* **39**, 63 (1962).
26. C. Litchfield, M. Farquhar, and R. Reiser, *J. Am. Oil Chemists' Soc.,* **41**, 588 (1964).
27. F. D. Gunstone and F. B. Padley, *J. Am. Oil Chemists' Soc.,* **42**, 957 (1965).
28. M. L. Blank, B. Verdino, and O. S. Privett, *J. Am. Oil Chemists' Soc.,* **42**, 87 (1965).
29. S. J. Purdy and E. V. Truter, *Chem. Ind. (London),* **1962**, 506.
30. A. Seher, *Mikrochim. Acta,* **1961**, 308.
31. O. S. Privett, M. L. Blank, and W. O. Lundberg, *J. Am. Oil Chemists' Soc.,* **38**, 312 (1961).
32. O. S. Privett and M. L. Blank, *J. Lipid Res.,* **2**, 37 (1961).
33. O. S. Privett and M. L. Blank, *J. Am. Oil Chemists' Soc.,* **39**, 520 (1962).
34. K. Krell and S. A. Hashim, *J. Lipid Res.,* **4**, 407 (1963).
35. F. Snyder and N. Stephens, *Anal. Biochem.,* **4**, 128 (1962).
36. J. L. Brown and J. M. Johnston, *J. Lipid Res.,* **7**, 480 (1966).
37. M. L. Blank, J. A. Schmit, and O. S. Privett, *J. Am. Oil Chemists' Soc.,* **41**, 371 (1964).
38. H. H. O. Schmid, W. J. Baumann, J. M. Cubero, and H. K. Mangold, *Biochem. Biophys. Acta,* **125**, 189 (1966).
39. R. Maier and R. T. Holman, *Biochemistry,* **3**, 270 (1964).
40. M. E. McKillikau and R. P. A. Sims, *J. Am. Oil Chemists' Soc.,* **40**, 108 (1963).
41. H. K. Mangold and D. C. Malins, *J. Am. Oil Chemists' Soc.,* **37**, 383 (1960).
42. R. G. Jensen, T. Sampugna, and G. W. Gander, *J. Dairy Sci.,* **44**, 1983 (1961).
43. M. Dobiasova, P. Hahn, and O. Koldovsky, *Biochem. Biophys. Acta,* **70**, 713 (1963).
44. A. Vacikova, V. Felt, and J. Malikova, *J. Chromatog.,* **9**, 301 (1962).
45. A. Chalvardjian, *Biochem. J.,* **90**, 518 (1964).
46. H. Kaunitz, E. Gaughitz, Jr., and D. G. McKay, *Metabolism,* **12**, 371 (1963).
47. W. C. Vogel, W. M. Doizaki, and L. Zieve, *J. Lipid Res.,* **3**, 138 (1962).
48. Y. Stein and O. Stein, *J. Atherosclerosis Res.,* **3**, 189 (1963).
49. N. Tuna and H. K. Mangold, in *The Atherosclerotic Plaque,* (R. J. Jones, ed.), Univ. Chicago Press, Chicago, 1964.
50. C. Michalec, M. Sulc, and J. Mesten, *Nature,* **193**, 63 (1962).
51. J. H. Broadbent and G. Shone, *J. Sci. Food Agr.,* **14**, 524 (1963).
52. H. P. Kaufmann and H. Wessels, *Fette Seifen Anstrichmittel,* **66**, 81 (1964).
53. F. D. Gunstone, F. B. Padley, and M. I. Omreschi, *J. Sci. Food Agr.,* **12**, 115 (1961).

COLUMN CHROMATOGRAPHY OF NEUTRAL GLYCERIDES AND FATTY ACIDS

6

K. K. Carroll† and B. Serdarevich

COLLIP MEDICAL RESEARCH LABORATORY
UNIVERSITY OF WESTERN ONTARIO
LONDON, ONTARIO, CANADA

I. Introduction

Column chromatographic methods for separation of neutral lipids and fatty acids may be divided into two main types: solid–liquid adsorption chromatography and liquid–liquid partition chromatography. Adsorption chromatography is based primarily on differences in the relative affinity of compounds for the solid adsorbent used as a stationary phase. Affinity is determined by polar and ionic forces and to a lesser extent by nonpolar van der Waals forces, and this means that polar groupings in the molecules to be separated (e.g., ester, ether, and hydroxyl groups) exert a much greater effect than nonpolar hydro-

† Medical Research Associate of the Medical Research Council of Canada.

carbon chains. Adsorption chromatography therefore tends to separate lipid mixtures into classes characterized by number and type of polar groups. Triglycerides from naturally occurring sources, for example, are normally eluted together as a class although they contain a variety of fatty acid substituents. This is not necessarily a disadvantage, because a preliminary separation into discrete lipid classes is often desirable and other methods may be used for subsequent analysis of the individual molecular species. It should be noted, however, that some separation of individual members of a lipid class may occur during adsorption chromatography as a result of differences in chain length or unsaturation of fatty acid substituents, and this has been used on occasion to separate lipids within a given class. More recently, the use of adduct-forming compounds and of impregnated adsorbents has also greatly facilitated the separation by adsorption chromatography of lipids differing in degree of unsaturation.

Adsorbents that have been used for adsorption chromatography of neutral lipids include charcoal (*1*), urea (*2*), magnesium oxide (*3*), magnesium sulfate (*4*), alumina (*5–8*), silicic acid (*8–12*), Florisil (*13,14*), and acid-treated Florisil (*15*). Of these adsorbents, silicic acid has been by far the most widely used in lipid research, but Florisil (a synthetic magnesia silica gel) has been used increasingly in recent years. Much of the chromatographic work with silicic acid has been carried out with fine-mesh material, and this has the disadvantage of slow flow rates, which prolong the time required for chromatographic separation. This problem has now been overcome, however, by the availability of coarse-mesh preparations of silicic acid and by the use of acid-treated Florisil, which has chromatographic properties similar to those of silicic acid but retains the coarse-mesh size of Florisil.

Liquid–liquid partition chromatography is based on the principles of countercurrent distribution, and separations are dependent on the relative solubility characteristics of compounds with respect to the stationary and moving phases of the column. There are two main types of liquid–liquid column: the partition type, in which the more polar liquid is the stationary phase (*16,17*), and reversed-phase partition, in which the less polar liquid is the stationary phase (*18*). Both types have been used for separation of fatty acids and their esters, but most of the recent work has been done on reversed-phase columns. Liquid–liquid partition chromatography permits separation of fatty acids according to chain length, degree of unsaturation, type of geometrical isomer, and presence or absence of polar substituents. Although gas–liquid chromatography has much wider application in analytical work on fatty acids, liquid–liquid partition chromatography is useful for preparative separations.

Theoretical and practical aspects of the above-mentioned techniques for column chromatography of neutral lipids have been described in a number of excellent books and review papers (*10,11,19–24*).

II. Adsorption

A. Silicic Acid

Silicic acid was first used as a chromatographic adsorbent for lipids by Kaufmann (*8,25*) and Trappe (*26,27*), and the method was further elaborated by Börgstrom (*9,28*) and by Fillerup and Mead (*29*), who demonstrated its high capability for separation of natural lipid mixtures. Hirsch and Ahrens (*10*) carried out a detailed study of the technique of silicic acid column chromatography of lipids, and much additional information on the method is available in review articles by Wren (*11*), Morris (*30*), and Hanahan (*31*).

Some attempts have been made to standardize column chromatography of lipids on silicic acid, but the method does not always give exactly reproducible results when used by different people in different laboratories. Preparations of silicic acid differ in adsorption characteristics, depending on pretreatment before use; and other factors such as shape and dimensions of the column, method of preparing the column, quality of eluting solvents, rate of elution, and temperature of column may also affect reproducibility. However, if attention is given to these details, an approach to reproducibility may be achieved.

Methods for pretreating silicic acid in order to obtain a standardized preparation have been described by several groups of workers (*10,32–34*). Hirsch and Ahrens (*10*) milled and sieved Mallinckrodt silicic acid to obtain material of uniform mesh size and then washed it with organic solvents. Horning et al. (*32*) and Hernandez et al. (*33*) pretreated with hydrochloric acid as well. The latter group investigated the adsorptive activity obtained by different conditions of heating and by addition of measured amounts of water. Hirsch and Ahrens, however, recommended the use of dehydrating solvents such as acetone and ether as a means of obtaining the desired degree of activation more reproducibly.

Column dimensions must be considered as well as amount of adsorbent if results obtained by column chromatography are to be reproducible. Longer, narrower columns give better separation than shorter, wider ones (35), and both the amount of adsorbent and column dimensions should be given when reporting chromatographic results. Although slower elution ensures better separation, a major disadvantage of silicic acid columns has been their slow flow rate, which increases the time required for chromatography and prolongs the exposure of lipids to adsorbent. Nearly all adsorbents are to some extent catalysts and will on longer contact with lipids induce changes in lipid structure, e.g., oxidation of unsaturated lipids and isomerization of mono- and diglycerides (9,36). These alterations can sometimes be avoided by techniques such as impregnation of the adsorbent with a complexing agent (37) or by working in an inert atmosphere. Wren and Szczepanowska (38) found that addition of small amounts of an antioxident to the eluting solvents protected unsaturated lipids from oxidation without affecting the chromatographic separation.

The size of sample used may also affect the reproducibility of the results obtained. In silicic acid chromatography, the adsorption isotherm normally has a convex shape and, during elution, the lipid classes separate in zones having a steep front and a long trailing edge. In this situation, a large sample tends to be eluted faster than a small one (24,35). Overloading should, of course, be avoided, but the permissible load depends to some extent on the separation desired. Larger loads can generally be used when the components to be separated differ more widely in chromatographic properties. Table 2 in the review by Stein and Slawson (24) gives a summary of column loads that have been used successfully by different investigators for separation of a variety of lipid mixtures. In general, amounts ranging from 10 to 30 mg/g of adsorbent have been used in separations of neutral lipid classes and somewhat similar loads (10 to 25 mg of lipid or 0.3–1.0 mg of phosphatide phosphorus/g of adsorbent) have been used in separation of polar lipids. For separation of neutral lipids from phosphatides, loads as high as 100 to 150 mg/g of adsorbent have been used. The adsorption of lipids on silicic acid depends on structure, porosity, granulation, water content, and other characteristics that are discussed in detail by Wren (11), Klein (39), and Snyder (34). Free electron pairs on the oxygen bridges and hydroxyl groups of silicic acid are responsible for the strong adsorption of lipids capable of hydrogen bonding. Thus, monoglycerides with two free hydroxyl groups are more strongly adsorbed than diglycerides, which have only one. Forces other than hydrogen bonding are also involved in binding lipids of different polarity to silicic acid, and these vary from the relatively weak van der Waals forces to stronger polar and ionic forces. Silicic acid, like other adsorbents, has least affinity for the nonpolar hydrocarbons. These are therefore eluted first from the column, followed by other lipid classes more or less in order of increasing polarity.

During chromatography, the adsorptive forces binding lipid to adsorbent are overcome and the lipids are moved progressively down the column by eluting with solvents in increasing order of polarity. The following list of solvents arranged in order of eluting power is an expanded and slightly modified version of Trappe's "eluotropic series" as given by Wren (11): Methanol > ethanol > 1-propanol > acetone > methyl acetate > ethyl acetate > ether > dichloromethane > benzene > toluene > 1,1-dichloroethane > 1,1,2,2-tetrachloroethane > chloroform > trichloroethylene > carbon tetrachloride > cyclohexane > ligroin. The solvents most frequently used for silicic acid chromatography are methanol, ether, benzene, chloroform, and various types of petroleum ether (ligroin).

During elution, the adsorptive bonds between lipid and silicic acid are normally broken before the polarity of the eluting solvent reaches the polarity of the adsorbed lipid. This is due to the fact that each adsorbed lipid molecule is surrounded by a larger number of eluting solvent molecules and is therefore more readily replaced by eluent at the adsorbent surface.

Wren (11) has also published a list of lipid classes in order of their elution from silicic acid (Table 1). This order, however, should only be considered as an approximation subject to modification. For example, Wren and Szczepanowska (40) in a later study found that monoglycerides were eluted between glycerophosphatidic acids and inositol-containing lipids. Factors such as the adsorption characteristics of the silicic acid used, the sequence of eluting solvents, and the characteristics of individual members of the lipid classes may modify the order of elution.

TABLE 1

Order of Elution of Lipids from Silicic Acid[a]

Hydrocarbons
Esters other than steryl esters and glycerides
Steryl esters
Fatty aldehydes (?)
Triglycerides
Long-chain alcohols
Fatty acids
Quinones
Sterols
Diglycerides
Monoglycerides
Glycolipids
Lipoamino acids
Bile acids
Glycerophosphatidic acids
Inositol-containing lipids
Phosphatidyl ethanolamines
Lysophosphatidyl ethanolamines
Lecithins
Sphingomyelins
Lysolecithins

[a] Reproduced from J. J. Wren, *J. Chromatog.*, **4**, 173 (1960), by permission of the author and the publisher.

Hirsch and Ahrens (*10*) chromatographed a mixture of lipid standards on an 18-g column of silicic acid using stepwise elution with petroleum ether–ethyl ether solvent mixtures and obtained the separation shown in Fig. 1. Silicic acid chromatography has been widely applied to separation of naturally occurring lipid mixtures, and similar separations of major lipid classes have usually been obtained. Some of this work is discussed in more detail below.

For best separation of natural lipids, mixtures should first be roughly separated into neutral and phosphatide fractions. This can be done conveniently on columns of coarse-mesh silicic acid or acid-treated Florisil, applying the mixture to the column in chloroform and eluting the neutral lipids with chloroform and the phosphatides subsequently with chloroform–methanol and/or methanol. The neutral lipid fraction may then be separated further into lipid classes by rechromatography, using an elution schedule similar to that shown in Fig. 1.

Although stepwise elution is simpler to carry out and is generally used for chromatography of neutral lipids, gradient elution is capable of giving better separation of lipid classes (*41,42*). In stepwise elution there is often some trailing of one lipid class into the next, and this may be avoided by gradient elution.

1. SEPARATION OF INDIVIDUAL LIPID CLASSES

a. Esters Other than Glycerides. Saturated hydrocarbons are normally eluted first from silicic acid columns with petroleum ether (light

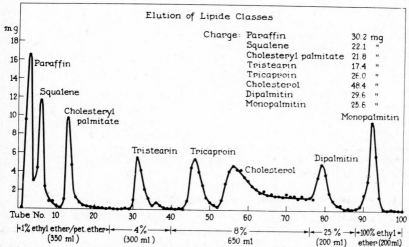

FIG. 1. Elution of representative substances of seven major classes of lipids from an 18-g column of silicic acid. [Reproduced from J. Hirsch and E. H. Ahrens, Jr., *J. Biol. Chem.*, **233**, 311 (1958), by permission of the authors and the American Society of Biological Chemists, Inc.]

petroleum, hexane, Skellysolve B, etc.) and are followed by unsaturated hydrocarbons such as squalene (*11,22*). Simple esters of fatty acids require a somewhat more polar solvent such as 1% ethyl ether or 10–18% benzene in petroleum ether (*10–12,22,32,43*). Wax esters, methyl esters, and cholesteryl esters are eluted more or less together, and the degree of separation may be determined more by the nature of the fatty acid than by the type of lipid class. Increasing unsaturation in the fatty acid chain increases the affinity for silicic acid suffi-

ciently to permit separation of fatty acid esters on this basis (44–47).
Esters containing hydroxy fatty acids are retained much more strongly
and require more polar solvents for their elution (48–49).

The separation of fatty acid esters differing in degree of unsaturation can be achieved more readily by preparing adducts of the unsaturated esters or by using silicic acid impregnated with silver nitrate
as adsorbent. Erwin and Bloch (50) separated fatty acids of protozoa
lipids according to degree of unsaturation by pretreating their methyl
esters with mercuric acetate and chromatographing the adducts on
silicic acid. Saturated esters were eluted with pentane–ether 95:5,
monounsaturated esters with pentane–ether 50:50, dienoic esters with
ethanol–acetic acid 99.8:0.2, trienoic esters with ethanol–acetic acid
99:1, and tetraenoic and polyenoic esters with ethanol–acetic acid
95:5. The same method was used by Korn (51) for separation of the
fatty acids of *Euglena gracilis*.

De Vries (52,53) used silicic acid impregnated with silver nitrate
and successfully separated methyl esters of saturated and unsaturated
fatty acids as well as geometrical isomers (Fig. 2). Cholesteryl esters
were separated by Haahti et al. (54), who used a continuous concentration gradient from 10 to 70% benzene in hexane for elution. Wagner
et al. (55) adapted de Vries' method to separate cerebroside fatty
acids in the form of their methyl esters. They used chloroform–ether
mixtures and eluted first the unsubstituted saturated esters, then unsubstituted unsaturated esters followed by hydroxy saturated esters,
and finally hydroxy unsaturated esters. Other workers have also used
silicic acid impregnated with silver nitrate for the separation of
methyl esters prepared from naturally occurring fatty acid mixtures
(50,56–59). Although degree of unsaturation is the primary basis of
separation, differences between positional isomers and fatty acid esters
of different chain length have been noted (56,57,59,60).

b. *Triglycerides*. Triglycerides are readily separated from wax esters
and steryl esters on silicic acid columns and are normally eluted with
3–5% ethyl ether in petroleum ether or with 60% benzene in hexane
(10–12,32). Shorter-chain and unsaturated triglycerides are somewhat
retarded and may overlap the free-fatty-acid fraction (10,11). Replacement of an ester linkage in triglycerides by an ether linkage does
not affect the rate of migration on silicic acid columns (10,61).

Using silicic acid impregnated with silver nitrate, de Vries (62)
separated triglycerides on the basis of unsaturation. The separation

depends on the total amount of unsaturation in the triglyceride molecule, which may be distributed among the fatty acid chains or may be concentrated to a greater extent in one or two of the chains (Fig. 3). Other workers have also used this method for separating triglycerides from natural sources on the basis of unsaturation (*63–65*).

    c. Fatty Acids and Fatty Alcohols. Free fatty acids and fatty alcohols are eluted from silicic acid columns soon after triglycerides

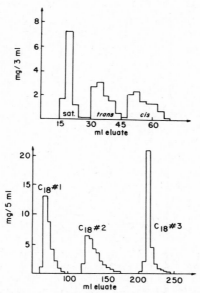

FIG. 2. *Upper chart:* Column chromatography of 10 mg each of a mixture of methyl stearate, elaidate, and oleate. Adsorbent 2 g AgNO₃–silica. Column height, 11 cm; diameter, 8 mm. Eluent, benzene–light petroleum 10:90–30:70. *Lower chart:* Column chromatography of 30 mg each of a mixture of methyl oleate, linoleate, and linolenate. Adsorbent 10 g AgNO₃–silica. Column height, 23 cm, diameter, 14 mm. Eluent, benzene–light petroleum 40:60–100:0. [Reproduced from B. de Vries, *Chem. Ind.* (*London*), **1962**, 1049, by permission of the author.]

and may not always be completely separated from the triglyceride fraction (*10,11,22*). As with other lipid classes, there is some separation on the basis of chain length and degree of unsaturation. McCarthy and Duthie (*66*) were able to separate free fatty acids quantitatively from other glycerides on silicic acid pretreated with isopropanol–KOH. Neutral glycerides were eluted with ether and free fatty acids with

2% formic acid in ether while phosphatides remained on the column. This method was used by Heald et al. (67) in studies of plasma free fatty acids. Preiss and Bloch (68) separated hydroxy acids from normal acids on silicic acid. The normal acids were eluted with pentane–ether 85:15 and the hydroxy acids with pentane–ether 1:1. Further details regarding chromatography of hydroxy fatty acids on silicic acid may be found in reviews by Downing (69), Applewhite (70), and Radin (49).

d. *Diglycerides.* Diglycerides are eluted from silicic acid columns with 20–60% of ethyl ether in petroleum ether, appearing between

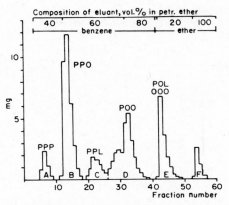

FIG. 3. Chromatography of palm oil (97 mg) on 10 g AgNO₃–silica. PPP, tripalmitin; PPO, dioleopalmitin; POL, palmitooleolinolein; OOO, triolein. [Reproduced from B. de Vries, *J. Am. Oil Chemists' Soc.,* **41,** 403 (1964), by permission of the author and the Society.]

sterols and monoglycerides (9–11). Some tailing of cholesterol into the diglyceride fraction may occur (Fig. 1). Diglycerides do not isomerize on silicic acid (36). De Haas and van Deenen (71) separated 1,3- and 1,2-glycerides, each containing one ester and one ether group, by eluting with 20% ethyl ether in hexane. Plasmalogenic diglyceride, in which one of the ester groups is replaced by an aldehydogenic group, was separated from conventional diglyceride by Kiyasu and Kennedy (72), using gradient elution with benzene–chloroform, and this separation was confirmed by Lands and Hart (73). Silicic acid chromatography was also used by Palameta and Kates (74) to purify synthetic diether analogs of diglycerides.

e. Monoglycerides. Monoglycerides are eluted from silicic acid columns with ethyl ether or chloroform and are well separated from most other neutral lipid classes (*9–11*). Monoglyceryl ethers, however, are eluted along with monoglycerides, and compounds such as lithocholic acid may also be eluted with ethyl ether (*10,11*). 1- and 2- Monoglycerides cannot be separated by chromatography on silicic acid. The reason for this may be intramolecular hydrogen bonding between the carboxyl oxygen of the ester group and a free hydroxyl on the glycerol, which brings both isomers to approximately the same degree of polarity (*75*). Both 1- and 2-monoglycerides isomerize during chromatography on silicic acid (*36*), but they can be chromatographed without isomerization on silicic acid impregnated with 10%, w/w, of boric acid (*76*).

2. COARSE-MESH SILICIC ACID

Most of the earlier chromatographic separations of lipids were done on fine-mesh silicic acid, but coarser-mesh preparations are now used more widely to avoid the disadvantage of slow flow rates and prolonged contact of lipids with adsorbent. In spite of their coarser mesh, these preparations may be used satisfactorily at the same load levels of lipid and give much the same separation of major lipid classes as the fine-mesh material. Examples of commercially available coarse-mesh silicic acid preparations for chromatography of lipids are Adsorbosil Cab, 60/100, 100/140, 140/200 mesh, Applied Science Laboratories, State College, Pennsylvania; Bio-Sil BH and Bio-Sil A, 100–200 mesh, Bio-Rad Laboratories, Richmond, California; and Unisil, 100/200 mesh, Clarkson Chemical Co., Williamsport, Pennsylvania. Another coarse-mesh preparation which gives separations of neutral lipid classes similar to those obtained on silicic acid is acid-treated Florisil, prepared by treatment of Florisil with strong mineral acid (*15*). This treatment removes most of the magnesium from Florisil and gives a preparation that is probably similar in structure to coarse-mesh silicic acid.

Separations of lipid standards on Unisil and on acid-treated Florisil are illustrated in Fig. 4. The same flow rate was used for each of these columns so that the results would be comparative, but the maximum flow rate with the acid-treated Florisil column is about three times that of the Unisil column, since acid-treated Florisil has a coarser-mesh size.

Although the pattern of elution of lipid classes was similar with both columns, some differences may be noted. Squalene was eluted soon after paraffin on the Unisil column but was retained more strongly on acid-treated Florisil and was eluted with steryl esters and methyl esters. Palmitic acid was cleanly separated from tripalmitin on the

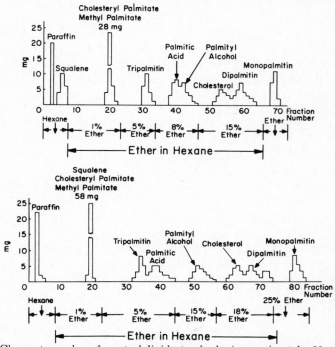

FIG. 4. Chromatography of neutral lipid standards (approximately 20 mg each) on Unisil (upper chart) and on acid-treated Florisil (lower chart). Weight of adsorbent was 20.4 g in each case. Column dimensions, 1.2 × 40.8 cm (Unisil) and 1.2 × 42 cm (acid-treated Florisil). Ten-milliliter fractions were collected. The term "hexane" refers to a distilled fraction of Skellysolve B boiling at 68–71°C.

Unisil column and overlapped with palmityl alcohol, whereas on acid-treated Florisil palmitic acid partially overlapped the tripalmitin peak and palmityl alcohol was eluted later as a separate peak. Cholesterol and dipalmitin overlapped to about the same extent on each chromatogram. The dipalmitin tailed somewhat on both columns, and the small extra peak on the acid-treated Florisil chromatogram is due to the change in solvent.

Most workers use fresh adsorbent for each chromatographic column, and this is good practice in careful analytical work, particularly when new or unknown mixtures are involved. We have found, however, that acid-treated Florisil can be regenerated and used repeatedly without any marked change in chromatographic properties. The usual procedure has been to wash the column with methanol at the end of the run to elute as much adsorbed lipid or other impurity as possible and then to replace the methanol with chloroform and finally with Skellysolve B in preparation for the next run. The washing has sometimes been done on the column and sometimes by unpacking the column, washing the adsorbent by the batch method with methanol and ether, and repacking with nonpolar solvent. Other workers have also informed us that they have successfully regenerated coarse-mesh silicic acid or acid-treated Florisil by similar techniques and reused them repeatedly.

The separation of neutral lipid classes is normally carried out in our laboratory on columns of Florisil (see Section II.B), and acid-treated Florisil was prepared primarily for chromatography of more polar lipids (e.g., phosphatides), which are not readily eluted from Florisil columns. Acid-treated Florisil may be useful, however, for chromatography of lipids that isomerize on Florisil (e.g., 1,2- and 1,3-diglycerides) and for certain specific separations of neutral lipids.

Acid-treated Florisil was used in our laboratory for purification of radioactively labeled fatty acids (77). The fatty acids were eluted with chloroform, and most of the radioactive impurities were retained more strongly and could be eluted later with methanol. Anderson and Hollenbach (78) used acid-treated Florisil impregnated with silver nitrate to separate labeled fatty acid esters obtained from *Penicillium javonicum*. Elution with hexane–benzene 9:1, 7:3, and 4:6 gave a clean separation of saturated, monounsaturated, and diunsaturated esters with quantitative recovery of both mass and radioactivity. Willner (79) also separated fatty acid methyl esters on acid-treated Florisil impregnated with silver nitrate. Myristic, palmitic, and stearic acids were eluted with 0.5% ethyl ether in hexane, palmitoleic, and oleic acids with 0.75–1% ether, linoleic acid with 5–7.5% ether, and linolenic acid with 10% ether in hexane.

Isomeric 1,3- and 1,2-diglycerides have been separated in our laboratory on acid-treated Florisil without isomerization (80). The 1,3-diglycerides were eluted with 20% ethyl ether in Skellysolve B and the 1,2-diglycerides with 25% of ether. Unsaturated diglycerides were retarded slightly on the column relative to their saturated analogs, and branched-chain diglycerides were eluted slightly ahead of diglycerides with straight-chain acids.

Monoglycerides isomerized on acid-treated Florisil in the same way as on silicic acid, but the isomerization was prevented by impregnating the acid-treated Florisil with 10%, w/w, of boric acid. The mono-glycerides were eluted with ether, and the 1- and 2- isomers were not separated on this impregnated adsorbent (37). Monoglyceryl ethers were also eluted from the column with ether and were not separated from monoglycerides (81).

B. Florisil

Florisil (Floridin Co., 2 Gateway Center, Pittsburgh, Pennsylvania) has been used increasingly in recent years for separation of lipids (13,14,82–93). It is available in standard mesh sizes ranging from 16/30 to 100/200, but material of 60/100 mesh activated at 1200°F is most commonly used for column chromatography. This material permits rapid flow rates, and in spite of its coarse-mesh size it gives good chromatographic separations at load levels similar to those used with silicic acid. Equilibration seems to occur rapidly on the column and the use of high flow rates does not appear to result in serious loss of resolving power, so that much time can be saved on routine separations (14). Resolution can be improved, if desired, by the use of multibore columns as described by Fischer and Kabara (87), but most of our separations have been performed on single-bore columns.

Lipids are, in general, adsorbed more strongly on Florisil than on silicic acid, but the adsorption isotherm on Florisil, as on silicic acid, has a convex shape in most cases (94,95). Snyder (42) studied the theoretical aspects of linear adsorption chromatography on Florisil in comparison with silicic acid and alumina. The adsorptive forces are probably similar for both Florisil and silicic acid, and their strength increases for lipids of increasing polarity. The elution pattern of neutral lipid classes on Florisil is similar to that on silicic acid, the main difference being that free fatty acids remain on the column until after other neutral lipids have been eluted (14). The acids are retained because of the basicity of Florisil and can be eluted with small amounts of acetic acid in ether. As in the case of silicic acid, rate of elution of lipids from Florisil is influenced to some extent by chain length, unsaturation, and chain branching *(82,95)*.

Florisil activated at 1200°F has the highest adsorption activity. Addition of measured amounts of water decreases this activity in a reproducible manner, and lipids may then be eluted with less polar solvents. Such treatment affects the elution volume of some lipids more than others and is one method by which desired separations may be

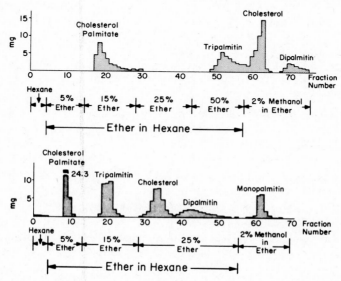

FIG. 5. Separation of neutral lipid classes on 12-g columns (1.2 × 15 cm) of Florisil activated at 1200°F (upper chart) and of Florisil hydrated with 7% water (lower chart). [Reproduced from K. K. Carroll, *J. Lipid Res.,* **2,** 135 (1961), by permission of the editor.]

improved *(14)*. Separations of lipid classes on highly active Florisil and on Florisil hydrated with 7% water are illustrated in Fig. 5. Chromatography of naturally occurring lipids from liver and blood on the hydrated Florisil gave similar elution patterns for the major neutral lipid classes.

In most of our chromatographic work, the Florisil has been discarded after running each column. Studies have shown, however, that it can be regenerated in the same way as acid-treated Florisil by washing with methanol, chloroform, and Skellysolve B. Such regeneration could be repeated several times without markedly affecting its chromatographic properties.

As in the case of silicic acid chromatography, cholesteryl esters are eluted from partially deactivated Florisil between the hydrocarbon and triglyceride fractions. Methyl esters and steryl esters overlap and are only partially separated on this column (*96*).

Chromatography on Florisil is a convenient method of isolating fatty acid methyl esters prepared from naturally occurring lipids. Kopaczyk and Radin (*97*) subjected brain lipids to alkaline transmethylation and chromatographed the products on Florisil. Methyl esters derived from ester-linked fatty acids were eluted with Skellysolve B–ethyl ether 99:1, cholesterol with the same solvents in 8:2 ratio, ceramides with chloroform–methanol 95:5, and the remaining glycolipids with chloroform–methanol 3:1. This method gave methyl esters uncontaminated with aldehydes or acetals, and the glycolipids were free of ethanolamine phosphatides. A somewhat similar procedure was used in our laboratory to isolate fatty acid methyl esters derived from adrenal cholesteryl esters (*98*). The methyl esters were eluted from a small Florisil column with Skellysolve B–ethyl ether 95:5 and were separated from the cholesterol liberated during the transesterification, which otherwise complicated analysis of the methyl esters by gas-liquid chromatography.

Radin et al. (*13,49,99*) separated saturated and unsaturated fatty acid esters by treatment with mercuric acetate followed by chromatography on Florisil. Saturated, nonhydroxylated esters were eluted with Skellysolve B and saturated hydroxy esters with Skellysolve B–ethyl ether 9:1. Unsaturated esters form a complex with mercuric acetate that is retained more strongly on the column. They were recovered in their original form by eluting with ethanol–chloroform–HCl 10:8:1.

Triglycerides are usually eluted from Florisil with 15% ethyl ether in Skellysolve B and are well separated from cholesteryl esters, cholesterol, and diglycerides (*14,100–102*). Long-chain fatty alcohols, however, are eluted along with triglycerides (*14*). Isopropylidene and benzylidene derivatives prepared as intermediates in the synthesis of monoglycerides and monoglyceryl ethers were also eluted from Florisil columns with 15% ether in Skellysolve B (*37,81*). The *cis* and *trans* conformational forms of the 1,3-benzylidene intermediate were separated by eluting with 10–15% of ether in Skellysolve B, the *trans* form being eluted before the *cis* form.

Free fatty acids, which may contaminate the triglyceride fraction from silicic acid columns, are retained more strongly on Florisil and may be eluted after other neutral lipid classes with 4% acetic acid in ether (*14*). Diglycerides are eluted from Florisil between cholesterol and monoglyceride fractions with 25–50% ether in Skellysolve B. Florisil causes isomerization of 1,2- and 1,3-diglycerides (*36*), but isomerization can be prevented by using Florisil impregnated with 10%, w/w, boric acid (*80*).

Monoglycerides and monoglyceryl ethers are eluted from Florisil with ether or 2% methanol in ether (*14,37,81*). Monoglycerides, like diglycerides, isomerize on Florisil (*36*), but this can be prevented by using Florisil impregnated with boric acid (*37*). It was possible to obtain a partial separation of 150 mg of a mixture of 1- and 2-mono-glycerides on a 1.5 × 43 cm column of impregnated adsorbent. Both isomers were eluted with 2% methanol in ether, the 2-monoglyceride being eluted slightly ahead of the 1-monoglyceride with about 20% overlap in the middle fractions (*80*).

C. Alumina

Early comparative studies of alumina and silicic acid as adsorbents for chromatography of lipids were carried out by Trappe (*26*), who found that they gave similar separation patterns of lipid classes. Trappe (*26*), Bergström (*103*), and others found, however, that alumina caused alterations in lipid structure during chromatography, namely, alumina partially hydrolyzed glycerides and other fatty acid esters, isomerized mono- and diglycerides, promoted autooxidation and isomerization of double bonds, and caused dehydration of peroxidized lipids. For this reason, alumina has not been used as widely as silicic acid for chromatography of lipids, although it may still be preferable for some separations.

As in the case of silicic acid and Florisil, hydrocarbons are eluted first from alumina columns by petroleum ether, hexane, or Skelly-solve B. By addition of 1–5% of ether to these solvents or with some other solvent mixtures, waxes, fatty acid esters, and cholesteryl esters are eluted (*26,104–106*). Haahti (*22*) separated wax esters from cholesteryl palmitate on a column of alumina pretreated with ethyl acetate. The wax esters were eluted with hexane and cholesteryl palmitate with 10% benzene in hexane. Cole and Brown (*107*) separated naturally occurring waxes by chromatography on an alumina column, eluting with heptane, ethyl ether, propanol, and acetic acid mixtures. Hydrocarbons were eluted with heptane, simple waxes by heptane–ether 99:1, waxes with free hydroxyl groups by heptane–propanol 99:1, and free fatty acids by heptane–acetic acid 99:1. Howard and Hamer (108) succeeded in fractionating peat wax on alumina (Woelm) of pH 4 and containing 2% added water by eluting with benzene, chloroform, and methanol mixtures. Leikola et al. (109) used column chromatography on alumina to separate different classes of pancreatic lipids and isolated substantial amounts of fatty acid methyl esters. Other workers have shown that separations of individual fatty acid esters may be obtained by chromatography on alumina (30,49). Kuemmel (110) separated fatty acid esters according to degree of unsaturation by treatment with mercuric acetate followed by chromatography on alumina.

Triglycerides are eluted with hexane (or petroleum ether)–ethyl ether mixtures after the steryl ester fraction (26,109), and some separation of triglycerides according to degree of unsaturation has been reported (30). Dobinson and Foster (111) separated cis and trans conformational isomers of benzylidene derivatives of monoglycerides and monoglyceryl ethers by column chromatography on alumina.

D. Other Adsorbents

Charcoal was used in a number of early studies as an adsorbent for separation of fatty acids according to chain length and degree of unsaturation (30,112). Different chromatographic techniques such as frontal analysis (113), displacement chromatography (114), and carrier displacement chromatography (115) were employed, but these techniques are now seldom used, since better methods are available.

Magnesium oxide and magnesium sulfate were used as adsorbents for separation of higher fatty acids and some unsaturated fatty acid derivatives (3,4). Urea has been used for separation of straight-chain from branched-chain fatty acids (2,116). This separation is based on preferential inclusion of straight-chain fatty acids, rather than adsorption.

Ion-exchange columns have been used in a number of instances to separate free fatty acids from neutral lipids (28,117). Ion-exchange column chromatography has also been used for separation of fatty acid mixtures. Wurster et al. (*118*) separated methyl esters of oleic, linoleic, and linolenic acids on a cation-exchange resin containing silver ion. Separation was based on the ability of silver ions to form complexes with unsaturated fatty acids. Methyl oleate was eluted with methanol–water 3:1, methyl linoleate with methanol, and methyl linolenate with ethanol saturated with butane-1. Emken et al. (*119*) used a similar type of column to separate methyl elaidate and methyl oleate. *Tran-trans, cis-trans,* and *cis-cis* dienes were also separated, but the *cis-cis* isomer was not recovered from the column.

III. Liquid–Liquid Partition

Liquid–liquid partition column chromatography, which is mainly used for separation of fatty acids and their derivatives, was developed by Martin and Synge (*120,121*) and later extended by Ramsey and Patterson (*16*) and Howard and Martin (*18*). As described above, separation of neutral lipid classes is normally carried out by adsorption column chromatography, whereas liquid–liquid partition chromatography offers advantages for separation of individual fatty acids and their esters. The latter type of chromatography is based on the principles of countercurrent distribution, where lipids are partitioned between two immiscible (or partly miscible) liquid phases moving in opposite directions. Liquid–liquid partition column chromatography always has one stationary liquid phase supported by some solid material (silica gel, celite, cellulose powder, rubber, or other polymers) and one immiscible liquid mobile phase. Stationary and mobile phases are normally equilibrated with one another prior to preparation of the column. The supporting material of the column is then impregnated with the phase that will serve as the stationary phase, and the other phase is used as the eluting solvent. The stationary phase forms a thin layer on the surface of the supporting material, and the lipid mixture is partitioned between this stationary phase and the mobile phase during elution of the column. The elution pattern in liquid–liquid partition column chromatography corresponds approximately to a linear adsorption isotherm in most cases (*35*). Temperature control is of importance in this type of chromatography to maintain equilibrium between the liquid phases.

More details on the technique and theoretical aspects of liquid–liquid partition chromatography may be found in papers and review articles by Lederer and Lederer (*19*), Giddings and Keller (*122*), Morris (*30*), Schlenk and Gellerman (*123*), Stein and Slawson (*24*), Vereshchagin (*124*), and Vink (*125*).

A. Partition

As noted earlier, liquid–liquid partition chromatography is carried out with a polar stationary phase and a nonpolar mobile phase. Ramsey and Patterson (*16*) used silicic acid as supporting material and impregnated it with water, which served as the stationary phase. With chloroform–butanol as mobile phase, they succeeded in separating C_1–C_4 acids. By partitioning between methanol as stationary phase and isooctane as mobile phase, they separated C_5–C_{10} acids; and by partitioning between furfuryl alcohol–2-aminopyridine and hexane, they separated C_{11}–C_{19} fatty acids (*126*). This procedure was extended and improved by Nijkamp (*17*) and by Vandenheuvel and Hayes (*127*). Fatty acids from various rat lipids were separated in a similar way by Lossow and Chaikoff (*128*). Davenport (*129*) separated hydroxamates of C_{14}–C_{20} saturated and unsaturated fatty acids with similar solvent systems and powdered cellulose as supporting medium. Other solvents such as nitromethane, acetonitrile, and phosphate, citrate, and glycine buffers have been used as stationary phases for separation of mono- and dibasic fatty acids and derivatives (*30,130*).

Youngs (*131*) separated triglycerides according to degree of unsaturation on a silicic acid column with aqueous ethanol as stationary phase and Skellysolve B as mobile phase. Frankel et al. (*132–134*) used a liquid–liquid partition system with methanol–benzene on silicic acid as the stationary phase to separate dimeric and polymeric products, fatty acid hydroperoxides, hydroxy fatty acids and esters, and other components of oxidized oils.

B. Reversed-Phase Partition

In reversed-phase liquid–liquid partition chromatography the less-polar solvent is used as stationary phase and the more-polar solvent as mobile phase. In this case it is necessary to have a lipophilic supporting material that will retain the nonpolar stationary liquid phase. In general, reversed-phase chromatography is the preferred system for lipids, since they are retained more strongly by the nonpolar stationary phase and better separations can usually be achieved.

Howard and Martin (*18*) treated Kieselgur (Hyflo Super Cel) with dichlorodimethylsilane to make it lipophilic so that it could be easily impregnated with nonpolar solvents. They used liquid paraffin as stationary phase and aqueous methanol or aqueous acetone as mobile phase and separated normal saturated fatty acids from C_{12} to C_{18}. Silk and Hahn (*135*) extended the method to separation of longer-chain fatty acids (C_{16}–C_{24}) and applied it to analysis of South African pilchard oil. Saturated fatty acids differing by two carbon atoms are well separated in this system, but unsaturated fatty acids tend to overlap with shorter-chain saturated acids. For example, palmitoleic, linoleic, and arachidonic acids overlap myristic acid and oleic acid overlaps palmitic acid (*136,137*). This difficulty can be overcome by preliminary separation into classes of one chain length or one degree of unsaturation followed by chromatography of the individual fractions. For smaller samples, where this may not be possible, Morris (*30*) suggested the following analytical schemes:

1. Total acids after hydrogenation and saturated acids after oxidative removal of unsaturated acids (*138*).

2. Total acids, saturated and unsaturated, and saturated acids after oxidative removal of unsaturated acids (*139*).

3. Total acids, saturated and unsaturated, and total acids of inseparable groups after hydrogenation (*135,137,140,141*).

4. Total acids, saturated and unsaturated, and total acids after conversion of unsaturated acids to polyhydroxy derivatives (*136*). The polyhydroxy derivatives may then be separated by reversed-phase chromatography on a stationary phase of castor oil.

Crombie et al. (*139*) have summarized the effects of some functional groups on elution rate of fatty acids in this type of reversed-phase system. A double bond increases the elution rate to a greater or lesser extent according to its position. When adjacent to the carboxyl group, it has little effect, but when further removed, it increases the elution rate to approximately that of the saturated acid having a chain length of two CH_2 fewer. Additional double bonds cause a progressive increase in elution rate. *Cis* and *trans* isomers are inseparable. [Howard and Martin (*18*) found, however, that elaidic acid was slightly retarded in relation to oleic acid.] One triple bond increases the rate of elution more than two double bonds. The presence of a keto group or a hydroxyl group markedly increases the rate of elution. The presence of two hydroxyl groups appears to reduce the solubility of the acids in paraffin so much that bad tailing results.

Gunstone and Sykes (*142*) used liquid paraffin, castor oil, and acetylated castor oil as stationary phases and acetone–water mixtures as mobile phases for analysis of complex mixtures of saturated, unsaturated, and oxygenated acids. Similar systems were also used by Elovson (*143*) for separation of normal and hydroxy fatty acids.

Reversed-phase chromatography has also been used for separation of fatty acid esters and glycerides. Morris et al. (*144*) isolated epoxy and hydroxy esters with isooctane as stationary phase and acetonitrile as mobile phase. Privett et al. (*145,146*) used heptane as stationary phase and acetonitrile–methanol 85:15 as mobile phase for separation and purification of fatty acid methyl esters. Fatty acid esters differing in chain length by two carbon atoms were cleanly separated on a 2.5×120 cm column, but, as in the systems described above, unsaturated acids were eluted before the corresponding saturated acid and overlapped with shorter-chain acids (Fig. 6). This method has been used successfully in our laboratory for separation of branched-chain fatty acids from bacterial lipids (*37*). The column is readily regenerated and may be used repeatedly without repacking. Black and Hammond (*147*) used heptane as stationary phase and aqueous acetone as mobile phase for separation of triglycerides. Using trilaurin and trimyristin as model substances, they calculated that under the best conditions on a 1.8×150 cm column it would be possible to separate glycerides differing by two double bonds or four carbon atoms. By adapting their own system and extending the elution time, Privett et al. have been able to separate triglycerides differing in some cases by only one carbon atom (*148*). Reversed-phase chromatography was used by Savary and Desnuelle (*149,150*) to separate products of hydrolysis of triglycerides. They first removed free fatty acids by passing the mixture through an ion-exchange resin (IRA-400) and then separated mono-, di-, and triglycerides on a column of siliconized Kieselgur with cyclohexane as stationary phase and aqueous ethanol or aqueous acetone as mobile phase.

Schlenk et al. (*123,151,152*) used a column of powdered Kieselgur impregnated with silicone oil as stationary phase and aqueous acetonitrile as mobile phase for separation of the methyl esters of odd- and even-numbered straight-chain fatty acids from fish oil (Fig. 7). Other techniques such as distillation and crystallization were used to obtain fractions enriched in particular fatty acids prior to the chromatographic separations. In additional studies, this reversed-phase sys-

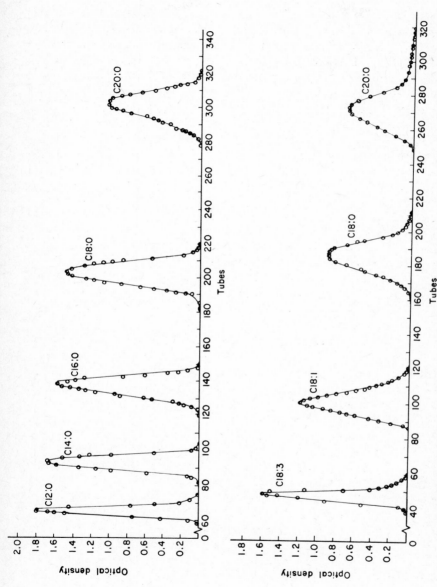

FIG. 6. Separation of 2-g samples of equimolar model mixtures of methyl esters by reversed-phase liquid–liquid chromatography using acetonitrile–methanol 85:15, v/v, as the mobile phase and heptane as the stationary phase on a 2.5 × 120 cm column of nonwetting celite. [Reproduced from O. S. Privett and E. C. Nickell, *J. Am. Oil Chemists' Soc.*, **40**, 189 (1963), by permission of the authors and of the Society.]

tem was used for separation and identification of fatty acids from both plant and animal lipids (*153–156*). One disadvantage of the method for preparative work is that the isolated fatty acid esters

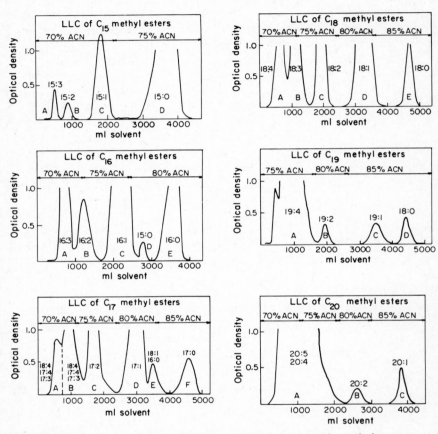

FIG. 7. Liquid–liquid chromatography of mullet fatty acid methyl esters on 2.7 × 90 cm columns of powdered Kieselgur impregnated with silicone oil as stationary phase and aqueous acetonitrile as mobile phase. [Reproduced from N. Sen and H. Schlenk, *J. Am. Oil Chemists' Soc.,* **41,** 241 (1964), by permission of the authors and of the Society.]

contain small amounts of silicone oil, but this contaminant can be removed by alembic distillation (*156*).

In all of the above-mentioned methods, Celite (Kieselgur, Hyflo

Super Cel, diatomaceous earth) was employed as a support for the nonpolar stationary phase. Schlenk et al. impregnated the Celite directly with silicone oil, whereas in other chromatographic procedures the Celite was first siliconized to render it lipophilic. A number of other supporting materials such as rubber powder, polyethylene powder, and methylated Sephadex have also been utilized for reversed-phase column chromatography.

Boldingh (*157,158*) was the first to use powdered rubber as a column-supporting material. He impregnated the rubber with hydrocarbons or triglycerides and used mixtures of methanol, acetone, and water as mobile phases to separate saturated fatty acids from C_6 to C_{22} as well as products formed by oxidation and isomerization of unsaturated fatty acids. Hofmann et al. (*159*) used a somewhat modified system to separate monounsaturated, saturated, and branched-chain fatty acids from bacterial lipids, and hydroxy acids were separated on a rubber column by Desnuelle and Burnet (*160*). Reversed-phase chromatography on rubber columns has also been used in a number of other laboratories for the separation of fatty acids and glycerides (*30,161*). Green et al. (*162*) used polyethylene powder as stationary phase and aqueous acetone as mobile phase and obtained good separations of C_6–C_{20} acids. This method has also been used by a number of other workers (*30*).

Hirsch (*163*) and Knittle and Hirsch (*164*) have described another new method of separating fatty acid esters and glycerides by reversed-phase chromatography, using Factice, a product obtained by polymerization of soybean oil with sulfur monochloride (Carter-Bell Manufacturing Co., Springfield, New Jersey). Factice is a lipophilic polymer that swells in nonpolar solvents and shrinks in polar solvents. Columns should therefore be packed in the presence of the solvent that is to be used as the mobile phase. The best separations were obtained by eluting with mixtures of acetone–water. It is suggested that the stationary phase in such columns is an acetone–Factice gel. Factice was found to be inert and did not induce chemical changes in lipids during chromatography.

The separation of some major lipid classes is illustrated in Fig. 8. The more polar lipids are eluted first, beginning at the right-hand side of the diagram. Separations may also be influenced by the nature of the fatty acid substituents, shorter-chain acids being eluted before longer-chain acids and unsaturated acids before saturated acids.

Phosphatides and free fatty acids tend to interfere and should be removed from lipid mixtures before chromatography on Factice. This may be accomplished by passing a chloroform solution of the lipids through a small Florisil column.

The proportions of acetone and water in the eluting solvent may be varied depending on the type of separation desired. The results shown in Fig. 8 were obtained by elution with 2% water in acetone. Individual cholesteryl esters may also be separated in this system, but for separation of triglycerides elution with 5% water in acetone is

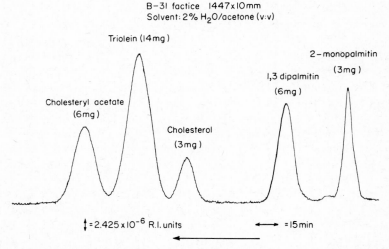

FIG. 8. Lipid class separation by Factice chromatography. The record was obtained by monitoring changes in the refractive index of the column effluent by means of a Phoenix Refractometer. [Reproduced from J. Hirsch, *J. Lipid Res.*, **4**, 1 (1963), by permission of the author and of the editor.]

most effective. For long-chain methyl esters 12% water and for short-chain methyl esters 15% water in acetone are recommended. The separation of short-chain methyl esters is illustrated in Fig. 9. As in other reversed-phase systems, the effect of a double bond on elution time is the same as shortening the fatty acid chain by two carbon atoms. Thus, methyl palmitate and oleate are inseparable in this system, wheras methyl linoleate is eluted with methyl myristate and methyl linolenate with methyl laurate.

Methylated Sephadex has recently been introduced by Nyström and

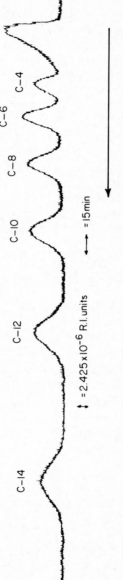

FIG. 9. Refractometric recording of a separation of methyl esters of fatty acids by Factice chromatography. The large initial peak at the right is the solvent front disturbance of refractive index. [Reproduced from J. Hirsch, *J. Lipid Res.*, **4**, 1 (1963), by permission of the author and of the editor.]

Sjövall (*165*) as a supporting material for reversed-phase column chromatography. They studied the chromatographic behavior of a variety of lipids (phosphatides, cerebrosides, triglycerides, cholesteryl esters, fatty acids, and bile acids) and concluded that the separations were due to several processes, with the major factors being liquid–liquid partition, gel filtration, and in some cases adsorption of polar compounds to the gel structure (*166*). The results indicated that either reversed-phase or "straight phase" partition could occur, depending on the composition of the eluting solvent. It was suggested that in chloroform–methanol systems with a low percentage of methanol, e.g., 8:1, there may be proportionately more methanol in the gel phase than in the mobile phase, so that straight partition chromatography occurs, whereas at higher percentages of methanol, e.g., 1:1, the mobile phase may contain more methanol, giving a reversed-phase system. Since the hydroxyl groups in the Sephadex were not all methylated, Sephadex could not be used satisfactorily with phases as nonpolar as benzene or heptane.

Most of the experiments were carried out with columns of 20-mm diameter containing 20–25 g of methylated Sephadex, which gave column heights of 25–36 cm. Sephadex G-25 and G-50, which differ in degree of cross-linking, were used as starting materials, and methylation was carried out by treatment with dimethyl sulfate. The methylated products had good stability, and columns were used continuously for over a year without change in properties.

It is difficult to assess the potential value of this new method until it has been explored further and the factors that affect the chromatographic separations are better understood. It may, however, prove to be a useful addition to methods based primarily on adsorption or partition, that are currently used for most column chromatographic separations of lipids.

Acknowledgment

The financial support of the Medical Research Council of Canada and technical assistance by Mr. H. E. Pedersen are gratefully acknowledged.

REFERENCES

1. H. G. Cassidy, *J. Am. Chem. Soc.*, **62**, 3073 (1940).
2. J. Cason, G. Sumrell, C. F. Allen, G. A. Gillies, and S. Elberg, *J. Biol. Chem.*, **205**, 435 (1953).

3. M. M. Graff and E. L. Skou, *Ind. Eng. Chem. Anal. Ed.*, **15**, 340 (1943).
4. R. O. Simmons and F. W. Quackenbush, *J. Am. Oil Chemists' Soc.*, **30**, 614 (1953).
5. F. T. Walker and M. R. Mills, *J. Soc. Chem. Ind.*, **61**, 125 (1942); **62**, 106 (1943).
6. C. L. Reinbold and H. J. Dutton, *J. Am. Oil Chemists' Soc.*, **25**, 117, 120 (1948).
7. J. Asselineau and J. Moron, *Bull. Soc. Chim. Biol.*, **40**, 899 (1958).
8. H. P. Kaufmann, *Fette Seifen Anstrichmittel*, **46**, 268 (1939).
9. B. Börgstrom, *Acta Physiol. Scand.*, **30**, 231 (1954).
10. J. Hirsch and E. H. Ahrens, Jr., *J. Biol. Chem.*, **233**, 311 (1958).
11. J. J. Wren, *J. Chromatog.*, **4**, 173 (1960); *Chromatog. Rev.*, **3**, 111, 177 (1961).
12. E. J. Barron and D. J. Hanahan, *J. Biol. Chem.*, **231**, 493 (1958).
13. Y. Kishimoto and N. S. Radin, *J. Lipid Res.*, **1**, 72 (1959).
14. K. K. Carroll, *J. Lipid Res.*, **2**, 135 (1961).
15. K. K. Carroll, *J. Am. Oil Chemists' Soc.*, **40**, 413 (1963).
16. L. L. Ramsey and W. I. Patterson, *J. Assoc. Offic. Agr. Chemists*, **28**, 644 (1945).
17. H. J. Nijkamp, *Anal. Chim. Acta*, **5**, 325 (1951); **10**, 448 (1954).
18. G. A. Howard and A. J. P. Martin, *Biochem. J.*, **46**, 532 (1950).
19. E. Lederer and M. Lederer, *Chromatography*, 2nd ed., Elsevier, Amsterdam, 1957.
20. L. J. Morris, *J. Lipid Res.*, **7**, 717 (1966).
21. E. Heftmann, *Chromatography*, Reinhold, New York, 1961; *Anal. Chem.*, **38**, 31R (1966).
22. E. Haahti, *Scand. J. Clin. Lab. Invest. Suppl.*, **13**, 59 (1961).
23. H. Schlenk, in *Fatty Acids*, 2nd ed., Pt. 3 (K. S. Markley, ed.), Wiley (Interscience), New York, 1964, p. 2125.
24. R. A. Stein and V. Slawson, in *Progress in the Chemistry of Fats and Other Lipids*, Vol. 8, Pt. 3 (R. T. Holman, ed.), Pergamon Press, London, 1966, p. 375.
25. H. P. Kaufmann, *Angew. Chem.*, **53**, 98 (1940).
26. W. Trappe, *Biochem. Z.*, **306**, 316 (1940).
27. W. Trappe, *Biochem. Z.* **307**, 97 (1941).
28. B. Börgstrom, *Acta Physiol. Scand.*, **25**, 101, 111 (1952).
29. D. L. Fillerup and J. F. Mead, *Proc. Soc. Exptl. Biol. Med.*, **83**, 574 (1953); **86**, 449 (1954).
30. L. J. Morris, in *Chromatography* (E. Heftmann, ed.), Reinhold, New York, 1961, p. 428.
31. D. J. Hanahan, *Lipide Chemistry*, Wiley, New York, 1960.
32. M. G. Horning, E. A. Williams, and E. C. Horning, *J. Lipid Res.*, **1**, 482 (1960).
33. R. Hernandez, R. Hernandez, Jr., and L. R. Axelrod, *Anal. Chem.*, **33**, 370 (1961).
34. L. R. Snyder, *J. Chromatog.*, **11**, 195 (1963).
35. L. Hagdahl, in *Chromatography* (E. Heftmann, ed.), Reinhold, New York, 1961, p. 56.

36. F. H. Mattson and R. A. Volpenhein, *J. Lipid Res.*, **3**, 281 (1962).
37. B. Serdarevich and K. K. Carroll, *J. Lipid Res.*, **7**, 277 (1966).
38. J. J. Wren and A. D. Szczepanowska, *J. Chromatog.*, **14**, 405 (1964).
39. P. D. Klein, *Anal. Chem.*, **33**, 1737 (1961) ; **34**, 733 (1962).
40. J. J. Wren and A. D. Szczepanowska, *J. Sci. Food Agr.*, **16**, 161 (1965).
41. R. S. Alm, R. J. P. Williams, and A. Tiselius, *Acta Chem. Scand.*, **6**, 826 (1952).
42. L. R. Snyder, *J. Chromatog.*, **12**, 488 (1963) ; **13**, 415 (1964) ; *Chromatog. Rev.*, **7**, 1 (1965).
43. A. Kuksis and J. M. R. Beveridge, *J. Lipid Res.*, **1**, 311 (1960).
44. R. W. Riemenschneider, S. F. Herb, and P. L. Nicholas, Jr., *J. Am. Oil Chemists' Soc.*, **26**, 371 (1949).
45. F. .E. Kurtz, *J. Am. Chem. Soc.*, **74**, 1902 (1952).
46. O. S. Privett, R. P. Weber, and E. C. Nickell, *J. Am. Oil Chemists' Soc.*, **36**, 443 (1959).
47. P. D. Klein and E. T. Janssen, *J. Biol. Chem.*, **234**, 1417 (1959).
48. S. Bergström and K. Pääbo, *Acta Chem. Scand.*, **8**, 1486 (1954).
49. N. S. Radin, *J. Am. Oil Chemists' Soc.*, **42**, 569 (1965).
50. J. Erwin and K. Bloch, *J. Biol. Chem.*, **238**, 1618 (1963).
51. E. D. Korn, *J. Lipid Res.*, **5**, 352 (1964).
52. B. de Vries, *Chem. Ind. (London)*, **1962**, 1049.
53. B. de Vries, *J. Am. Oil Chemists' Soc.*, **40**, 184 (1963).
54. E. Haahti, T. Nikkari, and K. Juva, *Acta Chem. Scand.*, **17**, 538 (1963).
55. H. Wagner, J.-D. Goetschel, and P. Lesch, *Helv. Chim. Acta*, **46**, 2986 (1963).
56. M. K. Bhatty and B. M. Craig, *J. Am. Oil Chemists' Soc.*, **41**, 508 (1964).
57. Y. Kishimoto and N. S. Radin, *J. Lipid Res.*, **4**, 437 (1963).
58. R. O. Weenink and F. B. Shorland, *Biochim. Biophys. Acta*, **84**, 613 (1964).
59. J. Elovson, *Biochim. Biophys. Acta*, **106**, 291 (1965).
60. Y. Kishimoto and N. S. Radin, *J. Lipid Res.*, **4**, 444 (1963).
61. N. Nicolaides and R. C. Foster, Jr., *J. Am. Oil Chemists' Soc.*, **33**, 404 (1956).
62. B. de Vries, *J. Am. Oil Chemists' Soc.*, **41**, 403 (1964).
63. O. Renkonen, O.-V. Renkonen, and E. L. Hirvisalo, *Acta Chem. Scand.*, **17**, 1465 (1963).
64. M. R. Subbaram and C. G. Youngs, *J. Am. Oil Chemists' Soc.*, **41**, 445 (1964).
65. A. Dolev and H. S. Olcott, *J. Am. Oil Chemists' Soc.*, **42**, 624 (1965).
66. R. D. McCarthy and A. H. Duthie, *J. Lipid Res.*, **3**, 117 (1962).
67. P. J. Heald, H. G. Badman, J. Wharton, C. M. Wulwik, and P. I. Hooper, *Biochim. Biophys. Acta*, **84**, 1 (1964).
68. B. Preiss and K. Bloch, *J. Biol. Chem.*, **239**, 85 (1964).
69. D. T. Downing, *Rev. Pure Appl. Chem.*, **11**, 196 (1961).
70. T. H. Applewhite, *J. Am. Oil Chemists' Soc.*, **42**, 321 (1965).
71. G. H. de Haas and L. L. M. van Deenen, *Biochim. Biophys. Acta*, **106**, 315 (1965).
72. J. Y. Kiyasu and E. P. Kennedy, *J. Biol. Chem.*, **235**, 2590 (1960).
73. W. E. M. Lands and P. Hart, *Biochim. Biophys. Acta*, **98**, 532 (1965).

74. B. Palameta and M. Kates, *Biochemistry,* **5,** 618 (1966).
75. O. S. Privett and M. L. Blank, *J. Lipid Res.,* **2,** 37 (1961).
76. A. E. Thomas, III, J. E. Scharoun, and H. Ralston, *J. Am. Oil Chemists' Soc.,* **42,** 789 (1965).
77. K. K. Carroll, *J. Lipid Res.,* **3,** 388 (1962).
78. R. L. Anderson and E. J. Hollenbach, *J. Lipid Res.,* **6,** 577 (1965).
79. D. Willner, *Chem. Ind. (London),* **1965,** 1839.
80. B. Serdarevich, *J. Am. Oil Chemists' Soc.,* in press.
81. B. Serdarevich and K. K. Carroll, *Can. J. Biochem.,* **44,** 743 (1966).
82. F. B. O'Neal and J. Carlton, *Anal. Chem.,* **30,** 1051 (1958).
83. D. H. Blankenhorn, G. Rouser, and T. J. Weimer, *J. Lipid Res.,* **2,** 281 (1961).
84. B. G. Creech, *J. Am. Oil Chemists' Soc.,* **38,** 540 (1961).
85. M. M. Rapport, H. Schneider, and L. Graf, *J. Biol. Chem.,* **237,** 1056 (1962).
86. H. Chino and L. I. Gilbert, *Science,* **143,** 359 (1964).
87. G. A. Fischer and J. J. Kabara, *Anal. Biochem.,* **9,** 303 (1964).
88. J. S. O'Brien and A. A. Benson, *J. Lipid Res.,* **5,** 432 (1964).
89. D. T. Downing, *J. Lipid Res.,* **5,** 210 (1964).
90. K. C. Kopaczyk and N. S. Radin, *J. Lipid Res.,* **6,** 140 (1965).
91. W. D. Suomi and B. W. Agranoff, *J. Lipid Res.,* **6,** 211 (1965).
92. S. M. Grundy, E. H. Ahrens, Jr., and T. A. Miettinen, *J. Lipid Res.,* **6,** 397 (1965).
93. A. Sheltawy, *Biochim. J.,* **95,** 561 (1965).
94. P. H. Monaghan, H. A. Suter, and A. L. Le Rosen, *Anal. Chem.,* **22,** 811 (1950).
95. E. D. Smith and A. L. Le Rosen, *Anal. Chem.,* **23,** 732 (1951).
96. K. K. Carroll and B. Serdarevich, unpublished work, 1966.
97. K. C. Kopaczyk and N. S. Radin, *J. Lipid Res.,* **6,** 140 (1965).
98. K. K. Carroll, *Can. J. Biochem. Physiol.,* **40,** 1115 (1962).
99. A. K. Hajra and N. S. Radin, *J. Lipid Res.,* **3,** 327 (1962).
100. A. Boudreau and J. M. Deman, *Biochim. Biophys. Acta,* **98,** 47 (1965).
101. H. Chino and L. I. Gilbert, *Biochim. Biophys. Acta,* **98,** 94 (1965).
102. D. Abramson and M. Blecher, *Biochim. Biophys. Acta,* **98,** 117 (1965).
103. S. Bergström, *Nature,* **156,** 717 (1945).
104. W. Fuchs and A. de Jong, *Fette Seifen Anstrichmittel,* **56,** 218 (1954).
105. V. R. Wheatley, *Biochem. J.,* **58,** 167 (1954).
106. V. R. Wheatley and A. T. James, *Biochem. J.,* **65,** 36 (1957).
107. L. J. N. Cole and J. B. Brown, *J. Am. Oil Chemists' Soc.,* **37,** 359 (1960).
108. A. J. Howard and D. Hamer, *J. Am. Oil Chemists' Soc.,* **39,** 250 (1962).
109. E. Leikola, E. Nieminen, and E. Salomaa, *J. Lipid Res.,* **6,** 490 (1965).
110. D. F. Kuemmel, *Anal. Chem.,* **34,** 1003 (1962).
111. B. Dobinson and A. B. Foster, *J. Chem. Soc.,* **1961,** 2338.
112. J. Cason and G. A. Gillies, *J. Org. Chem.,* **20,** 419 (1955).
113. S. Claesson, *Ann. N.Y. Acad. Sci.,* **49,** 183 (1948).
114. R. T. Holman and L. Hagdahl, *J. Biol. Chem.,* **182,** 421 (1950).
115. A. Tiselius and L. Hagdahl, *Acta Chem. Scand.,* **4,** 394 (1950).
116. H. J. Dutton, *J. Am. Oil Chemists' Soc.,* **36,** 513 (1959).

117. G. Lakshminarayana, F. A. Kruger, D. G. Cornwell, and J. B. Brown, *Arch. Biochem. Biophys.*, **88**, 318 (1960).
118. C. F. Wurster, Jr., J. H. Cupenhaver, Jr., and P. R. Shafer, *J. Am. Oil Chemists' Soc.*, **40**, 513 (1963).
119. E. A. Emken, C. R. Scholefield, and H. J. Dutton, *J. Am. Oil Chemists' Soc.*, **41**, 388 (1964).
120. A. J. P. Martin and R. L. M. Synge, *Biochem. J.*, **35**, 1358 (1941).
121. A. J. P. Martin, *Ann. N.Y. Acad. Sci.*, **49**, 249 (1948).
122. J. C. Giddings and R. A. Keller, in *Chromatography* (E. Heftmann, ed.), Reinhold, New York, 1961, p. 92.
123. H. Schlenk and J. L. Gellerman, *J. Am. Oil Chemists' Soc.*, **38**, 555 (1961).
124. A. G. Vereshchagin, *J. Chromatog.*, **14**, 184 (1964).
125. H. Vink, *J. Chromatog.*, **20**, 305 (1965).
126. L. L. Ramsey and W. I. Patterson, *J. Assoc. Offic. Agr. Chemists*, **31**, 441 (1948).
127. F. A. Vandenheuvel and E. R. Hayes, *Anal. Chem.*, **24**, 960 (1952).
128. W. J. Lossow and I. L. Chaikoff, *J. Biol. Chem.*, **230**, 149 (1958).
129. J. B. Davenport, *Chem. Ind. (London)*, **1955**, 705.
130. R. O. Feuge and E. R. Cousins, *J. Am. Oil Chemists' Soc.*, **37**, 267 (1960).
131. C. G. Youngs, *J. Am. Oil Chemists' Soc.*, **38**, 62 (1961).
132. E. N. Frankel, C. D. Evans, H. A. Moser, D. G. McConnell, and J. C. Cowan, *J. Am. Oil Chemists' Soc.*, **38**, 130 (1961).
133. E. N. Frankel, D. G. McConnell, and C. D. Evans, *J. Am. Oil Chemists' Soc.*, **39**, 297 (1962).
134. C. D. Evans, D. G. McConnell, E. N. Frankel, and J. C. Cowan, *J. Am. Oil Chemists' Soc.*, **42**, 764 (1965).
135. M. H. Silk and H. H. Hahn, *Biochem. J.*, **56**, 406 (1954); **57**, 577 (1954).
136. P. Savary and P. Desnuelle, *Bull. Soc. Chim. France*, **1953**, 939.
137. G. Popják and A. Tietz, *Biochem. J.*, **56**, 46 (1954).
138. F. A. Vandenheuvel and D. R. Vatcher, *Anal. Chem.*, **28**, 838 (1956).
139. W. M. L. Crombie, R. Comber, and S. G. Boatman, *Biochem. J.*, **59**, 309 (1955).
140. A. K. Lough and G. A. Garton, *Biochem. J.*, **67**, 345 (1957).
141. J. F. Mead and D. R. Howton, *J. Biol. Chem.*, **229**, 575 (1957).
142. F. D. Gunstone and P. Sykes, *J. Chem. Soc.*, **1960**, 5050.
143. J. Elovson, *Biochim. Biophys. Acta*, **84**, 275 (1964).
144. L. J. Morris, H. Hayes, and R. T. Holman, *J. Am. Oil Chemists' Soc.*, **38**, 316 (1961).
145. O. S. Privett, R. P. Weber, and E. C. Nickell, *J. Am. Oil Chemists' Soc.*, **36**, 443 (1959).
146. O. S. Privett and E. C. Nickell, *J. Am. Oil Chemists' Soc.*, **40**, 189 (1963).
147. B. C. Black and E. G. Hammond, *J. Am. Oil Chemists' Soc.*, **40**, 575 (1963).
148. O. S. Privett, private communication, 1966.
149. P. Savary and P. Desnuelle, *Bull. Soc. Chim. France*, **1954**, 936.
150. P. Savary and P. Desnuelle, *Biochim. Biophys. Acta*, **31**, 26 (1959).
151. J. L. Gellerman and H. Schlenk, *Experientia*, **15**, 387 (1959).
152. N. Sen and H. Schlenk, *J. Am. Oil Chemists' Soc.*, **41**, 241 (1964).

153. J. J. Rahm and R. T. Holman, *J. Lipid Res.,* **5,** 169 (1964).
154. H. Schlenk and J. L. Gellerman, *J. Am. Oil Chemists' Soc.,* **42,** 504 (1965).
155. D. Sand, N. Sen, and H. Schlenk, *J. Am. Oil Chemists' Soc.,* **42,** 511 (1965).
156. J. L. Gellerman and H. Schlenk, *J. Protozool.,* **12,** 178 (1965).
157. J. Boldingh, *Rec. Trav. Chim.,* **69,** 247 (1950).
158. J. Boldingh, *Koninkl. Vlaam. Acad. Wetenschap. Letter. Schone Kunsten Belg. Kl. Wetenschap. Intern. Colloq. Biochem. Probl. Lipiden, 1, Brussels,* **1953,** 64.
159. K. Hofmann, C. Y. Hsiao, D. B. Henis, and C. Panos, *J. Biol. Chem.,* **217,** 49 (1955).
160. P. Desnuelle and M. Burnet, *Bull. Soc. Chim. France,* **1956,** 268.
161. J. R. Trowbridge, A. B. Herrick, and R. A. Baumann, *J. Am. Oil Chemists' Soc.,* **41,** 306 (1964).
162. T. Green, F. O. Howitt, and R. Preston, *Chem. Ind. (London),* **1955,** 591.
163. J. Hirsch, *J. Lipid Res.,* **4,** 1 (1963).
164. J. L. Knittle and J. Hirsch, *J. Lipid Res.,* **6,** 565 (1965).
165. E. Nyström and J. Sjövall, *J. Chromatog.,* **17,** 574 (1965).
166. E. Nyström and J. Sjövall, *Anal. Biochem.,* **12,** 235 (1965).

GAS CHROMATOGRAPHY
OF NEUTRAL GLYCERIDES

7

A. Kuksis

BANTING AND BEST DEPARTMENT OF MEDICAL RESEARCH
UNIVERSITY OF TORONTO
TORONTO, CANADA

I. Introduction

Present methods for the analysis of neutral glycerides by gas–liquid chromatography (GLC) are based upon the use of short columns of relatively thin film packings prepared with highly thermostable liquid phases of low polarity and temperature programming. The preparation of packings suitable for glyceride separations requires procedures comparable with those developed for steroid work. The resolutions obtained depend on the boiling point or the molecular weight of the glycerides, but some fractionation due to the shape of the molecules may also be demonstrated. The resolving power of the gas chromatographic method has been greatly increased by the introduction of preparative thin-layer chromatography on silver-nitrate-impregnated plates, which allows a preliminary segregation of the compounds on the basis of unsaturation. Despite this, the GLC of neutral glycerides is still very much limited to exploratory work and research, although practical applications are becoming apparent. The lack of significant results in this area is largely a reflection of our ignorance of any importance the observed differences in neutral glyceride constitution might have upon the structure and function of physiological systems. This review discusses the technical advances to date and summarizes the more successful applications to natural mixtures of neutral glycerides.

II. Historical

The first practical demonstration of the separation of natural triglycerides by gas chromatography was described by Kuksis and McCarthy (1). Prior to that time, several investigators (2–5) had found that simple triglycerides could be recovered after gas chromatography under conditions involving extremely high temperatures and long retention times and that it might be possible to analyze natural fats containing glycerides as large as tristearin. Fryer et al. (*2*) obtained fingerprint chromatograms of various natural fats and oils. Martin et al. (*3*) obtained approximately twelve peaks when they chromatographed margarine stock. All separations were accompanied by degradation, and there was a general uncertainty as to the identity of any peaks. Huebner (*5*) recognized the carbon-number or molecular-weight basis of the GLC resolution of mono-, di-, and triglycerides, but his elution patterns also showed considerable degradation resulting, for example, in a complete distortion of the characteristic elution pattern of butterfat.

In all instances the temperatures of the flash evaporator at least approached or exceeded those of the cracking points of simple hydrocarbons. Although a number of investigators recognized at the time that the best way to reduce both temperature and retention times to useful levels was to decrease the amount of liquid phase and to shorten the column length, few trials were reported. This approach would have been investigated more widely had it not been for a general assumption that materials that have molecular weights over 900 could not have emerged from the column unchanged within the period of time indicated. This belief was supported both by authoritative opinion and by many laboratory results, although work at Applied Science Laboratories (*6*) had indicated that even triolein and trilinolein could be recovered from GLC columns as symmetrical peaks at rather low temperatures (290°C) using ¾% SE-30 coatings.

The problem was resolved by advances made in the GLC instrumentation, particularly the introduction of flame ionization detection, which permitted the handling of small samples, and the provision of means for on-column injection, which avoided the need for superheating the materials in the flash evaporator. As manufacturers have equipped their gas chromatographs with these facilities, most instruments are now suitable for the analysis of high-molecular-weight compounds including triglycerides. The temperature ranges for general analytical work usually can be restricted to 200–325°C, although with well-conditioned columns the elution of the highest-molecular-weight component may be completed below 300°C (*7*). The separation and elution of a complex triglyceride mixture may require 30–45 min, but much shorter periods of time may be sufficient for the resolution of simple mixtures. The sample size for analytical applications is at the microgram or submicrogram level. Column efficiencies ranging from about 500 to 800 theoretical plates per foot of ⅛-in. column are commonly employed.

III. Principles and Terminology

Although the principles of conventional gas chromatography are well known and the terminology is well established, high-temperature operation and temperature programming have introduced several new variables and have required the redefinition of certain terms and an alteration of the concepts of others. The complex nature of the neutral glyceride elution patterns has required the development of a simplified system of reference for the identification of the solutes or their mixtures.

A. Method of Chromatography

[Gas chromatography is a form of partition chromatography in which the stationary phase is a liquid film held in place on a solid support and the mobile phase is a carrier gas flowing over the surface of the liquid film in a controlled fashion.] Under conditions of high-temperature programming the gas phase also contains significant amounts of the vapor of the liquid phase, which also contributes to the partition behavior and may in fact be responsible for the effectiveness of the resolution and the protection of the solutes at the elevated temperatures.

Under isothermal operation, the distribution between mobile and stationary phases is about equal (8) and the peak appears as a sharp band. With time, the distribution, although remaining proportionately the same between the phases, begins to broaden so that longer times are required for the solutes to emerge from the column. The characteristic broadening continues as the higher-boiling solutes emerge, until the peaks effectively merge with the base line. Because of the wide range of boiling temperatures of the different glycerides commonly present in natural mixtures, the isothermal separation procedure is not ideally suited to the analysis of these compounds. It has nevertheless been used for glyceride analysis (9) and may offer advantages for the resolution of specific isomeric pairs.

With programmed temperature operation, the solute initially favors the stationary phase to a greater extent than the mobile phase. As the column temperature begins to rise, the distribution becomes more nearly equal and approaches the initial isothermal distribution. As the temperature increases, the distribution more and more favors the mobile phase and the solute bands remain sharp. This latter approach implies a limit to the useful column length for a specific heating rate.

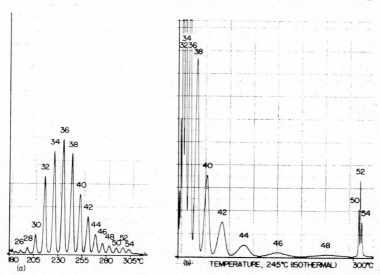

FIG. 1. GLC of triglycerides of coconut oil. Peaks identified by the total number of carbon atoms in the fatty acid moieties. (a) Linear temperature program; (b) isothermal. Instrument 1, Beckman GC-4 with special on-column inlet heater, dual columns, differential electrometer; columns, stainless-steel tubes, 20 in. × ⅛ in. o.d., packed with 3%, w/w, JXR on Gas Chrom Q (100–120 mesh). Nitrogen flow rate, 120 ml/min, 60-psi head pressure. On-column injectors, 280°C; detector, 340°C; detector line, 320°C. Column temperature as shown. Beckman 1-mv Potentiometric Recorder; chart speed, 0.2 in./min. Sample, 1 μl of 1%, w/v, solution in chloroform. Attenuation, 5 × 10³.

In Fig. 1 the elution patterns obtained with coconut oil isothermally and during linear temperature programming are compared.

The relationship between programmed and isothermal gas chromatography was graphically shown by Martin et al. (*10*). From the data of a plot of log retention volume at constant temperature with the reciprocal of temperature, the average linear velocity can be calculated. For linear temperature programming this can be plotted as a function of temperature and/or time.

If the area under the curves is accumulated until it equals the column length (time × velocity = distance), it is possible to predict the time and temperature at which the component will emerge from the column. From this it can be shown that a linear relationship exists between retention time and the number of carbon atoms in a homologous series for programmed operation, in contrast to the logarithmic relation for isothermal chromatography. Because of a continuously decreasing difference in volatility between higher-molecular-weight pairs in a homologous series, the glycerides are eluted at progressively closer intervals and the above derived linear relation exists only over a limited range of a glyceride series. This phenomenon severely limits the usefulness of linear temperature programs for the separation of complex glyceride mixtures. Nonlinear temperature programs give better resolution (7), but as yet have not found extensive application. The resolution obtained for coconut oil by using two different nonlinear temperature programs is shown in Fig. 2.

The ideas on retention volumes, retention times, and column performances all need modification in considering chromatograms from programmed temperature columns. The retention volume is extremely dependent upon the initial column temperature and the heating rate. Where these variables are not kept constant, repeatable results cannot be obtained and comparison of results from different origins is meaningless. Therefore, it has been proposed (11) to replace the retention volume by the retention temperature as the charactristic parameter of a peak.

Column efficiency in isothermal gas chromatography (12) is expressed numerically by the number of theoretical plates N.

$$N = 16(t_R/W)^2$$

The formula compares the width of the peak W to the time of residence t_R of the component in the column. An efficient column keeps peaks narrow. If this equation is applied directly to chromatograms obtained by programmed temperature chromatography, the number of plates so defined rises rapidly with retention time. Thus it has been shown (7) that 18 in. × ⅛ in. o.d. columns having a theoretical plate number of 800 under isothermal conditions exhibit apparent plate numbers of 3500–5000 under conditions of linear temperature programming. For

this reason Habgood and Harris (*13*) have proposed a modified
equation

$$N = 16(t_{RT}/W)^2$$

where t_{RT} is the residence time for the component in the same column
used isothermally at the temperature at which the peak emerges (re-

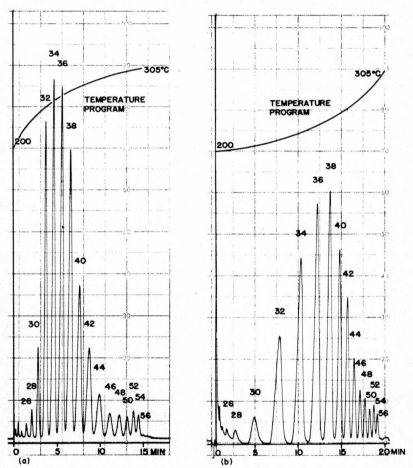

FIG. 2. GLC of triglycerides of coconut oil. Nonlinear temperature programming:
(a) concave up; (b) concave down. Peak identities, instrument, and other
operating conditions as in Fig. 1.

tention temperature). The number of plates calculated on this basis is more or less constant. However, since the latter equation, cannot conveniently be used for calculations, particularly with nonlinear temperature programs, it is best to describe the efficiency of the columns under selected isothermal conditions, which can be reproduced for future reference or for interlaboratory comparisons.

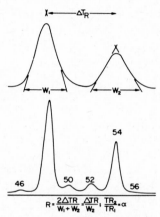

FIG. 3. Band-intercept resolution R and peak separation α. Peak identities as in Fig. 1. (a) 6-in. column; (b) 18-in. column. Instrument 2, Aerograph Hy-Fi Model 600 B, F, and M Linear Temperature Programmer Model 40. Columns, stainless-steel tubes, (a) 6 in. $\times$ ⅛ in. o.d., (b) 18 in. $\times$ ⅛ in. o.d., both packed with 3%, w/w, SE-30 on unsalinized Chromosorb W (60–80 mesh). Nitrogen flow rate, 200 ml/min, 12 psi head pressure. Flash evaporator, 325°C. Detector, in equilibrium with column. Column temperature programmed from 175° to 320°C at 3°C/min. Brown 1-mv recorder; chart speed, 1 in./3 min. Sample, 1 μl of 1%, w/v, solution in carbon disulfide. Attenuation, 16×.

The resolution of triglyceride peaks by GLC may be measured in terms of band-intercept resolution (*14*) or by determining the minimum carbon number difference between two adjacent triglycerides that could be separated with base-line resolution (*15*). These values are more useful guides than the number of theoretical plates in determining whether a specific GLC column would yield a desired separation under a given set of operating conditions. The band-intercept resolution between two peaks is given by the equation in Fig. 3. The chromatogram with the narrow peak widths offers the best resolution even though the separation as given by α is identical. Once the average base-line peak width is known for a homologous series, it is possible to calculate the number of peaks that can be separated in distance Δt_R. Equating Δt_R to the number of carbons by which the two reference peaks differ and dividing it by the base width yields directly the minimum carbon-number difference between two glycerides that can be completely separated.

Litchfield et al. (*15*) have defined (Fig. 4) an empirical ΔC value equal to the minimum carbon-number difference between two glycerides that could be separated with base-line resolution in the C_{42}–C_{48} region of the chromatogram. The authors point out that this value gives a better indication of resolution of triglyceride peaks that widely deviate from true Gaussian form upon which are based the standard formulas of peak resolution. Such a determination of ΔC, however, would be of greatest interest in the region C_{48}–C_{54} because most natural fats have the C_{48}, C_{50}, C_{52}, and C_{54} glycerides as major components. Furthermore, under conditions of linear temperature programming, the distance between adjacent peaks in this homologous series becomes progressively smaller as the carbon number increases, and a complete resolution in the lower-molecular-weight region does not guarantee a comparable resolution in the region where the higher-molecular-weight components emerge.

The van Deemter equation or some variation thereof (*16*) is a useful guide to optimizing parameters related to column design and operation. This relationship relates the theoretical plate height to the fundamental processes taking place as a gas migrates through the column. Detailed consideration of this relation is outside the scope of this discussion, but the practical considerations show that best resolution is obtained with thin liquid films on solid supports of small and uniform mesh. The minimum mesh size of the solid support is limited by the increased pressure drop, which is particularly critical in glyceride chromatography. Efficiency increases with decreasing column radius, as has been demonstrated with studies on capillary and preparative columns. Packing difficulties limit the column radius. In general, the resolution of analytical columns increases as the square root of the column length, but pressure drop, decreasing recoveries, and analysis time place a practical limit on column length. For highest efficiency the optimum flow rate should be determined for the components most difficult to separate. The usual practice is to err on the side of high flow rates; for this will cause less peak broadening and

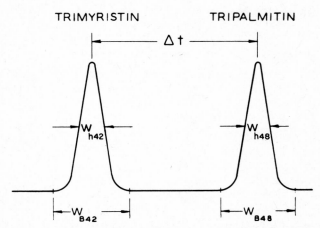

FIG. 4. Diagram of peak dimensions used to calculate peak resolution ΔC. Wh = peak width at half height; WB = base-line peak width. The formula ΔC was derived as follows:

$$WB42/48 = (WB42 + WB48)/2 = 1.6(Wh42 + Wh48)$$
$$M = \Delta t/(WB42/48) = \Delta t/1.6(Wh42 + Wh48)$$
$$\Delta C = 6/M = 9.6(Wh42 + Wh48)/\Delta t$$

where WB42/48 = average base-line width of triglyceride peaks in the C42–C48 region of the chromatogram

WB42 = base-line peak width for trimyristin peak
WB48 = base-line peak width for tripalmitin peak
Wh42 = peak width at half height for trimyristin
Wh48 = peak width at half height for tripalmitin
M = maximum number of peaks that could be separated with base-line resolution in the C_{42}–C_{48} region of the chromatogram
Δt = distance between the apexes of the trimyristin and tripalmitin peaks
ΔC = minimum carbon-number difference between two triglycerides that could be separated with base-line resolution in the C_{42}–C_{48} region of the chromatogram

Reproduced with the permission of the authors and the society. [From C. Litchfield, R. D. Harlow, and R. Reiser, *J. Am. Oil Chemists' Soc.*, **42**, 849 (1965).]

give the additional benefit of faster analyses and higher recoveries. Some column parameters optimized for analytical and preparative triglyceride separations are compared in Table 1.

Another useful relationship to consider in column design is the ex-

TABLE 1
Column Parameters for Analytical and Preparative GLC
of Glycerides[a]

	$\frac{1}{16}$ in. o.d.	$\frac{1}{8}$ in. o.d.	$\frac{1}{4}$ in. o.d.
Inside diameter, in.	0.047	0.065	1.155
Support, mesh size	100–120	80–100	60–80
Amount of liquid phase, %	3	3	5
Practical length, ft	2	1.5	2
Liquid film thickness, μ	5	5	5
Maximum plates per foot	1000	800	500
Maximum total plates	2000	1200	1000
Permeability, $\times 10^{-7}$ cm^2	1.5	2	
Optimum sample size, μl	1.0	2.0	20
Nitrogen flow rate, ml/min	100–150	150–200	200–250
Mode of injection	Flash	On-column	On-column

[a] Modified and extended from F. Baumann and J. M. Gill, *Aerograph Research Notes*, Spring, 1966. Published by Varian Aerograph, Walnut Creek, California.

pression for the number of plates required for a given resolution in terms of column and solute parameters (*17*).

$$N_{\text{req}} = R^2(a/a - 1)^2(b/K + 1)^2$$

where R is the resolution required as defined for Gaussian peaks, a is the relative retention of the two glycerides in the liquid phase, and b is the ratio of the volume of the carrier gas to the volume of the liquid phase. Factors that tend to increase K and decrease b lower the requirement for plates. All these values can be measured experimentally to a good approximation under the desired operating conditions and independently entered into the expression. The required number of plates can then be practically realized by adjusting the column length or changing the ratio of free gas volume to the volume of the liquid phase. In order to keep the film of the liquid phase thin, a solid support of higher porosity must be used.

In practice it is frequently impossible to work under optimum conditions and the theoretical requirements discussed must be adjusted to suit the sample and give reasonable recoveries and analysis times. Thus, suboptimal liquid films and column lengths may have to be used to complete at lower temperatures the separations of the higher boiling and thermally less stable glycerides.

B. Reference Conventions

Over the years a variety of abbreviations have been used for short-hand reference to simple triglycerides. Thus tripalmitin has been referred to as TP, PPP, or 16,16,16 glyceride and comparable systems of identification have been introduced for other simple and mixed glycerides (18). Specific mixed glycerides have been denoted by arranging the letters or numbers designating the constituent fatty acids in a specific order. Thus 1,3-dipalmito-2-monoolein has been abbreviated to POP or 16,18:1,16. Although this system of reference is very convenient for the description of individual glycerides and has been retained in this discussion, it is not suitable for the identification of the mixed glyceride peaks recovered on GLC. Because of the nature of the GLC resolution presently achieved, the only common and distinguishing characteristic of the glycerides found in a single GLC peak is the total number of carbon atoms in the fatty acid moieties. In keeping with the system established (19) for the shorthand designation of fatty acid methyl esters, tripalmitin can be referred to as C_{48} or simply a triglyceride of carbon number 48, and tristearin as C_{54} or 54. The great advantage of this type of reference system for triglycerides is the possibility for a universal application without prior knowledge of the actual composition of the fatty acids present in a glyceride peak recovered by GLC. Since all the common glycerides of identical number of acyl carbons are found in the same peak, the carbon number of any peak can be determined by reference to the retention times or temperatures of a pair of triglyceride standards.

The identification of the glyceride peaks by their acyl carbon number provides insight into the probable composition of the constituent glycerides on the basis of the over-all fatty acid composition of the oil or fat. Thus the C_{48} peak of a fat containing myristic, palmitic, and oleic acids would be anticipated to contain tripalmitin, monomyristomonopalmitomonoolein and dioleomonomyristin. Since GLC does not resolve saturated and unsaturated glycerides of identical carbon number, exact molecular weights cannot generally be used for the identification of individual peaks, but they may be useful in specific applications (20).

With increasing column efficiency, the GLC separations of triglycerides may be extended beyond their carbon numbers (21), and there is a need for a universally applicable reference system. The simplest example of this additional resolution is provided by the diacid glycerides containing two short- and one long-chain fatty acids. Thus the diacetates or dibutyrates of the symmetrical 2-monoglycerides are eluted ahead of their asymmetrical 1-monoglyceride isomers. This system of reference can be maintained also for long-chain diacid glycerides as long as the lone acid is used for the naming of the parent glyceride. Those diacid glycerides in which the lone acid is shorter than the other two may be referred to on the basis of a diglyceride of the long-chain acids. Assuming the fork type of orientation for the fatty acid chains in the glyceride molecule, such diacid glyceride derivatives can be divided into two groups on the basis of the molecular shape. The symmetrical derivatives, because of the shorter over-all length of their molecules, would form more compact structures and would be anticipated to be eluted ahead of the longer and less compact molecules of the asymmetrical derivatives. Most triacid glycerides can be accommodated into either the mono- or the diglyceride naming system by selecting appropriate reference chain lengths, and the predominating effects of symmetry or asymmetry usually can be recognized. On this basis, then, the ascending limb or the advance shoulder of a GLC elution peak can be attributed to the symmetrical derivatives of the glycerides of a given carbon number, while the descending limb or the trailing shoulder of the peak would be due to the asymmetrical derivatives.

It would appear that, in addition to symmetry considerations, the GLC elution of the glycerides is determined by the degree of interaction of the fatty acid chains in the 1 and 3 positions with the liquid phase. Thus diglyceride acetates and butyrates with long-chain fatty acids in the 1,3 positions are retained longer than diglyceride derivatives of comparable molecular weight with long-chain fatty acids in 1,2 positions. A lesser interaction of the fatty acid in the 2 position with the liquid phase has also been observed in thin-layer chromatography with silver nitrate, where the monounsaturated 2-oleotriglycerides migrate faster than comparable monounsaturated triglycerides with the oleic acid residue in 1 position. However, caution should be exercised in attempting to rationalize the GLC behavior of complex triglycerides, as the elution sequences have not yet been verified with a sufficient number of glycerides of known structure. Thus if the parent glycerides are named on the basis of the longest-chain fatty acid present, then the 1-monoglyceride and the 1,3-diglyceride derivatives are retained longer than the corresponding 2-monoglyceride and 1,2-diglyceride derivatives. The presence of unsaturated fatty acids in the glyceride molecule introduces further complications, and the contributions of the shapes of these molecules must be weighed against the differences in the molecular weight between the unsaturated and the corresponding saturated acids as well as differences in positional distribution. No clear-cut rules have yet emerged for the assignment of elution sequences in these glycerides.

The naming of the mono- and diglyceride derivatives other than fatty acid esters is based on similar considerations, and the carbon-number contributions of the parts of the molecules common to all members of a given homologous series are ignored. Thus the disilyl ether or the ditrifluoroacetate ester of monopalmitin is referred to as C_{16} or simply the 16 glyceride, and the monosilyl ether or monotrifluoroacetate of dipalmitin is named C_{32} or the 32 glyceride.

C. Nature of Separation

Basically, there are two kinds of separation obtained in gas chromatography with the commonly employed stationary phases. The so-called nonpolar or nonselective types effect separations based on molecular size, weight, and shape. The polar or selective stationary phases separate solute molecules on the basis of the functional groups also. Because of the prevalence of glycerides of different degrees of unsaturation, there is a demand for liquid phases that show selective retention for carbon–carbon double bonds. None of the presently available polar phases, however, have sufficient thermal stability to allow recoveries of glycerides below their own elution or decomposition points. With the choice of the liquid phases restricted to the thermostable nonpolar phases, practical separations of glycerides are limited to a segregation of the molecules on the basis of gross differences in their molecular weights. The presence of up to six double bonds per molecule (trilinolein) does not provide sufficient difference in molecular weight for a complete resolution from the saturated glyceride of corresponding chain length (tristearin), but definite shouldering can be noted. The peak broadening resulting from the presence of both saturated and unsaturated glycerides is reduced following hydrogenation.

The effect of the shape of the molecule upon the resolution is illustrated (*21*) by a run with the isomeric monopalmitin dibutyrates. Although a 5-ft-long column was used to achieve this separation, partial resolution of the isomers may also be obtained on short columns with the monoglyceride dibutyrates and diacetates of other long-chain fatty acids. The symmetrical derivative is consistently eluted first. The shape and/or degree of unsaturation of the glyceride molecule is responsible for the broadening and occasional shouldering of long-chain triglycerides observed when chromatographing natural fats and oils. Peaks of pure monoacid glycerides are always somewhat better resolved than equivalent peaks of natural mixtures. Furthermore, the triglycerides of unsaturated vegetable oils are frequently better resolved following hydrogenation.

Some difficulty arises when mixtures of oxidized and nonoxidized glycerides are chromatographed together. The azelaic acid ester residue in the azelaoglycerides derived from the unsaturated glycerides behaves on GLC as a residue of lauric acid (*22*). Thus the trimethyl ester of triazelain overlaps with trilaurin. Since most natural fats do not contain significant amounts of lauric acid, there is no real problem and the glycerides containing the dicarboxylic acid methyl ester can still be referred to as if they contained lauric acid residues. On this basis the diazelaomonopalmitin becomes a C_{40} or 40 glyceride, although it possesses only 34 fatty acid carbons. On more efficient columns, however, the triazelain is retained sufficiently longer to be recognized as a separate peak from trilaurin and the di- and monoazelaoglyceride derivatives form trailing shoulders of progressively diminishing resolution on the appropriate triglyceride peaks.

IV. Preparation of Derivatives and Preliminary Segregations

For the present discussion the neutral glycerides are defined to include the mono-, di-, and triglycerides and their neutral derivatives of both natural and synthetic origin. Although the triglycerides may be subjected to direct gas chromatographic examination, the mono- and diglycerides must first be converted into derivatives that increase their stability or volatility or both to ensure a more satisfactory resolution and recovery of these substances from the gas chromatograph. The preparation of partially oxidized glyceride derivatives permits the resolution of saturated and unsaturated glycerides of identical carbon number. In other cases the preparation of derivatives may help in the identification of GLC peaks, as may a preliminary resolution of the glyceride mixtures on the basis of functional groups or unsaturation.

A. Derivatives of Mono- and Diglycerides

The most suitable derivatives of mono- and diglycerides for GLC analysis are the acyl esters and the silyl ethers. Although the acyl esters give the advantage of relating the retention times and temperatures to those of the triglycerides, the preparation of the silyl ethers allows GLC at considerably lower temperatures.

1. ACYL ESTERS

The acetates may be prepared at room temperature with little or no isomerization by dissolving 1–5 mg of the glyceride in 1 ml of dry (BaO distilled) pyridine and adding 0.5 ml of acetic anhydride. The reaction mixture is allowed to stand in a tightly closed screw-cap vial overnight. The excess reagents are removed by evaporation under nitrogen, and the acetates are suitably diluted for injection into the gas chromatograph. Comparable techniques are used to prepare the trifluoroacetates except that shorter periods of time are sufficient and pyridine may be excluded. Higher-molecular-weight esters of mono- and diglycerides may be obtained by reacting the glycerides in chloroform solution with the fatty acid anhydrides in molar ratio of 1–1.2 in the presence of 0.03 mole of perchloric acid. The reaction is allowed to proceed for 3 hr at room temperature, when it is stopped by the addition of water, and the triglycerides are recovered by extraction with diethyl ether. The yields are nearly quantitative, and again there is little or no isomerization (*23*).

2. SILYL ETHERS

The silyl ethers of partial glycerides are prepared by dissolving 1–5 mg of the glyceride in 1 ml of dry pyridine and reacting it with 0.2 ml of hexamethyldisilazane and 0.1 ml of trimethylchlorosilane. The mixture is shaken for 15–30 sec and is allowed to stand for 5 min. The glyceride ethers are recovered by diluting the reaction mixture with hexane and partitioning with water. The combined hexane extracts are dried with calcium sulfate and evaporated to a small volume under nitrogen. The yields are quantitative and there is no isomerization (*24*). The silyl ethers may be stored for a limited period of time at −20°C.

3. OTHER DERIVATIVES

McInnes et al. (*25*) have converted monoglycerides to allyl esters of their corresponding fatty acids and found that these derivatives were suitable for separation by GLC. Under these conditions the 2-monoglycerides are isomerized to and measured as the 1-monoglyceride derivatives. Evidence has also been presented (*26*) indicating that the dehydration products of diglycerides obtained on pyrolysis of phosphatides can be adequately chromatographed by GLC. However, two apparently isomeric peaks are observed for each diglyceride derivative.

Monoglycerides may be chromatographed in the form of isopropylidene or benzylidene derivatives (*21,25*). The isopropylidene derivatives have proved particularly suitable for the GLC analysis of the glyceryl monoethers derived from naturally occurring diglyceride monoethers (*27,28*).

B. Oxidized Glycerides

Ordinary GLC methods do not distinguish unsaturated glycerides from the saturated ones having the same carbon numbers. The distinction can be achieved, however, by oxidizing the mixture. This converts the unsaturated glycerides into compounds with smaller carbon numbers, the GLC of which then permits the separation and estimation of the amounts of the unsaturated glycerides. Two general methods of oxidation are available.

1. PERMANGANATE-PERIODATE OXIDATION

This oxidation is accomplished essentially as given by Youngs and Subbaram (*22*) and von Rudloff (*29*). In a typical oxidation 10–20 mg of triolein, dissolved in 5 ml of tertiary butanol, is added to a mixture of 5 ml oxidant solution, 4 ml distilled water, 1 ml of a solution containing 8 mg potassium carbonate, and 10 ml tertiary butanol. The reaction mixture is stirred 2.5 hr at 65°C, cooled to room temperature, and decolorized by bubbling ethylene gas through the solution. The tertiary butanol is removed in a rotatory evaporator at 70–80°C. The aqueous mixture is acidified with dilute HCl, and the oxidized glycerides are recovered by extraction with chloroform. After evaporation of the chloroform the residue is taken up in methanol and esterified with diazomethane. The oxidized and methylated esters are suitably diluted with chloroform and analyzed in the GLC instrument. The proportions of reagents used for this oxidation are adjusted according to the fatty acid composition. Linoleic and linolenic acids respectively require 3 and 5 times, as much as that required for oleic acid (22).

2. REDUCTIVE OZONOLYSIS

Unsaturated glycerides may be split at the location of their double bonds by ozonization followed by catalytic reduction of the ozonides (30). The ozonization is virtually instantaneous at −65°C in a pentane solution saturated with ozone. The ozonides may be quantitatively reduced to the aldoesters in the presence of Lindlar catalyst. The aldehyde cores thus obtained have not as yet been subjected to GLC analysis, although simple aldehydes have been successfully separated and recovered by GLC methods (31).

C. Hydrogenated and Brominated Glycerides

The hydrogenation of glyceride mixtures prior to GLC examination serves two purposes. It increases the stability of the more unsaturated glycerides and reduces the peak spreading resulting from incomplete overlap of the saturated and unsaturated glycerides of the same carbon number. The hydrogenation of the glycerides can be conveniently accomplished in a microhydrogenation apparatus where the hydrogen uptake can be followed and measured with a microburette. Dioxane is a satisfactory solvent, and platinum black is an effective catalyst (19,32).

Saturated glyceride peaks free from contamination with unsaturated glycerides may be obtained by treating the glyceride mixture with a dilute solution of bromine in chloroform prior to GLC. The addition of bromine at the double bond takes place within a few minutes, and the excess reagent can be removed by evaporation (*33*).

D. Preliminary Separation of Glycerides by Thin-Layer Chromatography

The information obtained from the GLC of glycerides can be greatly enhanced by first segregating the glycerides on the basis of unsaturation. This can be effected readily by the method of Barrett et al. (*34*), who separated glycerides with 0–6 double bonds on silica gel G impregnated with 10–30% of silver nitrate. The plates may be developed with a variety of solvents, but mixtures of benzene and petroleum ether are preferred because their polarity is more easily controlled. The separated components are detected by spraying with 0.05% dichlorofluorescein in methanol and viewing the plates under UV light. The glycerides are recovered by eluting the silica gel scrapings from different bands with methanol–ether 50:50, v/v, or methanol–ether–water 50:50:10, v/v/v, mixtures.

Separations are also possible within certain double-bond numbers (*35,36*). Thus the two double bonds in linoleic acid form a stronger complex than the combined effect of two double bonds in two oleic acid chains, and the three double bonds in linolenic acid form a stronger complex than the four double bonds in two linoleic chains. In addition to the separation of the glycerides with the same number of double bonds, it is possible to separate, on the silver nitrate plates, certain positional and configurational isomers. Thus the 1- and 2-oleodistearins are separated, as are the oleodistearins and the elaidodistearins (*37*). The resolution of the bands on these plates is also affected by the chain length of the component fatty acids. The shorter-chain glycerides migrate more slowly than the longer-chain derivatives of the same number of double bonds. This effect can also be observed on silica gel G alone with the saturated glycerides of mixed chain length from such natural fats as coconut oil and butterfat.

Thin-layer chromatography on ordinary silica gel is also well suited for the resolution of the mixtures of saturated glycerides and the azelaoglyceride methyl esters obtained on partial oxidation of vegetable oils. Thin-layer chromatography is necessary for the resolution of the small amounts of the azelaoglyceride mixtures obtained on partial oxidation of the unsaturated glycerides recovered from individual bands of preparative silver nitrate plates.

V. Analytical Separations

Gas chromatography has proved to be a powerful tool for the analysis of glyceride mixtures. Even though higher-molecular-weight compounds are inherently more difficult to analyze by this means than lower-molecular-weight compounds, useful qualitative and quantitative data can be obtained with all glyceride mixtures if the operating conditions are carefully chosen.

A. Detection and Injection Systems

Second only to a satisfactory column, the detector and injector units of the gas chromatograph are critical for a successful triglyceride separation. Both should be located directly at their respective ends of the chromatographic column and should not be separated from it by unnecessary inserts of metal tubing of various lengths.

1. DETECTORS

In order to use thin-film columns and small samples for the analysis of complex glyceride mixtures, the ionization detectors are best suited to the detection of individual components. Although both the argon and the hydrogen flame ionization detectors have about the same sensitivity for the detection of carbon-rich compounds (0.01 μg for a single component), because of the self-purging nature of the flame ionization system, the latter is preferred. The injection of crude extracts of natural products may lead to decomposition and eventual fouling of the argon ionization cell. Since the bleed of the silicone liquid phase may be considerable at the elevated temperatures, it would tend to contaminate the argon detector to a much greater extent than the flame detector, where the substances are combusted in the burner and are lost to the atmosphere as volatile material.

Some hydrogen flame detectors show unusual effects that may be traced to the formation of silicone deposits on the flame jets or the collector assembly. These deposits are particularly liable to accumulate during high-temperature operation when the base-line bleed is significant. Their presence is demonstrated in the loss of signal below a certain level of sensitivity. The base line appears unusually quiet and stable, as the background noise and the minor components are not recorded. This loss in sensitivity cannot be regained by lowering the attenuation. The instrument shows poor linearity. Contamination and eventual impaired response also take place with the thermal conductivity detectors used in some of the early studies on glyceride separation by GLC. Furthermore, the hydrogen flame detector is relatively insensitive to small changes in flow and moderate variations in temperature and is thus extremely well suited to an operation with temperature programming. Great fluctuations in the rapid flow rates through the short columns, however, must be avoided, as the concentration of the sample in the detector will change with the rate of flow of the carrier gas, which in turn will change the apparent detector sensitivity.

The extremely wide linear range of the detector, which can be obtained with careful control of hydrogen and air flow rates, permits the use of a wide range of sample concentrations and a quantitative estimation of both major and minor components in the same run. Thus a plot of peak area versus amount of trilaurin injected has given linear response from 0 to 45 μg of triglyceride. Even when the 0 to 5-μg region was expanded to a larger scale, the plot was linear and passed through the origin. A similar plot for tristearin was linear in the range 0–20 μg, but showed a different slope in the 23 to 45-μg range *(15)*. Since under normal operating conditions a full-scale tristearin peak represents only 10–20 μg of triglyceride, this nonlinearity occurs only in overloaded columns and represents no problem in normal analyses. A dual, compensating flame detector is recommended. The flame should be designed for operation at high carrier gas flow rates.

Of great interest for metabolic studies with glycerides is the successful adaptation and testing of systems for simultaneous measurement of mass and radioactivity of such high-molecular-weight compounds as the naturally occurring cholesteryl esters. The elution conditions employed in these studies *(38)* were those commonly chosen for the GLC analyses of glycerides, although temperature programming was not used. There would appear to be no reason why this system should not work also with glycerides and under conditions of programmed temperature operation. The ultimate sensitivity of the instrument, which employs a copper oxide furnace to convert the organic material in the effluent stream into CO_2 and water prior to counting, is about 500 cpm per component.

2. INJECTORS

The construction of the inlet system and the type of sampling device used for triglyceride injection are particularly important, as the high flow rates required, through columns with fine mesh packings, lead to high inlet pressures. Under these conditions the usual microsyringes fail to give reproducible delivery, since the sample will blow back during injection. Better results have been claimed for the blind-end microinjectors and specially developed microsyringes. When working with internal standards, however, an absolute reproducibility of the sample size is not necessary and satisfactory results may be obtained with a Hamilton 10-μl syringe.

One of the major requirements of isothermal gas chromatography is that the sample must flash-vaporize onto the column. This prevents band spreading, and most instruments have been designed to operate primarily at isothermal conditions. With the programmed techniques, flash vaporization is not necessary if the sample can be deposited directly on the column. Since the column temperature can be maintained at a low independent level, the solute is prevented from traveling along the column until all the sample has been accommodated and the programming has been started. As the column temperature is increased, the components are eluted as sharp, well-resolved peaks. The elimination of the need for flash vaporization greatly facilitates the analyses of temperature-sensitive materials that decompose at the temperatures of flash vaporization. Among these substances must be included the high-molecular-weight triglycerides, the vaporization temperatures of which are above their cracking points. Either on-column injection or blowing of the sample onto the column in a partially evaporated state, when performed at low enough temperature, ensures that the sample enters the column as a narrow band. The delivery of the sample to the column is facilitated by the presence of solvent. Thus when a sample of tripalmitin was injected in CS_2 solution, the recovered area of the tripalmitin peak remained constant per given amount of weight over the temperature range 285–385°C. Injecting the triglyceride sample in solid form gave lower recoveries than solution injection at all flash heater temperatures, indicating substantial losses with solid sample injection (15).

In Fig. 5 is shown a simple but highly effective injector assembly (39) that allows an on-column sample injection with an ordinary

10-μl Hamilton syringe equipped with a 2-in. needle. The column is fitted with Swagelok fittings in such a manner as to allow the insertion of the injector end of the column into the flash evaporator up to 6 mm from the silicone septum. There is an absolute minimum of space between the outside of the column and the inside of the flash evaporator. The flash evaporator is maintained at a uniform temperature throughout by means of a hot-finger heater. For analyses of common natural glyceride mixtures, the injector part of the column may be kept at about 280°C, and good recoveries are obtained not only for tripalmitin but also for tristearin and triarachidin. A temperature of 300°C gave better than 75% recovery of trierucin. Ideally, the injector part

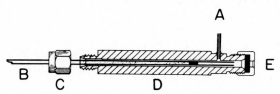

FIG. 5. Positioning of column in the injector assembly. A, carrier gas inlet; B, packed column; C, Swagelok nut (front and back ferrules not shown); D, injector barrel; E, injector and silicone septum. [Reprinted by permission, from the *Journal of Lipid Research*, **7**, **576** (1966).]

of the column should also be temperature-programmed together with the rest of the column. No decomposition of sample or other ill effects have been observed to result from the presence of column packing in the injector.

Flash heater temperatures up to 450°C are clearly unnecessary and hazardous for the gas chromatography of glycerides. Furthermore, in view of the excellent recoveries experienced with the on-column injection techniques, the use of flash evaporation should be avoided altogether and the solutes given the protection provided by the liquid phase right from the moment of injection.

B. Columns and Packings

The column has been called the heart of the gas chromatograph. It provides the resolution and governs the recoveries of the solutes. Although in a well-designed instrument even a poor column will perform reasonably well, the best column may give no resolution at all in a bad instrument. For optimum performance, therefore, certain features of its physical makeup and the operating conditions are important and deserve consideration.

1. COLUMNS

The dimensions of the columns employed for triglyceride GLC have ranged from 6 in. to 8 ft and have been of both ⅛ and ¼ in. o.d. Capillary columns do not appear to have been tried to date for the GLC of triglycerides or other compounds of comparable molecular weight, although they have been employed successfully for the resolution of the silyl ethers of monoglycerides (24). Glass, stainless-steel, and aluminum tubes have been used as the column material. Comparisons of the effect of glass and steel columns upon the recovery and resolution of triglycerides have shown (15) that, in the instruments tested, the columns were equivalent when either helium or nitrogen was used as the carrier gas at several rates. Under comparable operating conditions, the glass columns gave better peak resolution, but the recoveries of tripalmitin, for example, were identical. Having a more inert surface, the glass tubes might be better suited for the analysis of the more unsaturated triglycerides. Less adsorption of the material on the glass than on steel was also apparent in the more rapid equilibration of the solute between the gas and the liquid phase allowing rapid rates of temperature programming.

The great disadvantage of glass columns is the lack of adequate means of obtaining a reliable seal between the glass column and the metal inlet and outlet points. Because of this serious shortcoming and the greater mechanical durability of the metal column, stainless steel or aluminum is the preferred column material. Some of the adsorption on the metal surface may be reduced by siliconizing the tubes prior to filling (39). A potentially satisfactory seal between metal and glass in GLC applications has been obtained (40) with a graded series of glasses (Kovar seal).

The length of the column has a pronounced effect upon the resolution and recovery of high-molecular-weight triglycerides. While equal weights of tripalmitin and tristearin, for example, yield about the same peak areas on a 6-in.-long column and show only small losses of tristearin on an 18-in. column, large losses of the tristearin are observed on columns of 5 ft or more. With increased length of the column, however, there is increased column efficiency and better peak resolution. The decrease in the recovery appears at least partly due to a rapid drop in the flow rate of the carrier gas with increasing column length. Columns 18–24 in. long appear to provide optimum resolution combined with nearly quantitative recoveries.

2. PACKINGS

Of the large variety of supports recommended for gas chromatographic work, only a few have been tested for their suitability in triglyceride separations. Most studies have been limited to the Chromosorb W (Wilkens Instrument and Research, Inc.) and Gas Chrom A, P and Q (Applied Science Laboratories, Inc.). All of these supports in mesh sizes of 60–80 and 100–120 have given satisfactory columns at one time or another. However, the silanized supports have yielded high-quality columns with somewhat greater frequency. In order to obtain maximum efficiency with the short columns, it is desirable to use a relatively narrow range of rather fine mesh size. This greatly reduces the flow rate and the recovery of the higher-molecular-weight compounds. For quantitative work it is often better to use a narrow range of rather coarse particles (30–40 mesh). Although the neutral glycerides are much less polar and more stable compounds than most steroids, at the high temperatures necessary for successful GLC, the glycerides require about as high quality columns as the steroids, which can be chromatographed at considerably lower temperatures and frequently in the form of more volatile and less polar derivatives. It has been observed that the silanizing and support-coating techniques that produce good columns for the resolution of steroids are also satisfactory for the resolution of triglycerides. Silvering of support (*18*) appears less desirable.

On account of temperature and polarity limitations, not all liquid phases can be used for the TLC of glycerides. At the present time only three types of liquid phases have given adequate support coatings. All of these are silicone polymers, SE-30 (General Electric silicone gum), QF-1 (Dow-Corning fluoroalkyl silicone gum), and JXR (a polysiloxane polymer introduced by the Applied Science Laboratories). Because of a lower vapor pressure, the JXR phase has a lower tendency to bleed, and coatings prepared with this liquid phase last longer than coatings of comparable thickness obtained with other phases. QF-1 is the least stable of the three. All have about the same separation characteristics for the simple glycerides, but QF-1 may offer special advantages where selective columns are required. With the introduction of dual-column instruments, however, liquid phases of relatively high vapor pressures have also become amenable to exploration of their suitability for GLC of glycerides.

The amount of liquid phase is usually given in per cent values on a weight/weight basis. Coatings of 1–3% have been most useful. Because of a high bleed at the elevated temperatures, the actual thickness of the coating is usually not known. The given values refer to the ratios of the packing to the liquid phase at the time of application. Experimental techniques satisfactory for the preparation of the support, silanizing, and the application of the coatings have been described by Horning et al. (*41*). However, excellent column packings can be obtained at a relatively low cost from reliable manufacturers of gas chromatographic supplies.

C. Column Filling and Conditioning

The requirement for high efficiency demands that the short columns be uniformly packed and that they retain their structure at the time of conditioning and during the heating and cooling cycles of normal operation. This requirement is most easily met by a tightly packed column prepared with a relatively coarse mesh support (60–80). It is possible, however, to increase the column efficiency by using a finer mesh size but filling the column loosely. Columns tightly packed with fine-mesh packings require extra conditioning and at the end may give low recoveries of high-molecular-weight glycerides.

1. FILLING THE COLUMNS

The most popular method of filling narrow-diameter ($\frac{1}{16}$–$\frac{1}{8}$ in. o.d.) steel coils with packing is the suction technique, but column loading under positive pressure (15–25 psi) of inert gas may be preferable. Before it is filled, the outlet end of a suitably shaped steel tube is closed with a compact plug of siliconized glass wool extending about 6 mm from the end of the tube into the column interior. When the suction technique is used, the plugged end is attached to a water pump and the column is filled under suction with the help of a small funnel connected to the column inlet by Tygon tubing. During packing, the tube is mechanically vibrated (Vibro-Graver, Burgess Vibrocrafters, Inc., Grayslake, Illinois). Violent vibration of the column should be absolutely avoided, as this fractures the support particles and exposes adsorptive sites. The column is uniformly packed to about 35 mm from the inlet end, and a small siliconized-glass-wool plug is pushed down the tube to rest against the packing.

When a dual-column instrument is to be used, two columns are packed simultaneously under closely similar conditions. With fine-mesh packings some experience is necessary before too-tight packing can be consistently avoided. Metal tubes of ¼-in. diameter, or glass tubes, can usually be uniformly packed by gravity settling without suction. Such columns yield extremely high flow rates and apparently maintain optimum contact between the stationary and the mobile phases. Although they permit high recoveries of all glycerides, they do not provide the optimum resolution of peaks. The plugging of the column ends with the glass wool appears to be the least satisfactory and frequently the critical step in the column packing. There is a definite need for a better method of closing columns. Either a porous-glass or metal disc might be preferable.

2. CONDITIONING OF COLUMNS

Next to closing the ends of the column with glass-wool plugs, the conditioning of the column is the least scientific step, although empirical conditioning schedules have been worked out for specific applications (*42,43*). This is so because the processes of curing polymers at high temperatures are incompletely investigated and poorly understood. The conditioning step is intended primarily as a means of removing the low-molecular-weight residues from the stationary phase so as to maintain low and uniform bleeds of liquid phase throughout the working range of temperatures. This is normally accomplished by heating the column to a temperature some 50–75°C above the maximum operating temperature and maintaining it at this temperature for a few hours. However because of the danger of excessive losses of liquid phase, triglyceride columns are conditioned at the maximum working temperature. During conditioning, the column outlet end is free and the effluent is vented without passing through the detector.

At least two processes are believed (*44*) to take place during conditioning. There is an increase in cross-linking of the polymer, resulting from silicone–silicone bond formation accompanied by a loss of methyl groups, and there is a loss of low-molecular-weight polymeric fragments. Because of the desirability of obtaining as highly polymerized coating as possible, it has been suggested (45) that full-flow conditioning or low-molecular-weight-fragment venting should be preceded by a no-flow conditioning, and special benefits have been claimed for this process. The best method and the exact time required for optimum conditioning depend upon the actual batch of liquid phase polymer and possibly upon the characteristics of the support.

The early packings prepared with SE-30 required considerably longer periods of conditioning or thermal stripping (7) before satisfactory base lines could be obtained. With the higher-molecular-weight (or at least lower-vapor-pressure) materials now available, it is possible to apply relatively thin coatings of the liquid phase to start out with and to obtain satisfactory columns with a minimum of conditioning. Thus 1% JXR columns may be conditioned in 1 hr at 350°C and 3% columns in 4–6 hr. The first few runs, however, may not give completely satisfactory recoveries of all glycerides, and brief periods of additional conditioning may be necessary.

As a result of this conditioning, the base-line elevation experienced when single JXR columns are programmed from 200° to 350°C averages 10–20% of full scale at the usual working sensitivity. On further aging the column bleed may decrease to about 10% of full scale before the useful life of the column is exhausted. The losses of the liquid phase are usually somewhat higher with SE-30 columns. In the dual-column mode, the base-line bleed can be compensated for and an overall deviation of 1–2% can be obtained readily with both JXR and SE-30 columns (39). Instead of a reference column, the base-line compensation may be obtained by entirely electrical means, called base-line-drift anticipators (Wilkens Instrument and Research, Inc.).

D. Operating Conditions

Glyceride analyses require temperature-programmed operation with temperatures to 350°C. The ideal apparatus should be capable of dual, compensating column operation to minimize the effects resulting from liquid phase bleeding that occurs at these elevated temperatures. A dual, compensating flame detector is recommended. The general operating conditions, however, are the same regardless of whether a single- or a dual-column instrument is used.

1. CARRIER GASES AND FLOW RATES

It has been suggested (*46*) on a theoretical basis that the heavy gases, nitrogen and argon, give much smaller plate heights at slow flow rates than do hydrogen and, presumably, helium, whereas at faster flows the performances approach each other. Since the meaning of fast and slow flow rates is relative and the effect might vary with the shape and diameter of the column, the effect upon the resolution of specific components cannot be ascertained easily. Thus it has been shown (*15*) that, at least in one instrument, nitrogen at flow rates of 100 ml/min gave higher recoveries of tripalmitin and tristearin than did helium, whereas in another instrument helium and nitrogen gave equivalent recoveries for tripalmitin and tristearin. Helium carrier gas, however, gave significantly better peak resolution than nitrogen with equivalent calibration factors. Because of economy considerations, nitrogen has been the most frequently used carrier gas. It should be dry and free of oxygen.

The nitrogen flow rates have varied from about 100 to 300 ml/min depending on the molecular weight of the material to be analyzed and the amount of the liquid phase and column diameter. Heavier coatings require higher flow rates. Coatings of less than 1% liquid phase have given good triglyceride recoveries at flow rates of 30 ml/min. The flow of the carrier gas through the column is usually measured at atmospheric pressure and room temperature by means of a soap-bubble flowmeter. Most columns, however, tend to swell when the temperature is increased and the carrier gas flow rate decreases. For this reason it is necessary to measure the flow rate at the maximum operating temperature also and adjust it to the desired rate if necessary. Inlet pressures of 10–30 psi are usually sufficient, but for the operation of automatic flow controls the readout on the pressure gage may have to be set at 50 psi or higher. The outlet pressure is atmospheric. The maintenance of a constant carrier gas flow throughout the program is very important, as the rate of the flow through the detector determines the apparent detector response. It should be noted that the effect of the changes in the flow rate of carrier gas varies with the design of the instrument and the geometry of the detector. Thus, it has been shown (*15*) that changes in flow rate from 50 to 100, 150, and 200 ml/min nitrogen flow in two different gas chromatographs resulted in a sharply rising recovery of tristearin in one gas chromatograph and no apparent change in the other. As a result of the longer residence time of the solute in stationary phase and more complete equilibration, the resolution of the peaks at lower flow rates is improved, but higher elution temperatures are necessary. The optimum flow rate for maximum triglyceride recovery can readily be determined with a standard mixture of simple triglycerides.

2. TEMPERATURES

Temperatures are given in terms of column or oven starting temperatures, and the temperature increase is specified in degrees per minute. Either linear, exponential, or more complex rates of temperature increase have been used. The optimum temperature programs differ from instrument to instrument and vary with the glyceride mixture at hand. The great variability available in the program rates provided by the Thermotrack programmer (Beckman Instruments, Inc., Fullerton, California) offers the greatest flexibility in the selection of the optimum program.

Studies on the recoveries of selected triglycerides at different rates of programming have shown (*15*) that over the range of common operating conditions (1–4°C/min) varying the program rate had no effect on the quantitative recovery of tristearin. Slow program rates gave better peak resolution with steel columns, but the peak resolution was independent of program rates over the range of conditions studied with glass columns. Work with siliconized steel columns has shown (*47*) that programming rates of up to 20°C/min can be utilized for glyceride resolution without seriously affecting the proportional recovery of the individual components, provided that the starting temperature is low enough to allow a complete deposition of the lipid on the column prior to the start of the programming. For this reason, it may be necessary to maintain the column at isothermal temperature during the first few minutes after injection.

The column or oven starting temperatures depend on the molecular weight of the glyceride mixture. For glycerides of 24–54 fatty acid carbons, the most suitable starting temperatures are 175–200°C.

Higher starting temperatures apparently produce band spreading during the initial adsorption of the sample, and lower temperatures result in an incomplete transfer of the material from the injection chamber to the column packing. Consequently, on-column injection allows lower starting temperatures and higher initial programming rates than flash vaporization. For on-column injection, an injection port temperature of 280–300°C is adequate for work with the common glyceride mixtures. When using the flash injector, the injection port should be maintained at a temperature of 320–350°C. The detector base should be maintained at 300–340°C in both cases. When using the on-column type of injection, the injector temperature should not exceed the column temperature by more than 50–75°C; otherwise, substantial amounts of the liquid phase may be moving out of the injector part of the column and will produce false peaks in the detector. It should be noted that a clear-cut distinction may not always be made between on-column injection and flash vaporization. The upright mode of injection, using the empty or glass-wool-filled top part of the column as flash evaporator, represents a compromise between direct injection onto the packing and vaporization. On occasion it may prevent decomposition of the solute on the superheated support.

3. SAMPLE SIZE AND INSTRUMENT SENSITIVITY

The amount of sample required for the analysis depends upon the sensitivity of the instrument and the complexity of the glyceride mixture. With the common flame ionization detection systems, sample sizes range from 1 to 50 μg of total glyceride. The usual peak seen in triglyceride separations is 0.5–2 μg for a single component. The present limit of sensitivity for such work is about 0.01 μg. For most analyses, it is convenient to dissolve 10 mg of the glyceride mixture in 1 ml of solvent and to inject 0.5 to 1 μl of the solution. Chloroform is the best lipid solvent, but because of rapid column deterioration and detector fouling it should be used only when other solvents do not keep all components in solution. Hexane or petroleum ether is preferred whenever possible because of its tranquilizing effect upon the detector. Acetone, carbon disulfide, and diethyl ether are also well suited but would appear to be too volatile for reproducible sampling even over short periods of time.

Both the ultimate and the working sensitivity of the instrument refer to arbitrary instrument settings that provide electrical multiplication of the detector response. In comparisons of detector properties the term is usually used to indicate the lowest level of response of the system. The only meaningful comparisons of instrument sensitivity are those recorded with the complete systems under normal operating conditions using identical samples.

The introduction of dual-column systems results in a more effective background suppression and permits more accurate work with the smaller samples. Some practical sample sizes and common instrument settings with both single- and dual-column gas chromatographs are given in the legends to the figures in the text.

E. Separation of Standards

In the absence of other criteria, the best guide to the separations to be obtained on GLC of natural glycerides is the GLC behavior of appropriate standards. Because of the extremely complex nature of most natural mixtures the provision of suitable standards of adequate purity is a difficult task and the retention times and/or retention temperatures must frequently be extrapolated from the standards available. The following section represents a collection of GLC chromatograms of standard mixtures of mono-, di-, and triglycerides and their derivatives.

1. MONOGLYCERIDES

Huebner (5) reported that monoglycerides as such could not be eluted from a column containing 23% silicone grease but that acetylation assured their complete elution. McInnes et al. (25) had previously shown that it was practical to analyze monoglycerides in the form of their allyl esters and as the isopropylidene derivatives. The best separations of monoglycerides, however, have thus far been obtained by means of their trimethylsilyl ethers (24). These derivatives reduce the polarity and greatly increase the volatility of the diols and allow the completion of the separation at moderate temperatures (215°C or lower).

With thin films of liquid phase, as shown in Fig. 6, it is also possible to separate and recover unmodified monoglycerides from GLC columns in nearly correct weight proportions. Any β-monoglycerides present in the mixture were completely isomerized to the α form or else both isomers migrated with the same retention times under these GLC conditions, since only one peak was detected for each chain length. Because of the high polarity of these diols, their peaks show considerable tailing even on heavily silanized supports. Separation of saturated and unsaturated monoglycerides cannot be accomplished in the free form on nonpolar silicone columns. These compounds are also too polar

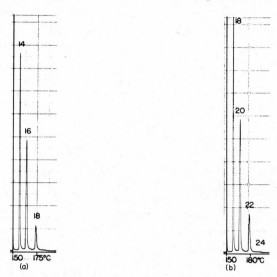

FIG. 6. GLC of mixed 1-monoglycerides. (a) Monoglycerides; (b) monoglyceride acetates. Peaks identified as in Fig. 1. Temperature programs as shown. Instrument and other operating conditions as in Fig. 1. (Synthetic 1-monoglycerides courtesy of Dr. E. Baer.)

for effective recovery from conventional polyester columns, which resolve saturated and unsaturated monoglycerides following conversion into the acetates or silyl ethers.

Also shown in Fig. 6 is the resolution obtained with the diacetates prepared from the same monoglycerides. The elution temperatures are comparable to those for the free monoglycerides, as are the peak heights and areas. The separations of the monoglycerides of the common long-chain fatty acids can be efficiently accomplished isothermally, but temperature programming improves the appearance of the peaks and is necessary for the analysis of more complex mixtures.

Huebner (20) has shown that the β-monoglyceride acetates emerge
ahead of the α-glyceride acetates. The resolution of the two isomers
can be best accomplished by employing longer columns (3 to 5 ft).
The saturated and unsaturated monoglyceride acetates can be resolved on short columns containing low concentrations of SE-30 or
JXR, but mono- and diunsaturated glycerides are not resolved. The
latter separations are effectively carried out on diethylene glycol

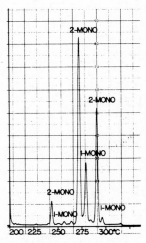

FIG. 7. GLC of isomeric monoglyceride dibutyrates. First pair, monomyristins;
Second pair, monopalmitins; Third pair, monostearins. The 2-monoglyceride
dibutyrates precede their 1-monoglyceride isomers. Column, 5 ft. × ⅛ in. o.d.
stainless-steel tube packed with 5%, w/w, SE-30 on silanized Chromosorb W
(60–80 mesh). Temperature program as shown. Instrument and other operating
conditions as in Fig. 1.

liquid phase (6%, w/w, on silanized Gas Chrom Q, 220°C) in a dual-
column instrument. Complications, however, arise when both α- and
β-monoglycerides of saturated and unsaturated fatty acids are present
in the mixture, because the unsaturated β-monoglycerides tend to
overlap with the saturated α-monoglycerides of the same chain length.

Figure 7 represents a chromatogram obtained with a mixture of
monoglyceride dibutyrates. The separation of these α and β isomers
requires a longer column (5 ft), as the reduced differences in the
molecular shapes of the higher-molecular-weight compounds are more

difficult to exploit. The resolution of the isomeric monoglyceride di-hexanoates required further increase in the column length, which could not be realized without great reductions in the carrier gas flow rates, which in turn resulted in impaired recoveries. Under the conditions of the dibutyrate resolution, the monoglyceride dihexanoates gave only shouldering, which nevertheless suggested that the order of elution established for the lower-molecular-weight derivatives was also main-tained for the isomers of the higher molecular weight. The progressive loss in the separability of these isomers, with increasing chain length of the derivatives, is of interest in assessing the relevance of factors responsible for the nearly complete overlap observed in the GLC of high-molecular-weight triglycerides of comparable molecular structure.

Wood et al. (*24*) illustrate the resolution obtained with the disilyl ethers of a mixture of saturated and unsaturated monoglycerides. Be-cause of the greatly increased volatility, these derivatives can be separated on DEGS columns at much lower temperatures than the monoglyceride diacetates. Furthermore, the columns may be much longer, and thereby provide more theoretical plates, before the reten-tion times become significantly increased and the recoveries greatly reduced. This increased column efficiency should allow a complete resolution of such monoglyceride pairs as α-monostearin and β-mono-olein and β-monolinolein and α-monoolein, which overlaps when chro-matographed as the diacetates. As yet, complete separations of this type have not been reported.

Separations of the monoglyceride isopropylidenes present no special problems (*25*). Under the conditions of the derivative formation, all the β-monoglycerides are isomerized to the α derivatives and only one peak is obtained for each fatty acid chain length. Comparable separa-tions are obtained with the monoglyceride benzylidenes. The latter derivatives are commonly obtained as intermediates in the synthesis of β-monoglycerides. Monoglyceride mixtures upon benzylidation yield only small amounts of the β-monoglyceride benzylidenes. Isopropyli-dene and benzylidene derivatives are also suitable for the resolution of saturated and unsaturated monoglycerides of the same chain length.

2. DIGLYCERIDES

Huebner (*5*) found that diglycerides, like monoglycerides, could not be eluted from a column containing 23% silicone grease and that acetylation of the free hydroxyl was necessary for satisfactory GLC. Thinner coatings of SE-30, however, have permitted analysis of diglycerides with free hydroxyl groups and the chromatograms have shown no signs of degradation or isomerization. The protection of the free hydroxyl by an acyl or a silyl group, nevertheless, may be desirable as a means of eliminating a potential source of isomerization, tailing, and dehydration. Acylation of the diglycerides also helps to

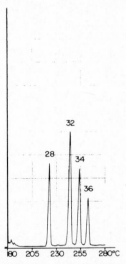

FIG. 8. GLC of mixed 1,2-diglycerides. 28, dimyristin; 32, dipalmitin; 34, monopalmitomonostearin; and 36, distearin. Temperature programs as shown. Instrument and other operating conditions as in Fig. 1. (Synthetic 1,2-diglycerides courtesy of Drs. E. Baer and D. Buchnea.)

bring their retention times in line with those of the natural triglycerides of comparable molecular weight.

The resolution obtained with a mixture of dimyristin, dipalmitin, monopalmitomonostearin, and distearin is shown in Fig. 8. On the basis of the method of preparation (*48*), all the diglycerides were thought to have been in the form of the 1,2 derivatives. At the elevated temperatures of chromatography, it was anticipated that they would isomerize at least partly to the 1,3 isomers. Since only one peak was observed for each diglyceride, it would seem that either the isomerization to the more stable 1,3 derivative was complete or the 1,2 and 1,3 derivatives had identical retention characteristics. It may be pertinent to note, that under comparable GLC conditions the 1,3 propane diol diester is well separated from its 1,2 isomer, which emerges first (*26*), whereas on thin-layer plates, the 1,3- diglycerides move faster than the 1,2-diglycerides (*49*). Apparently the greater affinity of the non-polar liquid phase for the fatty chains in the 1,3 positions is equally matched by the affinity of the support for the outer hydroxyl in the 1,2-diglyceride, resulting in equal retention times for both isomers.

FIG. 9. GLC of isomeric dimyristin acetates. The 1,2-dimyristin acetate precedes its 1,3 isomer. Temperature program as shown. Instrument and other operating conditions as in Fig. 1; column as in Fig. 7.

Following acetylation the 1,3-diglyceride acetate emerges last. The saturated and unsaturated diglycerides of the same chain lengths overlap on the SE-30 columns and as yet have not been recovered from polyester columns.

The resolution of the isomeric acetates of dimyristin is shown in Fig. 9. The spacings of the peaks are nearly comparable to those observed with true triglycerides of carbon number 30 and 32, with which they partially overlap. Huebner (*5*) had similarily observed that the 1,2-diglyceride acetates were eluted ahead of the 1,3 isomers. The diglyceride acetates can be recovered from polyester columns only with difficulty, but there is reason to believe that at least the saturated and the diunsaturated diglyceride acetates of comparable molecular weights are resolved. The diglyceride butyrates behave as true tri- glycerides, and their elution behavior is considered together with that of long-chain triglycerides.

The volatility of common fatty acid diglycerides can be greatly in- creased by conversion to their silyl ethers. The separation obtained with a mixture of diglyceride silyl ethers on an 18-in. SE-30 column is shown in Fig. 10. Under these conditions the isomeric dimyristins

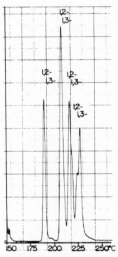

FIG. 10. GLC of mixed isomeric diglyceride silyl ethers. First pair, dimyristins; Second pair, dipalmitins; Third pair, monopalmitomonostearins; and Fourth pair, distearins. Temperature program as shown. Instrument and other operating condi- tions as in Fig. 1.

and dipalmitins overlap nearly completely. The greater volatility permits, however, the use of longer columns, and in other experiments complete separations between the silyl ethers of the 1,2- and 1,3- distearins have been observed. As noted for the diglyceride acetates (5), the 1,2-isomers are eluted ahead of the 1,3 isomers, but difficulties arise when both saturated and unsaturated derivatives are present. The silyl ethers of diglycerides may be suitable for use with polyester columns, but no experiments have yet been performed. The diglyceride elution patterns obtained following trifluoroacetylation are very similar to those resulting from silylation. In common with the silyl ethers, the trifluoroacetates are readily decomposed in the presence of traces of moisture. It is therefore best to inject the entire reaction mixture into the gas chromatograph.

As in the case of the monoglycerides, the diglycerides may be separated following their transformation into the trifluoroacetate esters. The trifluoroacetates are more volatile than the acetates but more polar than the silyl ethers. On occasion, they may provide suitable

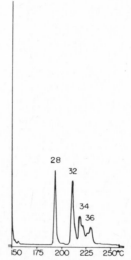

FIG. 11. GLC of mixed isomeric diglyceride trifluoroacetates. 28, dimyristins; 32, dipalmitins; 34, monopalmitomonostearins; 36, distearins. Temperature program as shown. Instrument and other operating conditions as in Fig. 1.

derivatives for the solution of a particular analytical problem. At high injector temperatures the trifluoroacetates appear to be decomposed to propene diol diesters (*26*). The latter products also are suitable for GLC. These derivatives may be comparable with the monoglyceride allyl esters that had been subjected to GLC earlier by McInnes et al. (*25*). The resolution obtained with mixed diglyceride trifluoroacetates under normal conditions of chromatography is shown in Fig. 11.

Dehydrated derivatives of diglycerides are also formed during pyrolysis of phosphatides. Thus the injection of dimyristoyl lecithin or distearoyl lecithin into the gas chromatograph, equipped for triglyceride resolution, results in the formation of peaks of retention times comparable to those in propylene glycol diesters. Surprisingly, each lecithin gives rise to a pair of peaks. Since the peak proportions vary from injection to injection, it appears that they represent isomeric products. The peak pairs resulting from the GLC of dimyristoyl and distearoyl lecithins are shown in Fig. 12. Although the exact identity of the peaks has not yet been established, it has been shown that they are free of phosphorus and yield two molecules of fatty acid upon saponification (*26*).

3. TRIGLYCERIDES

The separations of long-chain triglyceride standards have been largely confined to the simple monoacid derivatives. Thus the homologous series from triacetin (C_6) to trierucin (C_{66}) has been frequently examined and the retention-time characteristics evaluated. In addition, Litchfield et al. (*15*) have synthesized and determined the extent of separation of the C_{45}, C_{46}, C_{47}, and C_{48} triglycerides. They confirmed earlier expressed beliefs (*1*) that the odd-carbon-number triglycerides were eluted with retention times intermediate between those of their nearest even-carbon-number homologs. The separations of standard mixed fatty acid triglycerides have been limited to the examination of the GLC behavior of the various acyl esters of mono- and diglycerides.

A chromatogram obtained with the simple triglycerides trilaurin through tristearin is shown in Fig. 13. The spacing of the glyceride peaks is seen to decrease gradually under the influence of a linear temperature gradient. Under the selected experimental conditions some peaks exhibit nearly normal Gaussian curves, and others do not. Litchfield et al. (*15*) illustrate a separation of odd- and even-carbon-number triglycerides of C_{45}, C_{46}, C_{47}, and C_{48} triglycerides. This mixture was prepared by reacting palmitic and pentadecanoic acids in molar ratios in the presence of trifluoroacetic anhydride as suggested by Bourne et al. (*50*). A 5-ft column was necessary to effect a complete separation of these glycerides.

Figure 14 was obtained with a triglyceride mixture prepared by reacting equal proportions of palmitic and 12-methyl tridecanoic acid in the presence of trifluoroacetic acid anhydride (*50*). Before injection

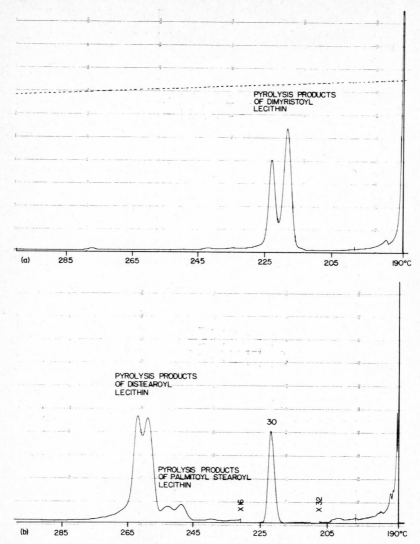

FIG. 12. GLC of the pyrolysis products of synthetic lecithins. (a) Dimyristoyl lecithin; (b) distearoyl lecithin, palmitoyl stearoyl lecithin, and tridecanoin. Peaks identified as in Fig. 1. Isomer sequence decided by reference to corresponding propenediol diesters. (Synthetic lecithins courtesy of Dr. E. Baer.) Instrument 3: Aerograph Hy-Fi Model 600 D, F and M Linear Temperature Programmer Model 240. Column, 21 in. × ⅛ in. o.d. stainless-steel tube packed with 3%, w/w, JXR on Gas Chrom Q (100–120 mesh). Nitrogen flow rate, 150 ml/min, 18 psi head pressure. On-column injector, 290°C. Detector, in equilibrium with column compartment. Temperature program as shown. Beckman 1-mv recorder. Sample, 1 μl of a 1%, w/v, solution in chloroform. Attenuation, 1 × 16.

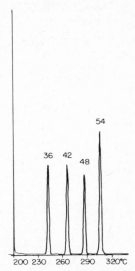

FIG. 13. GLC of simple triglycerides. Peaks identified as in Fig. 1. On-column injector, 325°C. Nitrogen flow rate, 200 ml/min. Temperature program as shown. Instrument and other operating conditions as in Fig. 1.

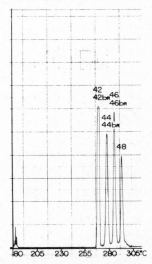

FIG. 14. GLC of triglycerides containing $C_{16:0}$ and iso-C_{14} fatty acids. Peaks identified as in Fig. 1. Peaks 42 and 48 contain approximately 50% of added trimyristin and tripalmitin, respectively. Temperature program as shown. Instrument and other operating conditions as in Fig. 1.

into the gas chromatograph the mixture was diluted with corresponding triglycerides prepared with the normal-chain acids only. The triglyceride with three branched chain acids (C_{42}br) is seen to be eluted ahead of the normal-chain C_{42} triglyceride. The effect of two branched-chain acids in the molecule (C_{44}br) is smaller, and that of one branched acid per molecule the smallest (C_{46}br), as both of these glycerides merge completely with the ascending limbs of the cor-

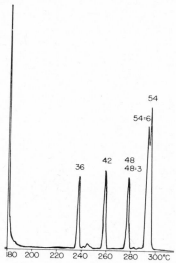

FIG. 15. GLC of homologous saturated and unsaturated triglycerides. 36, trilaurin; 42, trimyristin; 48, tripalmitin plus tripalmitolein; 54:6, trilinolein; 54, tristearin. Temperature program as shown. Instrument and other operating conditions as in Fig. 1.

responding normal-chain glycerides (C_{44} and C_{46}) in the elution pattern.

The effect of double bonds upon the resolution of triglycerides on a JXR column is shown in Fig. 15. The mixture contained approximately equal proportions of tripalmitin–tripalmitolein and tristearin–trilinolein. As with the unsaturated fatty acid methyl esters, the unsaturated triglycerides are eluted slightly earlier than their saturated counterparts from an SE-30 or a JXR column. Tripalmitolein with three double bonds does not form a shoulder on the ascending limb of the tripalmitin peak as trilinolein with six double bonds forms on the ascending limb of tristearin. Greater separations would be anticipated on DEGS columns, but to date it has not been possible to recover long-chain triglycerides from polyester columns.

In the determination of the triglyceride structure of natural fats by means of the permanganate–periodate degradation technique, it is necessary to assess the distribution of the fatty acids among the oxidized and nonoxidized glycerides. This can be readily done by GLC on the basis of molecular weight. In Fig. 16 are shown the GLC elution

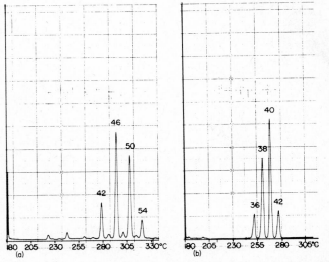

FIG. 16. GLC of a randomized mixture of trimyristin and triolein. (a) Before oxidation; (b) after oxidation. Peak identification as given in Fig. 1 and as discussed in the text. Temperature programs as shown. Instrument and other conditions of operation as in Fig. 1. (Original randomized mixture courtesy of Dr. F. H. Mattson.)

patterns obtained for a randomized triglyceride mixture of equal proportions of myristic and oleic acids before and after oxidation. Before oxidation, peaks are seen with carbon numbers of 42, 46, 50, and 54, representing trimyristin, dimyristomonoolein, monomyristodiolein, and triolein. After oxidation, peaks are seen with carbon numbers of 36, 38, 40, and 42, which represent triazelain, monomyristodiazelain, dimyristomonoazelain, and trimyristin, respectively. In terms of retention time and temperature the methylated azelayl residue in a glyceride molecule is equivalent to a residue of lauric acid (*22*). Tripalmitin and distearoazelain, for example, emerge at about the same time. On this basis relative carbon numbers, comparable to the carbon numbers of undegraded triglycerides, have been assigned to the oxidized glycerides and the peak identities established (*22*).

An alternative procedure for the partial oxidation of glycerides is the reductive ozonolysis method of Privett and Nickell (*30*). Thus far the products of this reaction sequence do not appear to have been tested in gas chromatographic separations.

VI. Preparative Separations

The separation of natural glycerides by GLC results in a grouping of the compounds on the basis of their molecular weight. Although each peak contains only glycerides of the same carbon number, these glycerides usually contain an assortment of fatty acids esterified to glycerol. To identify the glycerides it is necessary to isolate the individual peaks and to determine the constituent fatty acids. For this reason attempts have been made to develop gas chromatographic systems that would permit the separation and recovery of glyceride mixtures in quantities large enough for subsequent GLC analysis of the fatty acids.

A. Selection of Preparative Column

It has been shown earlier (*1*) that short, narrow-bore packed columns produce the most effective separations of natural glyceride mixtures. Because of the thin film of the liquid phase, however, the capacity of these columns is small, rendering them impractical for preparative separations even after many repeat injections. An increase in column diameter to ¼ in. allows the separation of considerably larger amounts of glycerides (up to 0.1 mg per peak), which can be further increased by extending the column length. However, since the longer columns require higher temperatures to complete the elutions, the advantages of the low-temperature operation of the thin-film columns are largely lost. A satisfactory compromise between the ideal temperature of elution and optimum column length to provide an acceptable sample load is reached in the utilization of a 5% liquid phase coating and a 2 ft × ¼-in.-o.d. column. It provides the desired glyceride yields with a minimum of inconvenience for all but the highest-molecular-weight components. The recovery of the C_{50}, C_{52}, and C_{54} triglycerides, however, is impaired and has to be corrected for by appropriate calibrations. Small amounts of the latter compounds can be quantitatively collected by using a 3% JXR packing, which also permits the extension of the column length to 3 ft. The columns are packed and conditioned as described for analytical separations.

B. GLC Apparatus and Method of Operation

Basically, the apparatus and the operation of the preparative system is similar to that used in analytical glyceride separations, except that

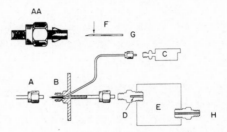

FIG. 17. Schematic representation of the post-column stream splitter showing its relation to column (A), flame detector (C), existing thermal conductivity detector (E), and the collector tube (H). AA, a magnified view of the outlet end of the packed column indicating the space inserted to accommodate the splitter needle (F). B, Swagelok fitting with $\frac{1}{16}$-in. bypass tube containing notched needle (F) and restrictor wire (G). D, $\frac{1}{16}$-in. restrictor in another Swagelok fitting. (Design and parts by Wilkens Instrument and Research, Inc.)

a post-column splitter is necessary to avoid the incineration of the whole sample. By means of the stream splitter, a small portion of the sample is fed to a hydrogen flame detector and the times of peak emergence are noted. The peaks can be collected manually or with the help of an automatic fraction collector activated by the recorder pen. Such an apparatus (51) was provided for the author by Wilkens Instrument and Research, Inc., by modification of the Aerograph Autoprep 700. A standard kit that is now available from Wilkens Instrument and Research, Inc., converts all A-90-P or A-700 instruments to flame units.

The critical part of the instrument is the splitter (Fig. 17), which consists of two stainless-steel tubes ($\frac{1}{8}$ in. and $\frac{1}{16}$ in. o.d.) silver-soldered into the male Swagelok fitting (B) accepting the outlet end of the $\frac{1}{4}$-in.-o.d. chromatography column (A). During chromatography, the column effluent was divided between the $\frac{1}{16}$-in. bypass tube to the flame detector (C) and the $\frac{1}{8}$-in. tube leading through the thermal conductivity detector (E) to the collector (H). With a $\frac{1}{16}$-in.-o.d. outlet hole (D) in the Swagelok fitting accepting the $\frac{1}{8}$-in. tube at the entrance to the old detector block, the approximate split ratio was 1:1 under the chosen working conditions. By placing a 21 gage needle (F) into the bypass tube at B, the gas flow to the flame detector was reduced and the split ratio became about 1:5 in favor of the collector. A further restriction resulting in a split ratio of approximately 1:10 was obtained by inserting a 0.007-in.-diameter wire (G) into the needle.

The splitter needle is embedded into the column packing material in the column. In order to prevent the solid support from plugging the splitter, the column exit (AA) is prepared for entrance of the needle by providing space by insertion of a 20-gage wire to penetrate the solid support $\frac{1}{4}$ to $\frac{1}{2}$ in. The restrictor wire (G), about 2 in. long, is completely inserted in the needle.

For optimum sensitivity the flow to the flame tip was kept at about 25–30 ml/min, which required an over-all flow of about 110–120 ml/min through the column giving a split ratio of 1:4.5. For a 1:10 split ratio, the flow through the column was set at about 300 ml/min. During chromatography, the detector oven housing, the splitter, and the flame ionization detector were maintained at 325°C. A specially designed elbow heater was used to obtain a temperature of about 325–350°C at the collector end of the tube. No special cooling devices were necessary for quantitative trapping of the eluted glycerides in the collection vials partially filled with clean glass wool.

The main steps involved in the operation of the preparative system consisted of the following:

1. Adjusting gas flows and injector, detector, and collector temperatures as required for the separation at hand.

2. Selecting a Powerstat setting that will provide a suitable temperature gradient for the resolution desired. (For resolution of high-molecular-weight triglycerides a temperature programmer is recommended because it provides more reliable control.)

3. Allowing the column temperature to reach the upper limit and

adjusting the cooling-cycle timer to the number of minutes necessary to cool the column down to the starting temperature or slightly below it.

4. Setting the precollection timer to bypass the solvent peak and the peaks of any low-boiling components emerging prior to the glycerides.

5. Injecting the sample into the flash evaporator or directly onto the column by using a Hamilton syringe with either a 2- or a 6-in. needle, respectively, and recording the time and temperature of injection. (For on-column injection, it is recommended that the injector barrel be drilled out to about ½ in. from the injector end so that the column can be inserted into the injector up to about 2 in. from the silicone septum, and a 2-in. Hamilton needle can also be used for on-column injection.)

6. Recording the temperature rise at regular time intervals and collecting the peaks automatically by setting the peak signal activator switch on the recorder to the desired level. (As each peak appears on the recorder, the signal switch is tripped and the collector table is automatically advanced.)

7. If necessary, using a post-collection timer to postpone the opening of the oven lid and the start of the cooling cycle until any remaining high-molecular-weight peaks have been vented or collected together in a separate collection bottle.

C. Recovery of Standards

The first preparative separations were performed with low-molecular-weight simple triglyceride standards ranging from tributyrin to tridecanoin. From these compounds were derived the correct split ratios and the optimum temperature gradients, which were applied to the recovery of the medium- and long-chain triglycerides. A successful demonstration of the recovery of simple triglycerides in pure form was necessary in order to justify attempts to recover pure fractions of the more complex mixed glycerides.

The elution pattern recorded for a mixture of equal proportions of trilaurin, trimyristin, tripalmitin, and triolein is shown in Fig. 18. As a result of the heavy coating of liquid phase (5% of support weight) and the flat temperature gradient, the recovery of triolein is low. Similar low recoveries were also experienced with tristearin under

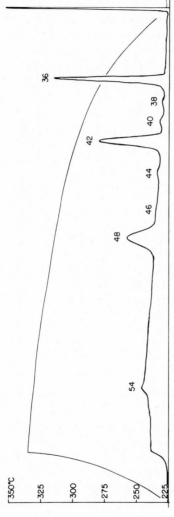

FIG. 18. Preparative GLC of simple triglycerides. Peaks identified as in Fig. 1. Instrument 4, Aerograph Autoprep 700 with flame ionization kit and splitter assembly as shown in Fig. 17. Column, 2 ft × ¼ in. o.d. aluminum tube packed with 5%, w/w, SE-30 on silanized chromosorb W (60–80 mesh). Nitrogen flow rate, 200 ml/min, 30 psi head pressure. Flash evaporator, 325°C; detector 330°C; temperature program as shown. L and N 1-mv recorder with a 6 in. scale and signal switch to activate fraction collector. Sample, 50 µl of a 25%, w/v, solution in petroleum. Split ratio 5:1.

these conditions. The proportions of the other triglycerides, however, are as calculated and correspond closely to those collected in the effluent, suggesting that the splitter system is satisfactory at least for the collection of widely spaced glyceride peaks. Thus in one experiment (51), 20 injections of 50-μl aliquots of a 5% solution of the mixed triglyceride in chloroform yielded about 10 mg each of trilaurin, trimyristin, and tripalmitin but only a trace (1 mg) of triolein. On rechromatography in an analytical GLC apparatus, the recovered fractions gave single peaks, indicating that the glycerides had not undergone interesterification or fragmentation during the preparative separation. Further tests of the preparative system with other triglyc-

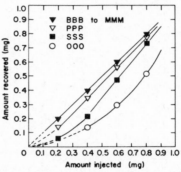

FIG. 19. Recoveries of simple triglycerides from a preparative GLC apparatus. The indicated quantities were injected directly onto the column as 1% solutions in chloroform. The triglycerides tributyrin through trimyristin (BBB to MMM) chromatographed under the conditions given in Fig. 18; other triglycerides were recovered by programming the oven temperature from 250° to 350°C. PPP, tripalmitin; SSS, tristearin; OOO, triolein.

eride mixtures using tridecanoin as an internal standard showed that the recoveries of the long-chain glycerides varied greatly with the operating conditions and that there occasionally were preferential losses of the unsaturated C_{54} triglycerides.

The approximate recoveries of various simple triglycerides from a 5% preparative column under optimum working conditions are indicated in Fig. 19. The low yields noted for triolein were not reflected in a general loss of all triglycerides containing unsaturated fatty acids. Mixed triglycerides containing oleic and linoleic acids in combination with saturated fatty acids of medium chain length could be recovered readily in apparently satisfactory condition. These glycerides gave typical triglyceride spectra in the infrared and could be completely digested by enzymatic and chemical methods and yielded the anticipated fatty acid patterns.

All the collected triglyceride peaks were contaminated with small amounts of the liquid phase, which was continuously lost from the column. This silicone bleed could be separated from the triglycerides on a thin layer of silica gel, where the less polar polysiloxane fragments migrate well ahead of the glycerides. An examination of the recovered triglycerides in the infrared (*51*), before and after the TLC purification, indicated that the minor contamination with silicone did not interfere with the identification of the absorption bands associated with the vibrations of the functional groups characteristic of the triglyceride molecules.

When working with triglyceride mixtures containing components differing by two carbons only, the time of the rotation of the fraction collector becomes more critical. Too early or too late indexing will result not only in incomplete recoveries of the desired peaks but also in cross-contamination. The difficulty arises from the relatively long distances that the triglyceride bands have to travel in separate tubes to reach the detector and collector ports simultaneously. Owing to differences in the tube diameter, minor variations in flow rate bring about changes in the split ratios and the rates of movement in the empty tubes. Proper adjustment of the temperature of the collector and detector housing is the factor contributing most to the collection of pure components. Unless these variables are adequately controlled, the recording of a good separation in the flame detector provides little assurance of peak purity.

With the apparatus described, the most complete collections are realized with the broader peaks, which may require about one minute for complete elution. Closely spaced sharp spikes present problems, as they may easily be missed completely or be recovered only partially as a result of an incomplete correspondence in the arrival times at the two exit ports (detector and collector). Therefore, the rather flat temperature gradients, obtained as the oven temperature asymptotically approaches a limiting value, may be more suitable for glyceride collection than the linear programs, which in addition produce progressively closer spacing of the higher-molecular-weight glycerides. Too flat peaks, however, cannot be reliably collected with this type of collector because any slight back pressure from the glass-wool plugs in the collection bottles will result in accidental tripping of the activator switch and a further rotation of the turn table. A convenient degree of spacing of glyceride peaks during preparative GLC is illustrated in Fig. 20. This run was obtained with the triglycerides of coconut oil (7). The collected components were shown by analytical GLC to be free of contamination with either preceding or following glyceride peaks. By using optimum conditions, quantitative collections of triglycerides of uniform molecular weight have been obtained with synthetic medium-chain triglycerides (51), molecular distillates of butteroil (21), and the medium-chain triglycerides of butterfat obtained from thin-layer chromatography on silver-nitrate-impregnated silica gel G (47).

VII. Quantification

The ultimate object of all analytical separations is the quantification of the separated components. In the GLC of glycerides as commonly conducted it is difficult to obtain direct measurements of either the components admitted to the gas chromatograph or those recovered from it. Samples are injected as solutions, and, although the concentrations and volumes can be controlled to some extent, it is not known how much of the measured sample actually reaches the column and appears in the detector in a concentration sufficiently high to be recorded as a peak. On the other hand, unless peak collections are made, the recoveries of the components from the column can be judged only by comparing the electrical signals recorded. Even when collections are made, there is uncertainty about the extent of condensation and location of all the components, as well as the variability of contamination with the liquid phase from the column. Despite these difficulties, ways and means have been found for obtaining reliable quantitative estimates of all glyceride components present in the sample.

A. Absolute Recoveries

The extent of actual recoveries of triglycerides from GLC columns can be ascertained by means of preparative runs and by runs using

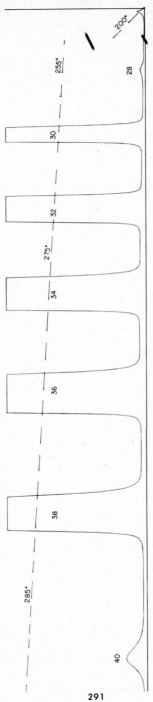

FIG. 20. Preparative GLC of coconut oil triglycerides. Peaks identified as in Fig. 1. Temperature program as shown. Instrument and other operating conditions as given in Fig. 18.

radioactive glycerides. By means of preparative GLC it has been shown that simple triglycerides up to trimyristin can be recovered in yields over 90% from well-conditioned 5% SE-30 columns. Comparable recoveries of tripalmitin (95%) and tristearin (85%) could be obtained from 3% JXR columns, and higher recoveries required still lower proportions of liquid phase and shorter columns. A total of about 10 mg of each glyceride was injected into the preparative GLC apparatus by using about 0.5 mg each of tripalmitin and tristearin and about 1 mg each of the lower-molecular-weight triglycerides per injection. The recovered components were purified by thin-layer chromatography and were weighed. The recoveries were calculated after correction for the amount burned in the detector, which was estimated from the split ratio. On the basis of the observations made, it is therefore correct to assume that trilaurin, which is frequently used as a reference standard in calculating glyceride recoveries, is quantitatively eluted from all columns that give at least a partial recovery of tristearin. The yields of triolein have been shown (51) to parallel those of tristearin, but trilinolein may be partially peroxidized during the collection in the open tube. The recoveries of standard triglycerides of molecular weight above that of tristearin have not been specifically determined by means of preparative GLC.

Although radioactive triglycerides themselves have not as yet been passed through GLC columns and the activity recoverable in the column effluents measured, measurements of radioactivity have been made with other compounds of comparable molecular weight and chemical stability. Thus Swell (38) used GLC columns suitable for triglyceride analysis for separation and simultaneous measurement of mass and radioactivity of cholesteryl esters. Both saturated and unsaturated fatty acid esters of cholesterol including the linoleate and the arachidonate were recovered with an over-all reproducibility of ±5%. The glass columns (24 in. × 4 mm o.d.) used in this demonstration were siliconized and contained 3% QF-1 on Gas Chrom Q (100–120 mesh). The system was operated isothermally at 290°C. The elution curves for the mass (flame ionization detection) and radioactivity (^{14}C and ^{3}H, modified Barber-Colman radioactive monitoring system) overlapped completely, although the return to the base line between peaks was not ideal. The latter problem, however, appears to have resulted from the use of the complex combustion train rather than from any irregularity of the band migration in the column or decomposition. Since it has previously been shown (*52,53*) that the gas chromatographic behavior of intact steryl esters is very similar to that of triglycerides, it may be assumed that labeled triglycerides would yield comparable elution patterns for mass and radioactivity resulting in quantitative recoveries of both. Comparable absolute recoveries of other neutral glycerides and their derivatives have not been specifically demonstrated.

B. Relative Response

Since all glycerides do not produce identical responses in the detector, equal areas on the chromatogram do not represent equal amounts of material. How much glyceride the areas do represent may be ascertained by careful introduction of measured amounts of known substances into the GLC apparatus. A comparison of these responses on an equal weight or mole basis gives the relative responses or correction factors for these compounds. Once these factors have been established, the injection of exact quantities of material becomes unnecessary and the concentrations of the unknowns present in the sample may be measured in relation to an added standard the exact amount and relative response of which is known.

1. CALIBRATION FACTORS

Theoretically, the flame ionization detector should give about the same response for the triglycerides as for the corresponding fatty acid methyl esters. A triglyceride molecule may be considered as three methyl ester molecules bound together. For example, Litchfield et al. (*15*) have pointed out that the molecule of tristearin has the same atomic composition and the same number of C—C, C—H, and C—O bonds as one molecule of methyl stearate plus two molecules of methyl oleate. In the hydrogen flame detector, the plot of the ratio of weight per cent area per cent versus the carbon number for long-chain fatty acid methyl esters approximates a horizontal line (*54,55*) with a slight negative slope (*56*). The negative slope has been attributed (*56*) to the decreasing weight per cent of oxygen as the carbon number increases. A similar plot for the triglycerides should therefore have the same slope, assuming that all injected sample landed onto the column and eventually reached the flame detector.

When the weight per cent area per cent is plotted against the carbon number of the simple triglycerides, it is seen that the plot approximates a horizontal line with a negative slope for all the triglycerides that are completely recovered from a given column. A positive slope in the region C_{42} to C_{54} obtained on some columns indicates that on those columns sample losses have occurred and that the greater the molecular weight the greater the loss. Although it has been estimated that the sample losses are proportional to the sample load and that the losses can be accurately compensated for with suitable correction factors, it may not be so with all columns and with all designs of the flame ionization detector. The bleed of the silicone liquid phase and its accumulation in the detector greatly affects the apparent component loss and the detector response, and may vary on different days even using the same detector. Comparisons of the calibration factors for saturated and unsaturated triglycerides of equal carbon number indicate that the double bond has a negligible effect. Litchfield et al. (15) have recorded quantitative weight response (weight per cent area per cent) and molar response (mole per cent area per cent) factors for a number of standard monoacid triglycerides under their working conditions. In Table 2 these factors are compared with those recorded in the author's laboratory under comparable conditions. The calibration factors are in the same range, and the variations may be attributed to differences in purity or stability of the substances.

In practice, triglyceride compositions are usually expressed in terms of mole per cent. All saturated glycerides of the same carbon number are assumed to have the same molar response. When the molar response values for saturated and unsaturated glycerides of the same carbon number are different, an average mole response is assigned to each peak, based on its estimated fatty acid composition. Since the calibration factors vary substantially with the operating conditions and may not be exactly reproducible from day to day, it may be convenient to equate the area percentages to weight percentages (1). Although the use of actual calibration factors should yield more accurate analytical results, for routine work with mixed long-chain triglycerides the weight–area approximation appears to be valid (7). In fact it may be shown that for certain common mixtures of triglycerides containing palmitic, palmitoleic, stearic, oleic, and linoleic acids, the ratio of weight per cent area per cent averages 1.

The above considerations emphasize that accurate calibration is essential for quantitative GLC of triglycerides. Since the response factors vary with the working conditions and the design of the instrument, the values given in Table 2 do not necessarily apply to other laboratories. The problems of quantification in triglyceride gas chromatography encountered with a thermal conductivity detector have

TABLE 2

CALIBRATION FACTORS FOR SIMPLE TRIGLYCERIDES UNDER OPTIMUM OPERATING CONDITIONS

| | | | Weight %/area % | |
| | | | Litchfield et al. (15) | Kuksis and Breckenridge (21) |
Triglyceride	C_n	Double bonds		
Trioctanoin	24	0	1.12	1.00
Tridecanoin	30	0	1.04	1.00
Trilaurin	36	0	1.00	1.00
Trimyristin	42	0	0.96	1.00
Tripalmitin	48	0	0.98	1.00
Tripalmitolein	48	3	1.01	
Trimargarin	51	0	1.05	
Tristearin	54	0	1.09	1.00
Triolein	54	3	1.03	1.00
Trielaidin	54	3	1.06	
Trilinolein	54	6	1.10	1.05
Trilinolenin	54	9	1.12	
Triarachidin	60	0	1.21	1.10
Tri-11-eicosenoin	60	3	1.10	
Tribehenin	66	0	1.43	
Trierucin	66	3	1.34	

been discussed by Huebner (5). Although an argon ionization detector has been used (4) for the detection of triglycerides in GLC column effluents, no attempt appears to have been made to ascertain the relative response of this detector to different glycerides.

Comparable calibration factors may be worked out for the homologous series of mono- and diglycerides and their derivatives. The acetic and butyric acid esters of the partial glycerides show responses comparable to those of low-molecular-weight triglycerides and present no special problem. The silyl and trifluoroacetyl derivatives require independent calibration.

2. USE OF INTERNAL STANDARDS

Once it has been established that standard glyceride mixtures of molecular weight comparable to that of the unknowns can be separated and recovered satisfactorily, the GLC system is ready for a quantitative examination of an unknown glyceride mixture. If this mixture is pure (contains only glycerides that are quantitatively measured in the detector) a determination of the peak proportions is sufficient to arrive at the per cent composition of the sample. If this mixture is impure (contains other volatile or nonvolatile substances), a determination of the peak proportions may still be useful in ascertaining the glyceride composition, but the amount of the total glyceride in the sample will remain uncertain. To determine the glyceride concentration from the detector response alone, it is necessary to know the calibration factors and the total weight or volume of the sample admitted to the column. The sample size, at the present time at least, cannot be conveniently controlled by injection with simple syringes.

It is possible, however, to obtain an accurate estimate of the total glyceride content in the sample by adding a known amount of a suitable internal standard to the unknown mixture and comparing the recorded area proportions. For the analysis of most natural triglyceride mixtures, it has proved to be convenient to add tridecanoin as the internal standard, but in special cases trioctanoin may be better suited. In Fig. 21 are shown a run with the saturated glyceride fraction of lard to which tridecanoin has been added as a standard and a run with the triglycerides of bovine milk fat to which trioctanoin has been added. The best standards are foreign triglycerides (containing fatty acids other than those present in the unknown) that run in a constant relation to the unknown glycerides, do not overlap them, and give about the same response in the detector. The glyceride/internal standard ratio can then be independently determined following transmethylation and gas chromatography of the fatty acid methyl esters. The total amount of the glyceride in the sample is calculated by the equations commonly employed in work with internal standards:

$$\text{mg TG}_{\text{total}} = \frac{\text{mg internal standard} \times \text{total triglyceride area}}{\text{area of internal standard}}$$

A similar expression may be used for estimating the total triglyceride content of the sample following transmethylation and fatty acid chromatography unless the sample and the internal standard contain common fatty acids, in which case a correction must be applied. A large selection of equations suitable for work with internal standards has been given in connection with steroid analyses by GLC (*57*) and

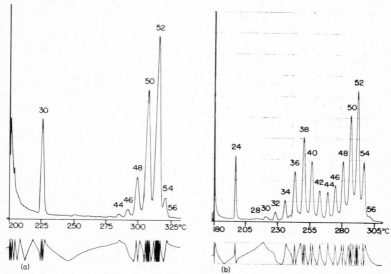

FIG. 21. GLC of natural triglyceride mixtures with internal standards. (a) Saturated triglycerides of lard plus tridecanoin; (b) bovine milk fat triglycerides plus trioctanoin. Peaks identified as in Fig. 1. Instrument 5, Aerograph 204-1B with dual columns and differential electrometer. Columns, 20 in. × ⅛ in. o.d. stainless-steel tubes packed with 3%, w/w, JXR on Gas Chrom Q (100–120 mesh) and assembled as shown in Fig. 5. Nitrogen flow rate, 120 ml/min at room temperature. On-column injector, 280°C. Detector, 340°C. Temperature program as shown. Brown Electronik 15 1-mv recorder; chart speed, 6 min/in. Sample, 1 μl of a 1%, w/v, solution in chloroform. Attenuation, 1 × 16.

is applicable also in glyceride determinations. For maximum accuracy the area response of the internal standard and the triglycerides should be carefully calibrated.

C. Calculation of Glyceride Recovery from Fatty Acid Data

When more than one glyceride peak is detected, the relative recoveries of glycerides may be calculated from a knowledge of the fatty acid composition of the sample (58). The calculation is based upon the recognition that when the triglyceride peaks are eluted in correct proportions, they fully account for the fatty acid distribution in the glycerides. If, for example, the longer-chain triglycerides are incompletely recovered, the mole proportions of the glycerides estimated from the GLC data will favor the shorter-chain fatty acids and will fail to account for all the carbon atoms available in the fatty acid mixture of the given acid proportions. Since the quantitative proportions of the fatty acids in the total glyceride sample are readily available from the GLC analyses of the fatty acid methyl esters, they constitute a convenient and informative check on both the individual and the over-all recoveries of glycerides.

The comparison may be expressed in terms of total fatty acid carbon yield (58), in the form of the average chain length of the fatty acids (15), or as the average chain length of the triglycerides (7). The theoretical fatty acid carbon number (TCN) is calculated (58) by summing the products of multiplication of the mole per cent of each fatty acid by its number of carbon atoms. This carbon number is compared to the experimentally determined carbon number (ECN) obtained by summing the products of multiplication of the mole per cent of each triglyceride by its fatty acid carbon number. Since it takes 3 moles of a fatty acid to make 1 mole of triglyceride, the total fatty acid carbon estimate derived from the fatty acid data must be multiplied by 3, or the estimate from the triglyceride data divided by 3, before the values can be matched.

The formulas and a sample calculation are given in Table 3. The average chain length of the fatty acids is calculated similarly from both the fatty acid methyl ester analyses and the analyses of the triglycerides, but the total carbon number is divided by the number of the fatty acids. Similar methods are used for the calculation of the average triglycerides, except that the total carbon number is divided by the number of different triglycerides (based on carbon number). The latter value can be correlated to the molecular weight of the average triglyceride estimated from the saponification value. Similar correlations can be made between the average fatty acid chain length estimated from GLC data and the data derived from saponification equivalents.

These estimates are most reliable when the fatty acid or glyceride mixture is very complex. Thus it has been shown (58) that the recoveries of butterfat triglycerides are nearly quantitative for every chain length and account completely for the fatty acid composition determined separately and independently. Excellent estimates of recovery have been obtained by this means also for coconut oil and other fats containing short- and long-chain fatty acids. Having fewer

TABLE 3

CALCULATION OF FATTY ACID CARBON RECOVERY (58)[a]

Theoretical fatty acid carbon number (TCN)	Fatty acid carbon number for experimentally determined triglyceride distribution (ECN)

Formula

$TCN = 3\Sigma_k$ (mole % FA × X)
  X = number of carbon atoms per fatty acid residue
  k = number of fatty acids

$ECN = \Sigma_n$ (mole % TG × X′)
  X' = number of fatty acid carbon atoms per triglyceride residue
  n = number of triglyceride types

Example

$TCN = 3\Sigma_{12}$ (7.8 × 4) + · · ·
  + (0.5 × 20) = 439,050

$ECN = \Sigma_{31}$ (4.00 × 54) + · · ·
  + (0.51 × 24) = 421,811

recovery of theoretical carbon yield = (ECN × 100)/TCN
= (421,811 × 100)/439,050 = 96%

[a] Dividing the TCN and ECN by 10,000 (the total number of triglycerides) gives the carbon number or the chain length of the average triglyceride; dividing by 30,000 gives the carbon number or the chain length of the average fatty acid. Both these values can be correlated to the respective molecular weight of the average triglyceride or average fatty acid estimated from the saponification value.

triglyceride peaks, the estimates of triglyceride recoveries for the common vegetable oils and animal fats have given data of the order of 95 to 105% of the theoretical carbon yield (7). Calculations of average fatty acid chain lengths from the fatty acid and triglyceride data for cocoa butter and rat adipose tissue fat have also yielded (15) closely agreeing values.

VIII. Applications

Although many laboratories have successfully recovered model mixtures of glycerides from GLC columns, the application of triglyceride GLC to specific problems in industrial, chemical, or medical research has been slow in realization. Exploratory studies have, however, been completed on virtually all types of glyceride mixtures, and in specific instances significant results have been obtained. The most promising applications have been made in the structural investigations, but the collection of glyceride elution patterns of natural fats and oils has also proved worthwhile. Many other avenues of exploration, such as those relating to the physical properties of glyceride molecules, have remained largely untouched.

A. Separations of Natural Glyceride Mixtures

Natural glyceride mixtures suitable for direct GLC analysis are usually obtained by TLC fractionation of a total lipid extract. In addition to the triglycerides, both mono- and diglycerides may be recovered and examined in the GLC apparatus, either as such or following the preparation of some suitable derivative. It is also possible to use the total neutral lipid fraction for direct GLC examination. All mono-, di-, and triglycerides are eluted in sequence from a single column provided the starting temperature is low enough. The steryl esters emerge between the diglyceride and the lower-molecular-weight triglyceride peaks without a serious overlap. In absence of significant amounts of diglycerides, the entire total lipid extract may be injected and excellent separations obtained for cholesterol, the pyrolysis products of phosphatides, steryl esters, and triglycerides. In fact, samples of plasma or concentrated solutions of lipoproteins may be directly injected into the gas chromatograph and effective resolutions obtained for all esters present. It should be recognized, however, that more detailed experimental data are obtained by an effective preliminary segregation of each ester class, on silver-nitrate-impregnated silica gel plates, preceding the GLC resolution. For a complete elucidation of the glyceride structures and the recognition of their organization at the cellular or subcellular level, there may be a need for an effective fractionation of the cellular organelles and their constituent lipoproteins prior to the glyceride analyses.

1. SEED OILS AND FATS

Oil seeds and nuts represent rich sources of natural glycerides for consumption and for industrial use. Since the palatability and industrial usefulness of these fats are determined by their chemical composition, a great deal of effort has been expended in the study of their glyceride structure. The direct analysis of natural fat triglycerides by GLC is of greatest value for fats containing a wide range of fatty acid chain lengths, but important data have also been obtained by GLC of fats of only a few fatty acid types.

One of the oils best suited for GLC, and the first one to be analyzed in any detail by the new technique, was coconut oil. GLC elution patterns recorded for a sample of refined coconut oil are shown in Figs. 1 and 2. Similar runs are obtained for samples of crude oil, except that the fronts of the chromatograms usually show small amounts of low-molecular-weight residues, which are eliminated in the refining process. Major peaks are seen for C_{36} and C_{38} triglycerides with only small contributions made by the C_{52} and C_{54} components, which are the main ones in most other plant oils. Collection of the major peaks (Fig. 20) by means of preparative GLC has shown (7) that the recovered fractions are triglycerides and that each peak contains most of the fatty acids of the original oil. The acid proportions are such that all of them can be recombined into triglycerides of the appropriate molecular weight. Palm kernel oil yields comparable runs and has been briefly reported upon by Huebner (5).

Litchfield et al. (59) have made a combined TLC–GLC study of the seed fat of an ornamental shrub (*Cuphea llavia*), a fat that contains 91.2% of decanoic acid. In this fat tridecanoin was recognized as a major component and most glyceride molecules contained at least two decanoic acid residues, but several minor triglycerides that did not conform to this pattern were also found. Application of the combined TLC–GLC technique to the ucuhuba kernel fat showed (60) that it contained at least 23 different triglyceride components. Irimyristin and laurodimyristin comprised over half the total triglycerides, which was expected, since the fat contained 20.0 mole % lauric and 71.4% myristic acids. Litchfield et al. (61) have further demonstrated the usefulness of this technique by analyzing 17 other seed fats containing C_8, C_{10}, C_{12}, C_{14}, C_{16}, C_{18}, and C_{20} fatty acids. The oils were obtained from eight species of the laurel (Lauraceae) family, six species of the Lythraceae (Cuphea) family, and three species of the elm (Ulmaceae) family, and the triglycerides were isolated by preparative thin-layer chromatography. From 10 to 18 peaks within the 22–58 carbon number range were resolved for each fat on a 24-in.

column packed with 3% JXR liquid phase of Gas Chrom Q. The most interesting finding was obtained with *Laurus nobilis*. This seed fat contained 58.4% lauric acid and 29.2–29.0% trilaurin. A maximum of 19.9% trilaurin would be predicted by a random, 1,3-random-2-random or 1-random-2-random-3-random distribution of lauric acid. This indicates a preference for the biosynthesis of a specific simple triglyceride by *Laurus nobilis* seed enzymes.

The highly unsaturated vegetable seed oils consisting primarily of the C_{18} unsaturated fatty acids yield only a few peaks on GLC, but even these separations can be very informative when preceded by a preliminary segregation of the oils by silver nitrate TLC. To safeguard against losses of the more highly unsaturated glycerides during the gas chromatography, the glycerides recovered from the TLC plates are first subjected to hydrogenation, and only then is the gas chromatography of the glycerides performed. With adequately prepared columns, however, there is no difficulty in recovering the unsaturated glycerides and there is no need for hydrogenation, at least of the triglycerides containing the C_{18} unsaturated acids. The elution patterns recorded for linseed oil and hydrogenated linseed oil are shown in Fig. 22. This oil contains a large proportion of linolenic acid, which is known to decompose and polymerize rapidly. In both runs major peaks are seen for C_{52} and C_{54} triglycerides. The peak proportions are about the same, indicating that no preferential loss of the unsaturated glycerides has taken place. There would have been a proportional increase in the more saturated glycerides, which contain two palmitic acid residues (C_{50}), had any of the highly unsaturated glycerides been lost.

In an early publication from the author's laboratory (*1*) the triglyceride elution patterns obtained for safflower, cottonseed, corn, and peanut oils were shown. Similar patterns were later recorded (*63*) for soybean, sunflower, olive, and wheat germ oils. In common with linseed oil, all these seed oils contain large amounts of C_{18} unsaturated fatty acids and therefore yield comparable triglyceride elution patterns. The run recorded for soybean oil triglycerides was similar to that obtained by Yates et al. (*62*) on a different instrument and under somewhat different working conditions. The chromatogram of olive oil was identical to that recorded by Schlater et al. (*8*). This suggests that even the runs with the unsaturated seed oils are reproducible from one laboratory to the other. The pattern recorded for cottonseed oil contained the C_{52} triglyceride as the major component, with relatively lower proportions for both the C_{50} and the C_{54} glycerides. Yates et al. (*62*) have published a comparable elution sequence for this oil.

Deviations from the general pattern of the vegetable seed oils were also noted for peanut oil. This oil is relatively rich in arachidic acid, and this is reflected in the glyceride elution pattern, which, in addition to the usual glyceride peaks found in the other vegetable seed oils, showed the presence of moderate concentrations of C_{56} and C_{58}

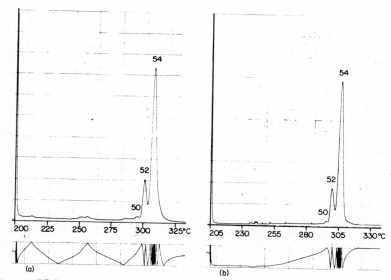

FIG. 22. GLC of linseed oil triglycerides. (a) before hydrogenation; (b) after complete hydrogenation. Peaks identified as in Fig. 1. Temperature program as shown. Instrument and other operating conditions as in Fig. 21.

triglycerides. In all cases examined, samples of hydrogenated oils gave glyceride elution patterns identical with those observed for the corresponding unsaturated parent oils. When both saturated and unsaturated glycerides were present in significant amounts, hydrogenation of the oils resulted in better peak resolution because of a more complete overlap among all the peaks of the same carbon number.

The seed oils containing large amounts of behenic and erucic acids are particularly rich in long-chain glycerides with major contributions made by peaks of carbon numbers in the range of C_{54}–C_{62}. Harlow et al. (*32*) have examined four erucic-acid-rich oils (from watercress, rapeseed, nasturtium, and *Lunaria annua*) and have established the pattern of incorporation of these long-chain acids into the triglycerides. Only when the long-chain acid content is above 66.7% are triglycerides with three long chain acids formed. On the other hand, when the long-chain acid content is below 33.3%, there is no comparable limitation to one long-chain acid per triglyceride molecule.

The elution patterns recorded in the author's laboratory for samples of refined rapeseed oil and rapeseed oil with "zero erucic acid" content are shown in Fig. 23. The common rapeseed oil contains somewhat

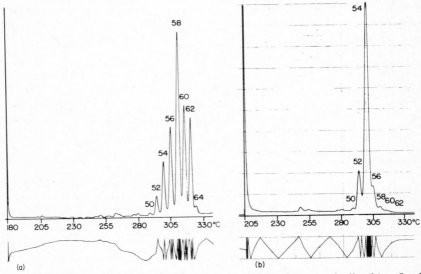

FIG. 23. GLC of rapeseed oil triglycerides. (a) Refined rapeseed oil; (b) refined rapeseed oil with "zero erucic acid." Peaks identified as in Fig. 1. Temperature program as shown. On-column injector, 325°C. Instrument and other operating conditions as in Fig. 21. (Rapeseed oil samples courtesy of Dr. B. Costigliola.)

over 33.3 mole per cent erucic acid. It gives a major peak for the C_{58} triglycerides, which suggests that a large proportion of this oil is made up of glycerides containing two C_{18} and one C_{22} fatty acids per molecule. The elimination of the bulk of the erucic acid from the "zero erucic acid" rapeseed oil has been accompanied by proportional increases in the content of the C_{18} fatty acids, which have given rise to a triglyceride composition much like that noted for the more common vegetable seed oils.

It is seen that the glyceride pattern recorded here for the industrially refined oil differs from that reported by Harlow et al. (*32*), which showed a major peak for the C_{62} component, apparently representing triglycerides of two C_{22} and one C_{18} fatty acids per molecule. Both oils had about the same erucic acid content. However, Harlow et al. (*32*) prepared their oil in the laboratory by extracting the crushed seeds with petroleum ether in a Soxhlet apparatus. It is possible that during the refining process preferential losses of the higher-molecular-weight triglycerides occurred and resulted in a distorted glyceride elution pattern. Furthermore, Harlow et al. (*64*) have shown that there are regional differences in the glyceride composition of seeds and that the oils commonly analyzed are mixtures of glycerides from various stages of maturity. For this reason the source of the oil should be carefully defined before attempting detailed analysis that can be reproduced by other workers. The recordings shown here are only examples intended to illustrate the scope of potential application. All of these oils could have been subjected to a preliminary separation on silver nitrate plates and further insight into their triglyceride structures thereby obtained. Such studies are in progress in several laboratories (*32,63*).

2. ANIMAL TISSUE LIPIDS

Only a few studies have been reported on the triglycerides of specific animal tissues. In most cases the analyses have been restricted to the examination of certain pooled glyceride mixtures that have been variously designated as lard, tallow, suet, or unspecified depot fat and have been mainly of industrial or nutritional interest. These mixtures are of little biochemical or physiological significance and do not usually justify the investment of time required for detailed analyses. Kuksis has shown (*65*) the triglyceride patterns recorded for samples of commercially rendered hog fat or lard, commercially rendered beef fat or tallow, and the hard fat, surrounding the kidneys of sheep, known as suet. The differences in the fatty acid composition of these fats give rise to characteristic triglyceride patterns by which the origin of the fat can be ascertained. Samples of lard, tallow, suet, and beef drippings from various commercial suppliers have shown remarkably similar patterns. As in the case of the vegetable oils, the information available from the small number of different glyceride groups resolved by GLC can be greatly upgraded by subjecting the fat to a thin-layer fractionation prior to the gas chromatographic analysis. Thus preliminary resolution of lard into the saturated and the mono-, di-, tri-, and polyunsaturated glycerides followed by GLC provides a sevenfold increase in the triglyceride resolution. As many as 30 distinct triglyceride groups may be distinguished in lard (*65*) if certain positional and fatty acid isomers also resolved on the silver nitrate plates are separately collected.

Triglyceride elution sequences recorded for selected animal organs are shown in Fig. 24. The triglyceride composition of beef liver, hog liver, and sheep kidney is remarkably similar to that of the respective adipose tissues of these species, but minor differences are discernible. The triglyceride patterns of bird tissue fat have also been examined (*65*). The adipose tissue fat of a hen is considerably different from that of a goose, and there are differences between the fats of adipose tissues and those of specific organs in the same bird. The C_{48}–C_{54} glycerides, however, remain the major components in all samples.

The triglycerides of a normal rat liver and those of a choline-deficient rat liver are given in Fig. 25. In addition to a tenfold increase in the glyceride content, this choline-deficient animal possessed liver triglycerides greatly enriched in tripalmitin and dipalmitomonoolein and dipalmitomonolinolein. The triglyceride elution sequences observed for normal rat liver were nearly identical with the glyceride patterns noted for the total lipid extracts of normal liver mitochondria, microsomes, and cell membranes. Comparable elution patterns have been recorded also for groups of lipoproteins prepared from these subcellular organelles. In many respects they are similar to the triglyceride patterns of rat adipose tissue and whole serum glycerides, but subtle differences may be noted following rechromatography of the glycerides after a preliminary resolution on TLC plates in the presence of silver nitrate.

A special group of animal fats is provided by the marine oils, which represent largely the liver oils of whales, seals, and salt water fish. These oils are rich in polyunsaturated long-chain fatty acids, which are derived to a considerable extent from the zooplankton of the sea. Despite the high molecular weight and great reactivity, these triglycerides survive the GLC conditions and emerge as symmetrical peaks of appropriate molecular weight. The resolution obtained for the triglycerides of raw and hydrogenated menhaden oil is shown in Fig. 26. Although the triglyceride patterns represent oils from two different sources, the over-all elution patterns are closely matched, suggesting

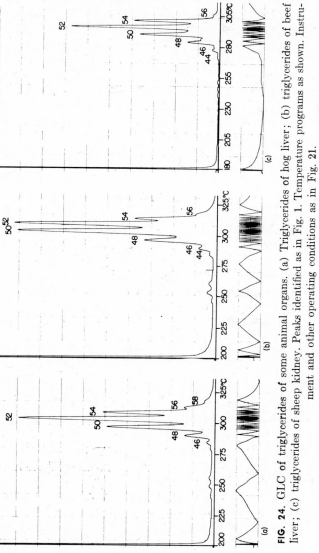

FIG. 24. GLC of triglycerides of some animal organs. (a) Triglycerides of hog liver; (b) triglycerides of beef liver; (c) triglycerides of sheep kidney. Peaks identified as in Fig. 1. Temperature programs as shown. Instrument and other operating conditions as in Fig. 21.

that both the unsaturated and the saturated oils were recovered to about the same extent. Calculations of recovery based on the fatty acid composition and the triglyceride peak distribution showed that essentially complete triglyceride elution was obtained for both oil samples. Harlow et al. (*32*) have recorded a nearly identical triglyceride gas chromatogram for another sample of menhaden oil following hydrogenation. In all samples, peaks were recognized for triglycerides of

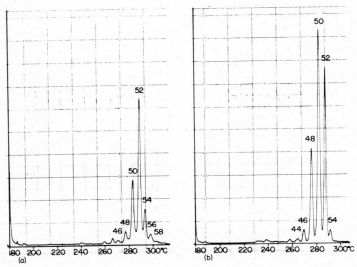

FIG. 25. GLC of triglycerides of rat liver. (a) Normal liver; (b) choline-deficient liver. Peaks identified as in Fig. 1. Temperature programs as shown. Instrument and other operating conditions as in Fig. 21. (Samples of floating fat courtesy of Dr. B. Rosenfeld.)

carbon number C_{50}–C_{58}. The highest-molecular-weight glycerides had a total of 62 carbons in their fatty acid chains. The hydrogenation of these oils is particularly advantageous if the fractions have to be stored, for instance, following a recovery from TLC plates after a preliminary resolution on the basis of unsaturation (*63*). Following the TLC separation an aliquot is transesterified and the fatty acids are identified and quantitatively measured; the rest of the eluate is hydrogenated and stored for further GLC examination. The GLC elution patterns recorded for herring oil (*65*) are similar to those shown in Fig. 26 for menhaden oil. Both types of fish apparently subsist on the same kind of fat and deposit triglycerides of related fatty acid composition. Comparable glyceride patterns have been recorded (*63*) for the triglyceride groups recovered following silver nitrate TLC. Although the latter technique did not resolve the glycerides containing more than four double bonds per molecule, some 60 triglyceride peaks could be identified and measured by GLC. Thus

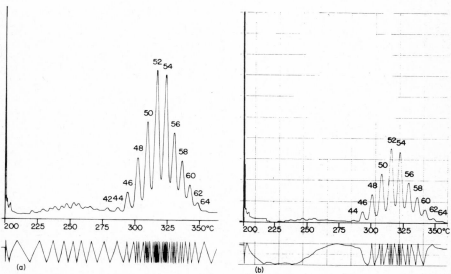

FIG. 26. GLC of triglycerides of menhaden oil. (a) Raw oil; (b) hydrogenated menhaden oil. Peaks identified as in Fig. 1. Temperature programs as shown. Instrument and other operating conditions as in Fig. 23. (Oil samples courtesy of Dr. W. G. Mertens.)

an examination based even on limited preliminary separation can greatly increase the over-all resolution of the oil and the amount of information obtained. In these analyses trilaurin was used as an internal standard.

The triglyceride elution patterns obtained with the body oils of two marine mammals are shown in Fig. 27. The whale oil was hydrogenated prior to gas chromatography, the seal oil was raw. Again the long-chain triglycerides are recovered in approximately the proportions anticipated both on the basis of the fatty acid compositions of these oils and the considerations advanced by Harlow et al. (*32*) in regard to erucic-acid-rich plant oils.

All the marine oils show very wide distributions of carbon numbers. Practically all possible di- and triacid glycerides appear in the chromatograms. These results indicate that GLC will be very helpful in the analysis of the triglycerides of these oils, the exceedingly complex structure of which has thus far defied quantitative fractionation by

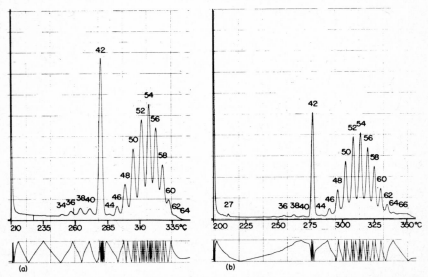

FIG. 27. GLC of triglycerides of seal and whale oils. (a) Raw seal oil; (b) hydrogenated whale oil. Peaks identified as in Fig. 1. Temperature programs as shown. Instrument and other operating conditions as in Fig. 23. Peak 42, trimyristin internal standard. (Oil samples courtesy of Dr. W. G. Mertens.)

chromatographic techniques. Quantitative GLC results for the marine oils have already been reported (*32*) for the even-carbon-number triglycerides, which represent the majority of triglycerides present in each oil. However, odd-carbon-number triglycerides are also present as indicated by the fatty acid composition and the appearance of small peaks between the C_{42}, C_{44}, C_{46}, and C_{48} peaks. The presence of odd-carbon-number and branched-chain fatty acids in the marine oils is also responsible for the poorer resolution of these triglyceride peaks. This effect has been most noticeable with tuna oil, which Harlow et al.

(*32*) have estimated to contain a total of 6.6% odd-carbon-number fatty acids. This is about twice the value recorded for menhaden oil. The complete resolution of all the peaks on one GLC column must await further improvements in the column efficiency.

3. SERUM, LYMPH, AND MILK LIPIDS

Although these lipids are also of animal origin, they have been selected for special consideration because of better defined origin, general availability, and characteristic composition. These lipids can be reproducibly sampled, and reliable interlaboratory comparisons can be obtained. Because of clinical interest, the serum triglycerides have been the most extensively studied of the three. Usually, however, the studies have been limited to a determination of the total triglyceride levels under a given dietary regimen or in a disease state. The GLC of serum or plasma glycerides provides a rapid qualitative and quantitative measurement and may be completed in the presence of other neutral lipids, such as cholesterol and cholesteryl esters, as well as phosphatides. In fact, preliminary studies have shown (*26*) that serum, lymph, and milk may be subjected to a direct GLC examination of their constituent lipids without a prior extraction with organic solvents.

The resolution obtained for a model mixture of neutral lipids of the type found in plasma, serum, lymph, or milk is shown in Fig. 28(a). Mixtures containing different proportions of free cholesterol, cholesteryl, myristate, palmitate, oleate and arachidonate, and lard were prepared. All of them were suitably diluted with tridecanoin, and the quantitative recoveries were determined. Such mixtures of standards had previously been used (*53*) to evaluate the elution patterns obtained for samples of neutral lipids in fasting plasma from a normal adult male and from a patient with hypertriglyceridemia. In addition to a greatly elevated level of the fasting plasma lipids, there was seen a distortion in the proportions of both the triglycerides and the steryl esters in hypertriglyceridemia. The neutral lipids of normal fasting human plasma are shown in Fig. 28(b).

In Fig. 29 is shown the elution pattern recorded for a total lipid extract of another sample of plasma and for the individual lipid classes recovered from a TLC plate of silica gel G developed with a neutral lipid system. It is seen that there is some overlapping between the steryl esters and the lower-molecular-weight triglycerides, but it is not sufficient to invalidate the quantitative measurements. The peaks eluted between the steryl esters and the tridecanoin apparently result from phosphatides and almost overlap those of the diglycerides of corresponding fatty acid composition. Subsequent studies, however, have shown (*26*) that these are not diglycerides but possibly consist of isomeric propenediol diesters of fatty acids formed on pyrolysis of the phosphatides. Work with synthetic dimyristoyl lecithin and

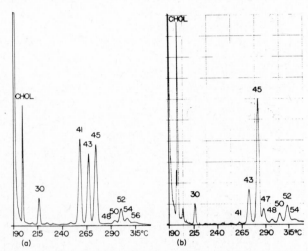

FIG. 28. GLC of neutral lipid mixtures. (a) Model mixture; (b) neutral lipids of fasting human plasma. Peaks 41, 43, 45, and 47 represent cholesteryl esters of C_{14}, C_{16}, C_{18}, and C_{20} fatty acids, respectively. Triglyceride peaks identified as in Fig. 1. Temperature programs as shown. Other conditions as shown for instrument 5 in Fig. 21.

distearoyl lecithin has shown that both of these lecithins yield two peaks, the proportions of which vary somewhat, but the yields average 60–70%. On this basis, it may be possible to estimate both the neutral lipids and the phosphatides, of a sample of total lipid extract, in one analysis. It has not been established, however, to what extent the phosphatides other than lecithin are pyrolyzed and whether or not they yield symmetrical peaks in the GLC apparatus. The diglyceride phosphatides are best examined by GLC as their diglyceride acetates (*66*).

The elution patterns recorded with the total plasma lipid extracts from a normal rat and from a rat with a tumor (Walker 256 carcinoma) are shown in Fig. 30. The sample from the tumor-bearing animal was hyperlipemic, and there was a marked elevation in the plasma triglycerides, steryl esters, and free cholesterol. Furthermore, the elevation in the peaks resulting from the pyrolysis of the phosphatides is proportional to that noted for the other plasma lipids, suggesting that the changes in the phosphatide levels have been correctly represented by changes in the peak heights of the pyrolysis products.

In Fig. 31 are shown the results of another preliminary study (*26*) in which 3 μl of dog plasma were injected directly into the GLC apparatus. Aside from the off-scale peak of cholesterol, the peaks are recorded in their correct proportions. All of them show symmetrical shapes and nearly complete return to base line. Apparently there has been no interference, by the water vapor or the smoldering protein remains, with the initial vaporization or the subsequent elution of the fatty esters. The increased levels of the phosphatide pyrolysis products would suggest that the original lipid extraction [see Fig. 29(a)] left some phosphatides unextracted.

In Fig. 32 are shown the triglyceride elution patterns recorded for dog lymph collected during the height of absorption of corn oil and butterfat (*67*). The lymph triglycerides are seen to resemble the dietary fats to considerable degree, but there has been a dilution of the dietary fat by endogenous fat during the chylomicron formation. Other studies (*68*) have shown that in the lymph it is possible to distinguish at least three different triglyceride pools. In addition to the largely dietary fat carried in the form of the chylomicrons, there is the lymph serum triglyceride pool, which resembles the fasting plasma triglycerides. These two triglyceride types are separated by the chylomicron membrane, which contains a characteristic high-melting triglyceride mixture and represents the third pool. The latter do not appear to change appreciably with the dietary fat carried in the chylomicron. The GLC patterns obtained for the high-melting triglyceride mixtures of the dog chylomicron membranes are shown in Fig. 33. Although these triglyceride mixtures have not been completely analyzed, there is sufficient superficial resemblence to suggest that they may have originated from triglycerides bound to the intracellular lipoproteins of the mucosal cells of the dog intestine.

Because of the occurrence of relatively high concentrations of both short- and long-chain fatty acids in most animal milks, the triglyc-

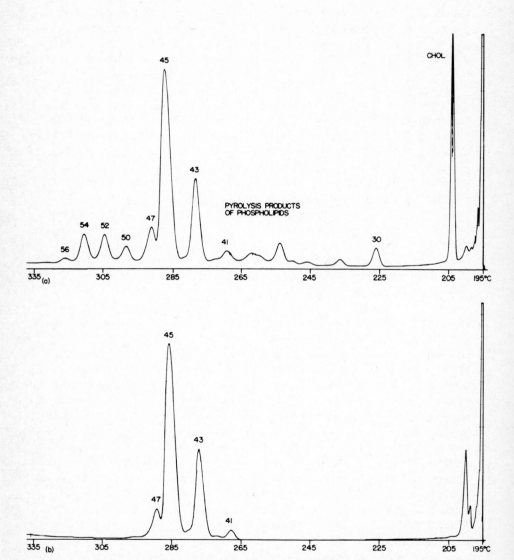

FIG. 29. GLC of total lipids of fasting plasma. (a) Total; (b) cholesteryl esters; (c) triglycerides; (d) pyrolysis products of phosphatides. Individual peaks identified as in Fig. 28. Temperature programs as shown. Instrument and other operating conditions as in Fig. 12.

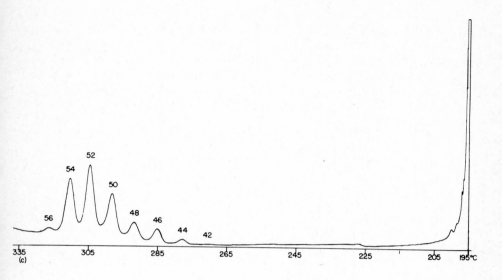

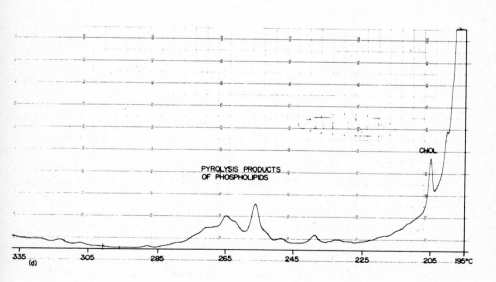

Fig. 29 (*Continued*).

eride mixtures of these fats exhibit a wide range of carbon number distributions, which is conducive to successful gas chromatographic analysis. The triglycerides of cow's milk have been most extensively studied, and the approximate proportions of both even- and odd-carbon-number glycerides have been determined (58). Molecular distillates of butteroil have been fractionated by preparative GLC, and many of the individual peaks have been collected and their fatty

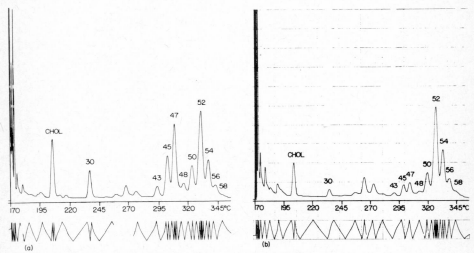

FIG. 30. GLC of total lipids of fasting rat plasma. (a) Normal rat; (b) rat with a tumor. Unmarked peaks emerging after peak 30 and preceding peak 43 represent pyrolysis products of phosphatides. Other peaks identified as in Fig. 32. Temperature programs as shown. Instrument and other conditions of operation as in Fig. 21. (Samples of rat plasma courtesy of Dr. K. Itiaba.)

acid compositions determined (21). The data obtained have permitted the assessment of the triglyceride distribution in butterfat and an assignment of structures. All bovine milk samples show a characteristic two-hump distribution in their glyceride populations. The ratios of the long- and the short-chain glycerides making up the two humps, however, undergo considerable variation from season to season and also show some differences according to the breed of cattle. The glyceride elution patterns recorded for the triglycerides of the milk fat of Holstein and Jersey breeds of dairy cattle are compared in Fig. 34. The Jersey milk fat contains proportionally more of the shorter-chain glycerides. Apparently, the higher fat content of this milk is largely due to a greater capacity for fatty acid synthesis by Jersey mammary gland tissue. The milk fat of the Holstein cow was obtained soon after parturition and is more like that of colostrum (*39*). As the milk production progresses, the proportion of the triglycerides forming the first hump of the pattern (short-chain glycerides) increases and approaches that of mature milk [Fig. 21(b)].

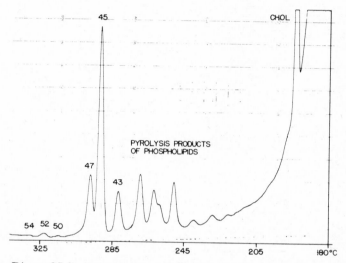

FIG. 31. Direct GLC of the lipids of fasting plasma of dog. Peaks identified as in Fig. 30. Temperature program as shown. Instrument and other operating conditions as in Fig. 12. Sample, 3 µl of fasting plasma.

Preliminary segregation of the milk fat triglycerides on thin-layer plates impregnated with silver nitrate results in a quantitative estimation of some 80 different triglyceride groups. Detailed calculations, based upon these separations, have shown (*70*) that the glyceride distribution of milk fat is highly specific, although randomness may be seen in pooled samples at the more superficial levels of examination.

The triglyceride elution patterns recorded for sheep and goat milk are shown in Fig. 35. Both of these triglyceride mixtures contain significant amounts of short-chain fatty acids and, in common with bovine milk, show the peculiar two-hump distribution. These glyceride mixtures appear to contain less of the odd-carbon-number glycerides

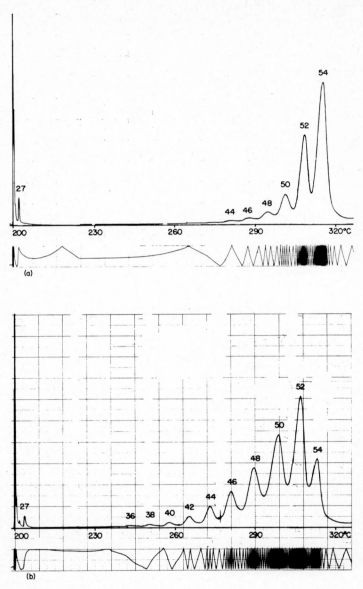

FIG. 32. GLC of dog lymph triglycerides. (a) During corn oil absorption; (b) during butterfat absorption. Peaks identified as in Fig. 1. Column, 18 in. × ⅛ in. o.d. stainless-steel tube packed with 2.25%, w/w, SE-30 on unsilanized Chromosorb W (60–80 mesh). Temperature program as shown. Instrument and other operating conditions as in Fig. 3.

than do those of bovine milk (Fig. 34) and consequently give better peak resolution on the GLC. Mare's milk (*65*) shows a particularly unusual distribution of triglycerides. In contrast with the sheep, cow, and goat, the mare produces milk that is especially rich in the glycerides of carbon number C_{40} to C_{46}, a range where the other ruminants show a minimum in their glyceride distribution. Further peculiarities are observed in the distribution of the glycerides in the milk of the

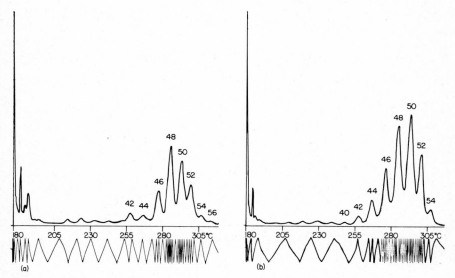

FIG. 33. GLC of high-melting triglycerides of chylomicron membranes. (a) From corn oil feeding; (b) from butterfat feeding. Peaks identified as in Fig. 1. Temperature programs as shown. Instrument and other operating conditions as in Fig. 1.

guinea pig and the dog (*47*). Both these milks are low in the short-chain fatty acids and contain mostly the long-chain glycerides resembling depot fat. Human milk is also relatively poor in the short-chain fatty acids and shows a milk triglyceride distribution that is greatly different from that of the cow, sheep, and mare. The triglyceride distributions of mother's milk 3 days and 2 weeks after parturition are compared in Fig. 36. As noted for the cow milk, the colostrum sample is richer in the longer-chain triglycerides than the milk produced on sustained nursing. The triglyceride distributions of the glyceride groups recovered from silver nitrate TLC are most interesting and provide new insight into the formation of milk fat triglycerides and the mechanism of milk secretion itself.

The milk fat globules appear to be surrounded by a lipoprotein envelope that contains a group of high-melting triglycerides of charac-

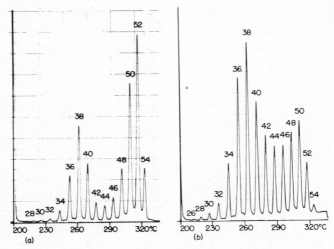

FIG. 34. GLC of milk fat triglycerides. (a) Holstein cow; (b) Jersey cow. Peaks identified as in Fig. 21. Temperature programs as shown. Instrument and other operating conditions as in Fig. 21.

teristic composition. The elution patterns observed for the glycerides of bovine milk globule fat and for the same milk fat globule membrane (*69*) are compared in Fig. 37. Although the chromatograms of the high-melting triglycerides of the globule membrane are somewhat similar to those of the globule fat triglycerides, the milk serum triglycerides show distribution patterns more comparable with those observed for bovine blood plasma triglycerides.

4. OTHER LIPIDS

Direct GLC of glycerides has also proved useful in the analysis of various industrial lipids, such as food fats, margarines, shortenings, and cooking oils. Many of these have been prepared from natural oils and fats by hydrogenation, interesterification, or other modifying and

refining processes. The recordings obtained for two margarines available on the Canadian market are shown in Fig. 38. The characteristic
triglyceride patterns indicate the origin of the oils by revealing the
vegetable or fish oil bases of these modified fats. Since the physical
properties of these fats depend upon the glyceride types and proportions, the analysis of the intact triglycerides is much more informative than a fatty acid determination. Interesterified fats, which possess a
more random placement of fatty acids, give GLC runs that differ from

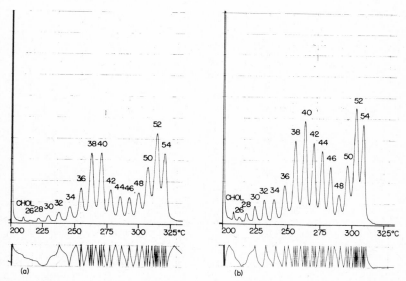

FIG. 35. GLC of milk fat triglycerides. (a) Sheep milk; (b) goat milk. Peaks
identified as in Fig. 21. Temperature programs as shown. Instrument and other
operating conditions as in Fig. 21.

those of the native oils. Changes in glyceride distribution following
this type of rearrangement have been recorded (*71*) in the past for
butterfat and coconut oil samples. Extensive GLC analyses have also
been made with molecular distillates of butteroil (*72*) and corn oil
(*53*).

Another application of direct GLC analysis of triglycerides has
been made in the detection of adulteration of certain premium food
fats by fats of lower food value. It has been shown (*73*) that GLC of

unknown butterfat samples can detect 5–10% of added lard or vegetable fat. The ease of detecting the adulterants varies with the type
of fat added. Thus mixtures of lard and coconut oil that closely
matched the gas chromatograms of butterfat could be made. Leegwater and van Gend (*9*) have described a GLC method for the
detection of coconut oil or palmkernel oil interesterified with other
fats. The concentration of C_{40}–C_{46} triglycerides increased with respect

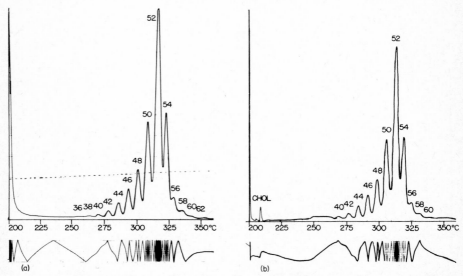

FIG. 36. GLC of milk fat triglycerides of man. (a) Colostrum; (b) mature milk.
Peaks identified as in Fig. 21. Temperature programs as shown. Instrument and
other operating conditions as in Fig. 21.

to these oils and was taken as the criterion of adulteration. The triglyceride GLC was carried out isothermally at 250°C. As little as 1%
adulterant can be detected if the GLC pattern of the original fat is
known and the added material is confined to a few peaks only (*73*).
Then distortion in triglyceride proportions is clearly visible, whereas
the fatty acid composition of the mixture deviates only slightly from
that of the unadulterated sample.

Direct analysis of glycerides is also applicable to the glyceride
mixtures recovered from partial hydrolysates of triglycerides by
lipases. In conventional determinations of fatty acid distributions

within the glyceride molecule, the fatty acids liberated by lipase are analysed and compared with the acids present in the monoglyceride fraction. In Fig. 39 is shown the glyceride pattern recorded for corn oil triglycerides at a selected stage of lipase hydrolysis and for a model mixture. As the enzymatic hydrolysis proceeds, more of the triglyceride is converted into the monoglyceride through an intermediate diglyceride stage. The distribution of the peaks among the di- and triglycerides is about the same, suggesting that the fatty acid

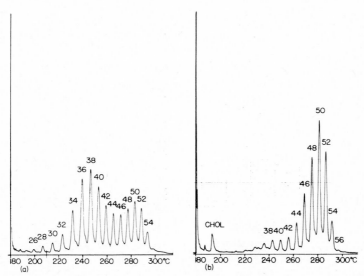

FIG. 37. GLC of triglycerides of bovine milk. (a) Globule fat; (b) high-melting glycerides of globule membrane. Peaks identified as in Fig. 21. Temperature programs as shown. Instrument and other operating conditions as in Fig. 1.

chain length distribution is similar within the two ester classes. The monoglycerides, however, are made up solely from the C_{18} fatty acid esters. Further insight into the mechanics of the enzyme hydrolysis and into a possible preferential release of fatty acids from certain triglyceride types may be obtained if the partial glycerides are first resolved on the basis of unsaturation on the silver-nitrate-impregnated silica gel plates. The mono- and diglycerides may be analyzed either in the free form or following acetylation, trifluoroacetylation, or silylation.

The monoglyceride mixtures isolated on partial lipase hydrolysis of natural fats have been effectively analyzed by Wood et al. (24) following the conversion of the monoglycerides into the silyl ethers. These derivatives can be readily resolved, on the basis of molecular weight and unsaturation, on polyester columns. Some difficulty arises, however, from the presence of the 1- and 2-monoglyceride derivatives that also are resolved under these conditions. Feldman and Feldman

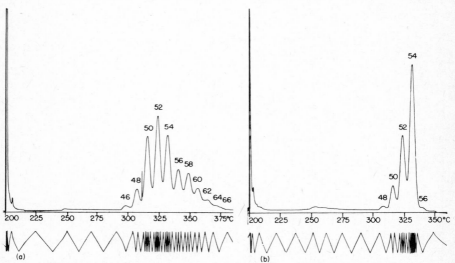

FIG. 38. GLC of triglycerides of some common Canadian margarines. (a) A partially hydrogenated fish oil margarine; (b) a partially hydrogenated vegetable oil margarine. Peaks identified as in Fig. 1. Temperature programs as shown. Instrument and other operating conditions as in Fig. 21.

(74) have reported the occurrence of short-chain triglycerides in human and cattle lenses following a GLC examination of the glyceride fraction. Although no specific short-chain glycerides were identified, their presence was suggested by the finding of appreciable quantities of fatty acids of fewer than 12 carbon atoms in this tissue.

GLC analysis of glycerides has also been performed on the unabsorbed lipids eliminated in the feces. Under normal conditions very little fat is excreted in the form of triglyceride because the lipases of the intestine hydrolyze most of it completely to free fatty acids. When diarrhea or dietary overloading occur, however, large quantities of undigested triglyceride may be found in the feces. Even under normal conditions, certain enzyme-resistant glycerides may accumulate in the feces. Thus characteristic GLC patterns have been recorded (*63*) for the triglycerides recovered from the feces of healthy young adults on hydrogenated corn oil and butterfat diets. The hydrogenation of the corn oil apparently resulted in impaired absorption of the long-chain glycerides. Similarly, the unabsorbed triglycerides from the butterfat ingestion were also primarily of the saturated long-chain type. In addition, there are relatively large amounts of triglycerides containing

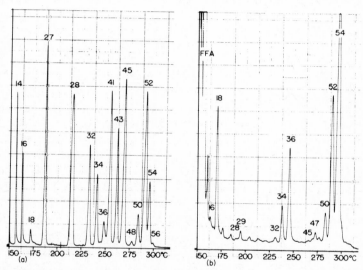

FIG. 39. GLC of corn oil glycerides during lipase hydrolysis. (a) Standard mixture; (b) 5 min after adding enzyme. Peaks 14 to 18, monoglycerides; 27 to 29, sterols; 28 to 36, diglycerides; 41 to 47, steryl esters; 48 to 56, triglycerides.

odd-carbon-number and branched-chain fatty acids in both fecal lipid samples, which may represent a bacterial contribution or a simple enrichment from the food fats.

Other triglyceride analyses have been carried out on the fats of insects. Yates et al. (*62*) have separated the triglycerides from the neutral fat fraction of four different types of boll weevils. Significant differences were observed in the patterns of the triglycerides obtained during reproducing and nonreproducing phases, between sexes, and between different species. The small amounts of sample required make this technique particularly apt for the study of insect fat, which has been neglected by the older methods.

B. Determination of Glyceride Structure

The structural analysis of glycerides has for many years been an intriguing but somewhat frustrating problem. Only TLC on silver-nitrate-impregnated plates and GLC have been capable of effective resolution of the large number of similar glycerides found in natural fats. Despite this, there still remains a large area of structural analysis where statistical speculation must be substituted for experimental measurement. The advantages of the combined approach utilizing both TLC on silver nitrate plates and GLC have been pointed out by McCarthy and Kuksis (77), who showed that this system was particularly fruitful in the analysis of the triglyceride structures of the common vegetable seed oils. The method can be extended to lard and other fats, all of which give characteristic GLC patterns following segregation on TLC plates on the basis of unsaturation (63,65,70). The glyceride bands recovered from the TLC plates may be subjected to specific lipase hydrolysis, and the hydrolysis products may be analyzed by GLC. An evaluation of the pooled data from the different analyses can yield more information than has thus far been obtained by any other methods or their combinations. Furthermore, the glyceride bands recovered from the TLC plates may also be subjected to permanganate–periodate oxidation, and the oxidation products may be diazomethylated and analyzed further by GLC. In addition, the methylated oxidation products may be digested by lipase and the digestion products identified. Each of these steps has been demonstrated to be practical, and the methods have become routine in several laboratories. Thus far, however, no natural fat has been analyzed by a combination of all of them.

A combination of permanganate–periodate oxidation and gas chromatography of glycerides has been used by Youngs and Subbaram (22) for the analysis of several natural fats, but in this case the total fat was oxidized and only the oxidation products were resolved on adsorption chromatography (silica columns) prior to GLC. An identical procedure was used by McCarthy and Kuksis (75) to investigate the triglyceride structure of lard. Oxidation and diazomethylation of triglycerides of uniform degree of unsaturation, which can be recovered from TLC on silver nitrate plates, followed by GLC is much more informative. Oxidation of total glyceride mixtures cannot distinguish between mono- and polyunsaturated acids, as all unsaturated acids yield the same azelaic acid residues. Nevertheless, a large number of vegetable and animal fats have been analyzed (*78*) by this method, and the results obtained have been compared with those derived from lipolysis of the fractionated azelaoglycerides. The results generally agreed with those predicted by lipolysis with the exception of lard, human fat, and bitter gourd seed fat (*Momordica charantia*). The latter contained approximately twice as much (80%) stearodiunsaturated glyceride as would have been calculated from lipase hydrolysis data (45%).

In all cases the value of the data could have been greatly enhanced by performing a silver nitrate–silica gel fractionation on the total fat prior to the oxidation step. Subbaram and Youngs (*79*) fractionated glycerides according to their degree of unsaturation by means of silica columns impregnated with silver nitrate and determined the composition of each fraction by GLC of the oxidized glycerides. All the chemically different glycerides (with the exception of positional isomers) of the common saturated and unsaturated fatty acids could be determined. Analyses of lard and coconut butter allowed the detection of 24 and 18 glycerides, respectively. Since TLC on silver-nitrate-impregnated silica gel permits more complete resolution, the glyceride fractions obtained by this means are better suited for oxidation and subsequent GLC analysis. Additional information regarding the constitution of the glycerides can be obtained by collecting the oxidized derivatives by preparative gas chromatography and determining the composition and positional disposition of the fatty acids for each group of molecular weights.

Analyses of natural fats by consecutive application of chromatographic techniques based on thin-layer and gas chromatography have also been performed by Litchfield et al. (*59*), Jurriens and Kroessen (*18*), and Culp et al. (*60*). Although no oxidations or collections of the triglyceride peaks separated by GLC were made, the application of the combined method to unmodified glycerides yielded much of the desired information.

The combined TLC–GLC approach has been most extensively utilized (*70*) to date for the analysis of butterfat. The relatively low molecular weight of a large part of the fat has permitted the collection of several of the triglyceride peaks from the gas chromatograph. Furthermore, this fat can be effectively segregated on silica gel alone to yield short-, medium-, and long-chain triglyceride classes. Molecular distillation yields comparable segregation (*72*). Each of the glyceride groups so obtained can be effectively resolved on the basis of unsaturation on silver-nitrate-treated silica gel plates. The elution patterns recorded for some of the triglycerides recovered from silver

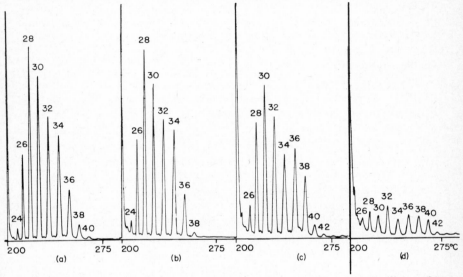

FIG. 40. GLC of a molecular distillate of butteroil. (a) Original distillate; (b) saturates; (c) monoenes; (d) dienes. Peaks identified as in Fig. 1. Temperature programs as shown. Instrument and other operating conditions as in Fig. 21.

nitrate TLC of the most volatile molecular distillate of butteroil (*47*) are represented in the combination print in Fig. 40. The information obtained by this means is effectively supplemented by the data obtained from the fatty acid analysis of the peaks collected from preparative triglyceride GLC. The results of such studies have agreed with those of Blank and Privett (*76*) in which silver nitrate TLC was combined with reductive ozonolysis. These studies have shown that butterfat is a nonrandom mixture of triglycerides and that the non-randomness, or specificity, increases as the resolution progresses. Further improvements in the GLC method of fractionation of butter-fat triglycerides have been described by Kuksis and Breckenridge

(*21*), who presented evidence for a fractionation of certain triglyceride types of this fat beyond that based on their carbon numbers. For a complete exploitation of these separations for analytical purposes, more efficient columns than those presently available are necessary.

In most instances, the results of the combined TLC–GLC separation techniques have been interpreted to confirm the predicted 1,3-random-2-random placement of fatty acids in the glyceride molecules. The significance of such observations, however, remains in doubt in view of the superficial nature of many of the resolutions actually realized. Had more elaborate experimental methods been applied, perhaps discrepancies would also have been revealed in the analyses of the more common vegetable oils. It has been shown for the more extensively investigated fats that the over-all distribution of fatty acids is an average of many individual nonrandom distributions. Furthermore, many of the triglyceride samples have been of rather poorly defined origin and have represented pooled fat, which may account for the apparent random nature. It would, therefore, appear unwise to conclude that any fat that shows a random or partially random triglyceride distribution on the basis of a statistical examination of superficial experimental data necessarily possesses such a distribution of glycerides in reality.

The determination of the structure of partial glycerides by GLC is considerably simpler than triglyceride analysis. Separation of the α- and β-monoglycerides of the various chain lengths can be readily accomplished following either acetylation or silylation. Although some overlap occurs (see Section V.E.1), unless the saturated and unsaturated derivatives are first resolved on silver nitrate plates, it may be avoided by selecting more specific liquid phases and by increasing the column efficiency. The determination of diglyceride structure by GLC is considered together with methods for the investigation of the structure of phosphatides.

C. Determination of Phosphatide Structure

The determination of phosphatide structure by means of GLC is based upon the resolution of the diglyceride parts of these complex lipids (*66*). Specific enzymes should be used to release the diglycerides, which can then be determined by GLC as such or following the preparation of suitable derivatives. In Fig. 41 are shown the GLC separations obtained for the diglycerides of rat liver lecithin prepared by phospholipase C hydrolysis. In one case the diglycerides were resolved as such; in the other case they were first acetylated. Prior resolution of the acetylated diglycerides on the basis of unsaturation by means of silver nitrate TLC produces triglyceride groups associated with each of the TLC bands. In Fig. 42 are shown the elution patterns recorded for some of the diglyceride acetate fractions obtained from egg yolk lecithins following silver nitrate TLC. A combination of these analyses

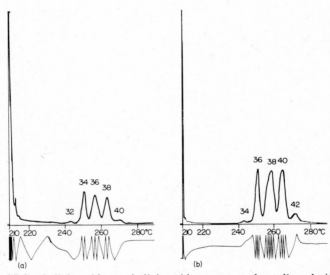

FIG. 41. GLC of diglycerides and diglyceride acetates of rat liver lecithins. (a) Diglycerides; (b) diglyceride acetates plus dimyristin, peak 28. Peaks identified as in Fig. 1. Temperature programs as shown. Instrument and other operating conditions as in Fig. 1.

with specific lipase hydrolyses of the diglyceride acetates and appropriate determination of the fatty acid composition produces sufficient information for an essentially complete description of the structure of such phosphatides as lecithins, phosphatidyl ethanolamines, phosphatidyl serines, and phosphatidyl inositides.

The diglyceride acetates of the common phosphatides may also be prepared by acetolysis as originally described by Bevan et al. (*80*) and subsequently modified by Renkonen (*81*) (see Chapter 2), but some isomerization takes place (*82,83*). The molecular weights of the diglyceride acetates are low enough for an effective collection of individual peaks by preparative GLC and an identification and quantitative estimation of the component fatty acids. The diglyceride acetates of the collected peaks may also be subjected to specific lipase hydrolysis, in which case a complete assignment of the structures of the component lecithins may be made. The method is applicable to other diglyceride phosphatides, but no extensive analyses have been made.

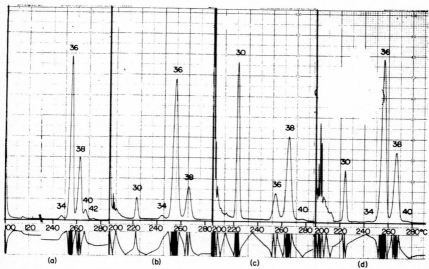

FIG. 42. GLC of diglyceride acetates of egg yolk lecithins. (a) Total diglycerides; (b) monoenes; (c) dienes containing two monounsaturated fatty acids; (d) dienes containing one diunsaturated fatty acid. Peaks identified as in Fig. 1. Temperature programs as shown. Instrument and other operating conditions as in Fig. 1.

There are other as yet incompletely explored approaches to a determination of the phosphatide structure by GLC. An examination of the products of heat fragmentation appears to be particularly promising. It has been shown by Kuksis and Marai (*26*) that lecithins undergo pyrolysis in the flash evaporator of the GLC apparatus at relatively low temperatures (280–300°C) and are converted, in a 60–70% yield, to isomeric propenediol diesters that migrate as symmetrical peaks in the GLC columns. Experiments with standard dimyristoyl and distearoyl lecithins have shown that each lecithin molecule yields two peaks the proportions of which vary somewhat with the exact conditions of pyrolysis. The isomers are resolved with a retention time difference of approximately two carbon atoms. For this reason the peaks produced on pyrolysis of mixed lecithins partially overlap. The overlap is such that the mixed lecithins of human plasma, for example, which contain significant amounts of C_{16}, C_{18}, and C_{20} fatty acids, show only three major peaks when their pyrolysis products are separated on GLC. Although the exact mechanism of the formation of these diglyceride derivatives has not been established, the potential of the method for the study of the structure of phosphatides and for their determination has been recognized.

D. Study of the Physical Properties of Glycerides

Although GLC of glycerides is now widely used for analysis, no recorded attempts have been made to appraise the potential of the system for physical measurements. Studies of column performance apparently can yield information about solubility, adsorption, diffusion, and thermodynamic quantities. Furthermore, specialized techniques have been suggested for the measurement of solid surface areas, boiling points, and vapor pressures, which may permit the determination of the structure of the molecules. The most attractive feature of this technique is the relative simplicity of the measurement and the extreme rapidity with which high-quality information may be obtained. An additional advantage is the possibility of performing many of the measurements on mixed samples, thus avoiding the need of absolutely pure starting materials. This is of particular importance in work with lipids, which are difficult to purify. Purnell (*84*) has summarized a number of the more obvious applications of GLC for the study of the physical properties of low-molecular-weight compounds. Many of the methods, however, are also applicable to the determination of the physical properties of glycerides.

IX. Summary and Conclusions

The development of techniques for direct gas chromatographic analysis of natural fat triglycerides has been a major breakthrough in the study of fats and oils. This technique, together with the gas chromatography of the fatty acids and the chromatography of glycerides on thin layers of silica gel impregnated with silver nitrate, now provides analytical means the power of which the lipid chemist has never before known. Although only a relatively small number of natural fats have been analyzed by the combined TLC–GLC system and further improvements in instrumentation and technology must be made, sufficient evidence has been obtained to assure general applicability of the method. Furthermore, the combined system is flexible enough to permit a rapid separation and quantification of a variety of chemical and enzymatic modification products, which have proved valuable in the determination of the constitution of natural fats by other methods.

Despite these advances, there still is considerable reluctance to be committed to an intensive application of the new techniques to the study of natural fats, because it is believed that it may yet be possible to obtain a fairly accurate estimate of the glyceride composition of a fat from a knowledge of its fatty acid composition and simple mathematical manipulations to suit a random or a partial random distribution. It does not appear to be realized that these methods of calculation have never been submitted to a critical experimental examination and that to establish their validity would require an analytical technique of the type provided by the combined TLC–GLC approach. The need for an experimental examination of the glyceride composition of all natural fats is particularly urgent because of the observation that the fats that have been analyzed in detail by the combined approach have failed to confirm the assumptions involved in the 1,3-random, 2-random theory, thus invalidating the calculations of the triglyceride structure deduced from enzyme hydrolysis results. Moreover, no evidence has been obtained for a common pattern of fatty acid distribution in vegetable fats in general. It may therefore be impossible to calculate accurately the triglyceride composition of any fat simply from a knowledge of fatty acid composition.

Before undertaking any detailed analyses of natural fat triglycerides, it should be recognized that the source of the fat should be completely defined and reproducibly sampled. There is little biochemical or physiological interest in collecting detailed data on glyceride mixtures pooled from different tissues, cells, or even different subcellular organelles. The discovery of a random distribution of fatty acids in the glycerides of a refined vegetable oil or in a commercially rendered pool of animal fats has no biological significance, and any inferences regarding any enzymatic specificity or lack of it during the synthesis of these fats are unwarranted. For maximum reliability and reproducibility, the lipids should be isolated from specific subcellular components and, if possible, from individual lipoproteins.

Analytical techniques based on the use of gas chromatography and thin-layer chromatography permit analyses with tissue samples of the size conveniently prepared in analytical centrifuges and by established methods of lipoprotein fractionation. A complete analysis of the lipids from well-defined sources appears mandatory if the design and purpose of the extreme complexity of natural lipid mixtures is to be understood.

For the GLC analysis of neutral glycerides, columns of higher efficiency are still required. With further improvements in column technology and methods of chromatography it should be possible to use columns considerably longer than 2 ft without serious impairment of sample recovery. Better resolution of glycerides within a selected carbon-number range may be obtained by high-temperature liquid phases of greater selectivity than the common silicone gums or by use of GLC columns providing a greater number of theoretical plates per foot of column length. The eventual separation of saturated and unsaturated glycerides of the same carbon number appears certain, and there is reason to believe that it will also be possible to resolve the positional isomers of most triglycerides by GLC. The latter accomplishment should give a more satisfactory assessment of triglyceride structure by eliminating the uncertainties of enzymatic positional analysis of glycerides.

The advances made in the separation of triglyceride mixtures are of practical interest in the determination of the complete structure of natural diglyceride phosphatides. The most successful analytical approaches to date have been those that depend on a conversion of the phosphatides into a triglyceride form by enzymatic dephosphorylation and acetylation of the resulting diglycerides. The diglyceride acetates may then be subjected to the entire array of techniques of triglyceride analyses including selective argentation and gas chromatography, and the whole spectrum of molecular species of phosphatides may then be obtained.

Acknowledgments

The author wishes to thank Dr. J. Gilbert Hill for constructive criticism of the manuscript.

The investigations by the author and his collaborators reported herein were supported by the Medical Research Council of Canada, the Ontario Heart Foundation, the Charles H. Best Foundation, the Eli Lilly Comany, Indianapolis, Indiana, and the Special Dairy Industry Board, Chicago, Illinois.

Appreciation is expressed to Miss Helen Christie for proofreading the copy and to Miss Patricia Arnold for accurate typing of it.

REFERENCES

1. A. Kuksis and M. J. McCarthy, *Can. J. Biochem. Physiol.*, **40,** 679 (1962).
2. F. H. Fryer, W. L. Ormand, and G. B. Crump, *J. Am. Oil Chemists' Soc.*, **37,** 589 (1960).
3. A. J. Martin, C. E. Bennett, and F. W. Martinez, Jr., in *Gas Chromatography 1960* (R. P. W. Scott, ed.), Butterworth, London, 1960.
4. N. Pelick, W. R. Supina, and A. Rose, *J. Am. Oil Chemists' Soc.*, **38,** 506 (1961).
5. V. R. Huebner, *J. Am. Oil Chemists' Soc.*, **38,** 628 (1961).
6. Anon, in *Gas-Chromatography Newsletter,* Vol. 2, No. 4, Applied Science Laboratories, State College, Penn., Sept. 1961.
7. A. Kuksis, *J. Am. Oil Chemists' Soc.*, **42,** 269 (1965).
8. J. M. Schlater, L. Mikkelsen, and M. G. Beck, in *Lectures in Gas Chromatography 1962* (H. A. Szymanski, ed.), Plenum Press, New York, 1963, p. 105.
9. D. C. Leegwater and H. W. van Gend, *Fette Seifen Anstrichmittel,* **67,** 1 (1965).
10. A. J. Martin, C. E. Bennett, and F. W. Martinez, Jr., in *Gas Chromatography* (H. J. Noebels, R. F. Wall, and N. Brenner, eds.), Academic Press, New York, 1961, p. 363.
11. H. G. Boettger, in *Lectures in Gas Chromatography 1962* (H. A. Szymanski, ed.), Plenum Press, New York, 1963. p. 133.
12. Committee recommendations, *Vapour Phase Chromatog. Proc. Symp. London, 1956,* **1957,** xi.
13. H. W. Habgood and W. E. Harris, *Anal. Chem.*, **32,** 450 (1960).
14. L. S. Ettre, *J. Gas Chromatog.*, **1**(2), 36 (1963).
15. C. Litchfield, R. D. Harlow, and R. Reiser, *J. Am. Oil. Chemists' Soc.*, **42,** 849 (1965).
16. A. B. Littlewood, in *Third International Gas Chromatography Symposium* (N. Brenner, J. E. Callen, and M. D. Weiss, eds.), Academic Press, New York, 1962, p. 141.
17. S. Dal Nogare and R. S. Juvet, Jr., *Gas-Liquid Chromatography,* Wiley (Interscience), 1962, p. 70.
18. G. Jurriens and A. C. J. Kroessen, *J. Am. Oil Chemists' Soc.*, **42,** 9 (1965).
19. J. W. Farqhuar, W. Insull, Jr., P. Rosen, W. Stoffel, and E. H. Ahrens, Jr., *Nutr. Rev. Suppl.*, **17,** 1 (1959).
20. V. R. Huebner, *J. Am. Oil Chemists' Soc.*, **36,** 262 (1959).
21. A. Kuksis and W. C. Breckenridge, *J. Am. Oil Chemists' Soc.*, **42,** 978 (1965).
22. C. G. Youngs and M. R. Subbaram, *J. Am. Oil Chemists' Soc.*, **41,** 218 (1964).
23. F. H. Mattson and J. B. Martin, *J. Lipid Res.*, **5,** 374 (1964).

24. R. D. Wood, P. K. Raju, and R. Reiser, *J. Am. Oil Chemists' Soc.*, **42**, 161 (1965).
25. A. G. McInnes, N. H. Tattrie, and M. Kates, *J. Am. Oil Chemists' Soc.*, **37**, 7 (1960).
26. A. Kuksis and L. Marai, submitted for publication.
27. D. C. Malins, J. C. Wekell, and C. R. Houle, *J. Lipid Res.*, **6**, 100 (1965).
28. D. J. Hanahan, J. Ekholm, and C. M. Jackson, *Biochemistry*, **2**, 630 (1963).
29. E. von Rudloff, *Can. J. Chem.*, **34**, 1413 (1956).
30. O. S. Privett and E. C. Nickell, *J. Am. Oil Chemists' Soc.*, **39**, 414 (1962).
31. O. S. Privett and E. C. Nickell, *Lipids*, **1**, 98 (1966).
32. R. D. Harlow, C. Litchfield, and R. Reiser, *Lipids*, **1**, 216 (1966).
33. A. T. James, in *Methods of Biochemical Analysis,* Vol. 8 (D. Glick, ed.), Wiley (Interscience), New York, 1960, p. 50.
34. C. B. Barrett, M. S. J. Dallas, and F. B. Padley, *J. Am. Oil Chemists' Soc.*, **40**, 580 (1963).
35. F. D. Gunstone and F. B. Padley, *J. Am. Oil Chemists' Soc.*, **42**, 957 (1965).
36. O. S. Privett, M. L. Blank, D. W. Codding, and E. C. Nickell, *J. Am. Oil Chemists' Soc.*, **42**, 381 (1965).
37. B. De Vries and G. Jurriens, *J. Chromatog.*, **14**, 525 (1964).
38. L. Swell, *Proc. Soc. Exptl. Biol. Med.*, **121**, 1290 (1966).
39. A. Kuksis and W. C. Breckenridge, *J. Lipid Res.*, **7**, 576 (1966).
40. N. S. Radin, *J. Chromatog.*, **20**, 392 (1965).
41. E. C. Horning, W. J. A. VandenHeuvel, and B. G. Creech, in *Methods of Biochemical Analysis,* Vol. 11 (D. Glick, ed.), Wiley (Interscience), New York, 1963, p. 69.
42. F. A. VandenHeuvel, G. J. Hinderks, and J. C. Nixon, *J. Am. Oil Chemists' Soc.*, **42**, 283 (1965).
43. G. L. Feldman and J. F. R. Kuck, Jr., *Lipids*, **1**, 158 (1966).
44. C. Chen and O. Gaeke, *Anal. Chem.*, **36**, 72 (1964).
45. J. A. Schmit and A. Mather, personal communication, 1964.
46. J. Bohemen and J. H. Purnell, in *Gas Chromatography 1958,* (D. H. Desty, ed.), Academic Press, New York, 1958, p. 33.
47. W. C. Breckenridge and A. Kuksis, *J. Lipid Res.*, in press, (1967).
48. J. C. Sowden and H. O. L. Fischer, *J. Am. Chem. Soc.*, **63**, 3244 (1941).
49. A. F. Hofmann, in *New Biochemical Separations* (A. T. James and L. J. Morris, eds.), Van Nostrand, Princeton, N.J., 1964, p. 283.
50. E. J. Bourne, M. Stacey, J. C. Tatlow, and J. M. Teder, *J. Chem. Soc.*, **1949**, 2976.
51. A. Kuksis and J. Ludwig, *Lipids*, **1**, 202 (1966).
52. A. Kuksis, *Can. J. Biochem.*, **42**, 407 (1964).
53. A. Kuksis, *Can. J. Biochem.*, **42**, 419 (1964).
54. L. S. Ettre and F. J. Kabot, *J. Chromatog.*, **11**, 114 (1963).
55. J. L. Moore, T. Richardson, and C. H. Amundson, *J. Gas Chromatog.*, **2**, 318 (1964).
56. R. G. Ackman and J. C. Sipos, *J. Am. Oil Chemists' Soc.*, **41**, 377 (1964).
57. A. Kuksis, in *Methods of Biochemical Analysis,* Vol. 14 (D. Glick, ed.), Wiley (Interscience), New York, 1966, p. 325.

58. A. Kuksis, M. J. McCarthy, and J. M. R. Beveridge, *J. Am. Oil Chemists' Soc.,* **40,** 530 (1963).

59. C. Litchfield, M. Farqhuar, and R. Reiser, *J. Am. Oil Chemists' Soc.,* **41,** 588 (1964).

60. T. W. Culp, R. D. Harlow, C. Litchfield, and R. Reiser, *J. Am. Oil Chemists' Soc.,* **42,** 974 (1965).

61. C. Litchfield, E. Miller, R. D. Harlow, and R. Reiser, *J. Am. Oil Chemists' Soc.,* **43,** 102A (1966).

62. M. L. Yates, H. L. Conrad, and J. S. Forrester, *Bulletin AB-509,* Micro Tek Instruments, Inc., Baton Rouge, La., 1965.

63. A. Kuksis, unpublished observations, 1966.

64. R. D. Harlow, C. Litchfield, H. C. Fu, and R. Reiser, *J. Am. Oil Chemists' Soc.,* **42,** 747 (1965).

65. A. Kuksis, in *Proceedings of the American Meat Science Association,* 19th annual meeting held at Cornell Univ., Ithaca, N.Y., June 1966.

66. A. Kuksis and L. Marai, in *Abstracts of Papers of the 40th Fall Meeting of the American Oil Chemists' Society, Philadelphia, Pa., 1966,* p. 42; *Lipids,* in press.

67. A. Kuksis and T. C. Huang, *Can. J. Physiol. Pharmacol.,* in press.

68. T. C. Huang and A. Kuksis, *J. Am. Oil Chemists' Soc.,* **42,** 148A (1965).

69. T. C. Huang and A. Kuksis, in *Abstracts of the 47th Canadian Chemical Conference, Kingston, Ont.,* June, 1964.

70. W. C. Breckenridge, *A Comparative Study of the Triglyceride Structure of Bovine and Human Milk Fat,* M.Sc. thesis. Univ., Toronto, 1966.

71. A. Kuksis, M. J. McCarthy, and J. M. R. Beveridge, *J. Am. Oil Chemists' Soc.,* **41,** 201 (1964).

72. M. J. McCarthy, A. Kuksis, and J. M. R. Beveridge, *Can. J. Biochem. Physiol.,* **40,** 1693 (1962).

73. A. Kuksis and M. J. McCarthy, *J. Am. Oil Chemists' Soc.,* **41,** 17 (1964).

74. G. L. Feldman and L. S. Feldman, *Lipids,* **1,** 86 (1966).

75. M. J. McCarthy and A. Kuksis, *Proc. Can. Fed. Biol. Soc.,* **6,** 42 (1963).

76. M. L. Blank and O. S. Privett, *J. Dairy Sci.,* **47,** 481 (1964).

77. M. J. McCarthy and A. Kuksis, *J. Am. Oil Chemists' Soc.,* **41,** 527 (1964).

78. M. R. Subbaram, M. M. Chakrabarty, C. G. Youngs, and B. M. Craig, *J. Am. Oil Chemists' Soc.,* **41,** 691 (1964).

79. M. R. Subbaram and C. G. Youngs, *J. Am. Oil Chemists' Soc.,* **41,** 445 (1964).

80. T. H. Bevan, R. A. Brown, G. I. Gregory, and T. Malkin, *J. Chem. Soc.,* **1953,** 127.

81. O. Renkonen, *J. Am. Oil Chemists' Soc.,* **42,** 298 (1965).

82. O. Renkonen, *Lipids,* **1,** 160 (1966).

83. L. J. Nutter and O. S. Privett, *Lipids,* **1,** 234 (1966).

84. J. H. Purnell, *Endeavour,* **23,** 142 (1964).

8

ISOLATION AND GAS-LIQUID CHROMATOGRAPHY OF ALKOXY LIPIDS

Helmut K. Mangold and Wolfgang J. Baumann
UNIVERSITY OF MINNESOTA
THE HORMEL INSTITUTE
AUSTIN, MINNESOTA

I. Introduction

Long-chain alkyl ethers and dialkyl ethers of glycerol and, most likely, also of other polyhydric short-chain alcohols, occur widespread in nature. Thus, alkyl glycerol-(1) ethers (**1**) have been found in hydrolyzates of lipid extracts from human (*1–10*), animal (*2,5,7,11–50*), and plant (*51*) tissues. It has been shown that they occur as *O*-alkyl diglycerides (**2**) (*5,6,9,10,14,26–28,37,41,44–46,48–50*), phosphatides (e.g., **3**) (*2,4,23,31,33–35,38–43*), and glycolipids (*36*).

$$
\begin{array}{ccc}
& H_2C\!-\!O\!-\!R & H_2C\!-\!O\!-\!R \\
& HC\!-\!O\!-\!\underset{\underset{O}{\|}}{C}\!-\!R' & HC\!-\!O\!-\!\underset{\underset{O}{\|}}{C}\!-\!R' \\
H_2C\!-\!O\!-\!R & & \\
HC\!-\!OH & & O \\
H_2C\!-\!OH & H_2C\!-\!O\!-\!\underset{\underset{O}{\|}}{C}\!-\!R'' & H_2C\!-\!O\!-\!\underset{\underset{O\ominus}{|}}{\overset{\overset{O}{\uparrow}}{P}}\!-\!O\!-\!\underset{\underset{H\ H}{|}}{\overset{\overset{H\ H}{|}}{C}}\!-\!\underset{}{C}\!-\!\overset{\oplus}{N}H_3 \\
(1) & (2) & (3)
\end{array}
$$

Alkyl glycerol-(2) ethers, i.e., compounds isomeric to (**1**), have not been conclusively identified in natural sources.

The presence of 1,2-dialkyl glycerol ethers (**4**) has been demonstrated in lipid hydrolysates from human (*52*) and animal tissues (*53*) as well as bacteria (*54–57*). It is highly probable that they also occur as *O,O*-dialkyl monoglycerides (**5**), but so far they have been obtained only from phosphatides (e.g., **6**) (*52–57*).

$$
\begin{array}{ccc}
& H_2C\!-\!O\!-\!R & H_2C\!-\!O\!-\!R \\
& HC\!-\!O\!-\!R' & HC\!-\!O\!-\!R' \\
H_2C\!-\!O\!-\!R & & \\
HC\!-\!O\!-\!R' & H_2C\!-\!O\!-\!\underset{\underset{O}{\|}}{C}\!-\!R'' & H_2C\!-\!O\!-\!\underset{\underset{O\ominus}{|}}{\overset{\overset{O}{\uparrow}}{P}}\!-\!O\!-\!\underset{\underset{H\ H}{|}}{\overset{\overset{H\ H}{|}}{C}}\!-\!\underset{}{C}\!-\!\overset{\oplus}{N}H_3 \\
H_2C\!-\!OH & & \\
(4) & (5) & (6)
\end{array}
$$

Evidence for the existence in nature of 1,3-dialkyl glycerol ethers, e.g., compounds isomeric to (**4**), has not been brought forward.

Lipids derived from alkyl glycerol ethers are often accompanied by the corresponding alk-1-enyl-1-glycerol ether derivatives. Thus, *O*-alkenyl-diglycerides ("neutral plasmalogens") (**7**) have been identified in human (*5,6,9,10*) and animal (*5,30,45,46,48,49*) tissues containing *O*-alkyl diglycerides (**2**). It has been shown that *O*-alkyl diglycerides, neutral plasmalogens, and triglycerides can be resolved by adsorption chromatography and that each of these classes of compounds can be recovered for an analysis of its long-chain moieties (*9,10,49*).

Methods for the separation of the corresponding three classes of phosphatides have not been described. The long-chain moieties of alkyl acyl phosphatides (e.g., **6**), alkenyl acyl phosphatides ("plasmalogens") (e.g., **8**), and diacyl phosphatides usually are determined after enzymatic removal of the phosphoryl choline or phosphoryl ethanolamine and acetylation of the remaining alkyl monoglycerides, alkenyl monoglycerides, and diglycerides (*58–60*).

Recently it has been shown that diesters and alk-1-enyl ether esters of 1,2-ethanediol and various other short-chain diols are constituents of many animal and plant tissues as well as microorganisms (e.g., *61*; for additional references see *62*). Probably owing to the procedures employed for the detection of "diol lipids," alkyl ethers of diols have not been found. However, considering the chemical and most probable metabolic relationship between alkyl ethers and alk-1-enyl ethers, the occurrence of alkyl ethers of diols in nature can be expected.

Alkyl ethers and alk-1-enyl ethers differ greatly in their chemical behavior, and therefore it is possible to analyze their derivatives in presence of each. The scheme on p. 342 presents a survey of reactions utilized in the analysis of the long-chain moieties in alkyl diglycerides and neutral plasmalogens (*9*). The same reactions are useful, of course, in work with the acetylated alkyl monoglycerides and alkenyl monoglycerides derived from phosphatides.

This chapter has been written as a guide to well-established and reliable methods applicable to the detection and isolation of various classes of alkoxy lipids, the fractionation of these classes into their

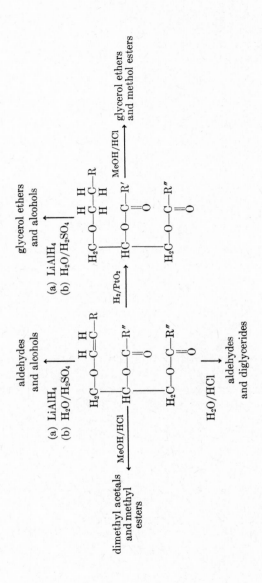

constituents, and the identification of pure compounds. Studies concerned with the distribution of alkoxy lipids in nature, their biosynthesis and catabolism, and their possible physiological functions have been reviewed elsewhere (*63,64*).

II. Reference Material

Synthetic methods are well worked out for the preparation of alkyl glycerol-(1) ethers (*65,66*) and alkyl glycerol-(2) ethers (*65*), dialkyl glycerol-(1,2) ethers (*67,68*), trialkyl glycerol-(1,2,3) ethers (*68*), alkyl diglycerides (*69*), dialkyl glycerides (*69*), as well as alkyl ethers and ether–esters of 1,2-ethanediol and 1,3-propanediol (*62*). A few of these compounds are available commercially.

The liver oils of dogfish (*Squalus acanthias*) and ratfish (*Hydrolagus colliei*) are good sources of natural alkoxy lipids. These oils can be obtained from the U.S. Bureau of Commercial Fisheries Technological Laboratory, 2725 Montlake Boulevard East, Seattle 2, Washington, and from Western Chemical Industries, Ltd., LaPointe Pier, Vancouver, British Columbia, Canada.

III. Detection of Alkoxy Lipids

Neutral alkoxy lipids can be separated from the corresponding ester lipids by adsorption chromatography. In Fig. 1, as an example, is shown the fractionation, by thin-layer chromatography, of representative pure compounds. Trialkyl glycerol ethers, dialkyl glycerides, and alkyl diglycerides are well separated from each other and from triglycerides. The clear resolution of the alkyl diglycerides and triglycerides of ratfish liver oil demonstrates that samples that contain a multiplicity of constituents differing in both chain length and number of double bonds of their aliphatic moieties are well resolved into classes of compounds having the same type and number of functional groups.

A. Analysis of Mixtures Containing Neutral Alkoxy Lipids

The sample and several reference compounds (50–300 μg) are spotted side by side on a plate coated with a 0.25-mm layer of activated silica gel G. The chromatoplate is developed, at room temperature, with petroleum hydrocarbon (b.p. 40–60°C)–diethyl ether–acetic acid 90:10:1, v/v/v, in a tank lined with filter paper (*69*). After 40–50 min, the plate is sprayed with a saturated solution of potassium dichromate in 70% sulfuric acid and the substances are made visible by charring them in an oven at 180°C.

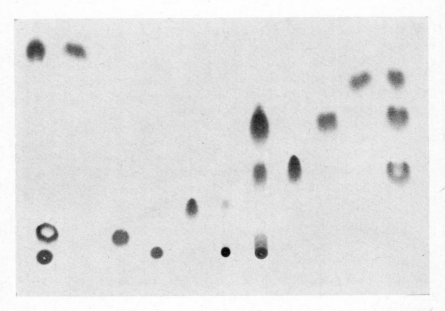

FIG. 1. Thin-layer chromatogram of alkoxy lipids (*69*). (a) Octadecyl glycerol-(1) ether, dioctadecyl glycerol-(1,2) ether, and trioctadecyl glycerol-(1,2,3) ether; (b) trioctadecyl glycerol-(1,2,3) ether; (c) dioctadecyl glycerol-(1,2) ether; (d) octadecyl glycerol-(1) ether; (e) octadecanoic acid; (f) nonsaponifiable lipids from ratfish (*Hydrolagus colliei*) liver oil; (g) ratfish liver oil; (h) tristearin; (i) *O*-octadecyl distearin; (j) *O,O*-dioctadecyl monostearin; (k) tristearin, *O*-octadecyl distearin, and *O,O*-dioctadecyl monostearin. For experimental conditions, see text.

Neutral plasmalogens migrate between alkyl diglycerides and dialkyl glycerides. On thin-layer chromatograms, neutral plasmalogens are recognized as yellow spots after spraying with a saturated solution of 2,4-dinitrophenylhydrazine in ethanol containing 10% conc. sulfuric acid. As free aldehydes yield the same color reaction, it is advisable to augment the identification of neutral plasmalogens by the following method.

B. Detection of Neutral Plasmalogens (*45*)

Between 200 and 500 μg of the sample is spotted in a corner of a plate, 20×20 cm, which is coated with a layer of silica gel G, 0.3 to 0.4 mm in thickness. First, the sample is separated into classes of compounds. The plate is developed twice in a jar lined with filter paper using petroleum hydrocarbon (b.p. 40–60°C)–diethyl ether 95:5, v/v, as the solvent. Thereafter, neutral plasmalogens are cleaved, in situ, by acid fumes. The plate is mounted, layer down, about 15 cm above a dish containing conc. hydrochloric acid which is preheated to 40–50°C. After having been exposed to the acid fumes for 5 min, the reaction products, i.e., free aldehydes and diglycerides, are resolved from each other and from unchanged compounds by chromatography in the second direction. The plate is developed with petroleum hydrocarbon (b.p. 40–60°C)–diethyl ether 80:20, v/v. All lipid fractions are made visible by spraying with a saturated solution of potassium dichromate in 70% sulfuric acid and charring in an oven at 180°C.

Diol lipids cannot be separated from the corresponding glycerol-derived lipids by adsorption chromatography (*61,70*). These compounds are easily separated, however, by reversed-phase partition chromatography (*62*) as well as by gas–liquid chromatography (*50*).

Methods for the complete resolution of alkyl acyl phosphatides, alkenyl acyl phosphatides, and diacyl phosphatides are not available. The acetylated alkyl monoglycerides, alkenyl monoglycerides, and diglycerides that can be derived from these lipids are separated, according to class, by thin-layer chromatography on silica gel G as described in Chapter 2.

Plasmalogens (alkenyl acyl phosphatides) can be visualized on chromatograms by spraying with 2,4-dinitrophenylhydrazine solution (Chapter 2). A chromatographic method for the detection and identification of plasmalogens has been published recently (*71*) (see Chapter 2).

IV. Isolation of Alkoxy Lipids

Alkyl diglycerides and neutral plasmalogens in lipid extracts can be enriched by chromatography on columns of silicic acid (*5,6,30*).

However, concentrates of these two lipid classes are obtained more conveniently by chromatography on "thick" layers of the same adsorbent (*9,48,49*).

For the identification and quantitative analysis of the long-chain moieties in alkyl diglycerides and neutral plasmalogens it is not necessary to isolate each of these two classes of compounds in pure form. Instead, the alkyl glycerol ether and fatty acids in alkyl diglycerides, the aldehydes and fatty acids in neutral plasmalogens, and the fatty acids in triglycerides are determined in the concentrate. The following procedure is applicable to the enrichment of alkoxy lipids from samples containing less than 1% of these compounds, such as human depot fat.

A. Isolation of a Concentrate of Alkyl Diglycerides and Neutral Plasmalogens (*48*)

A 10% solution of 0.75 g lipid in hexane is applied as a band along one edge of a glass plate, 20×20 cm, coated with a layer of silica gel H, 2 mm in thickness. The plate is developed twice, for about 40 min each, with petroleum hydrocarbon (b.p. 40–60°C)–diethyl ether 95:5, v/v, using an unlined tank. The triglyceride fraction is visible in transmitted light without the use of an indicator. A band 2 cm wide, including 2 mm of the upper edge of the triglyceride fraction, is scraped off. Lipids are eluted from the adsorbent with diethyl ether; the slurry is filtered through a jacketed sintered-glass funnel kept at about 30°C. The material isolated is rechromatographed once or twice under the same conditions.

The flow diagram shown in Fig. 2 illustrates the course of analysis of the long-chain moieties in alkyl diglycerides and neutral plasmalogens isolated by the above method. Procedure for the analysis of alkyl glycerol ethers by gas–liquid chromatography are described on pages 349–354, those for aldehydes and fatty acids in Chapters 10 and 9, respectively.

This scheme is applicable also the the analysis of acetylated alkyl monoglycerides and acetylated alkenyl monoglycerides that have been derived from mixtures of phosphatides (see Chapter 2).

Preparative thin-layer chromatography is used for the isolation of pure alkyl diglycerides and pure neutral plasmalogens from concentrates prepared by the above procedure. Lipid extracts containing more than 1–2% of these classes of compounds, such as dogfish and ratfish liver oils, are fractionated without prior enrichment.

B. Isolation of Pure Alkyl Diglycerides and Pure Neutral Plasmalogens (9,48)

A hexane solution of 0.07 g of a lipid mixture containing 2% or more of neutral alkoxy lipids is applied as a band along one edge of a glass plate, 20 × 20 cm, coated with a 0.5-mm layer of silica gel G. The plate is developed twice under the conditions stated in the pre-

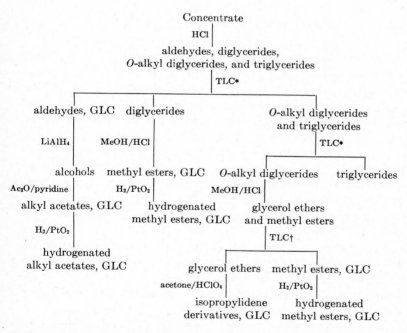

FIG. 2. Scheme for the analysis of *O*-alkyl diglycerides. Neutral plasmalogens and triglycerides in mixtures (*10*). *: adsorbent, silica gel G; solvent, hexane–diethyl ether 95:5, v/v, developed twice in tanks lined with filter paper. †: adsorbent, silica gel G; solvent, hexane–diethyl ether, 90:10 v/v, in tanks lined with filter paper.

ceding procedure; a band 3 cm wide is scraped off, starting *at* the upper edge of the triglyceride fraction; and the lipids are eluted with diethyl ether. The material is rechromatographed until pure.

Methods for the preparation of various classes of phosphatides, including alkenyl acyl phosphatides, are described in Chapter 2.

Alkyl glycerol ethers and dialkyl glycerol ethers are obtained by lithium aluminum hydride reduction (*72*) of total lipid extracts or enriched fractions and subsequent preparative chromatography on layers of silica gel G. Reductive cleavage does not lead to the formation of emulsions. This method is, therefore, more suitable than is saponification, especially for the isolation of alkyl glycerol ethers from lipid mixtures that contain only small amounts of alkoxy lipids (*9,41,49*).

V. Fractionation of Classes of Alkoxy Lipids into Their Constituent Compounds

Adsorption chromatography serves for the isolation of classes of alkoxy lipids having the same basic structure and the same type and number of functional groups per molecule. These classes can be separated further by one or several of the following methods: Argentation chromatography on thin layers of silica gel G impregnated with silver nitrate (see Chapters 2 and 5) resolves each class on the basis of the number and configuration of double bonds in its constituent compounds (*47*). Thin-layer chromatography of mercuric acetate adducts separates according to number of double bonds (*38*). Chromatography on layers of silica gel G impregnated with boric acid is suitable for separating symmetrical and asymmetrical alkyl glycerol ethers (*73*). Reversed-phase partition chromatography on hydrophobic paper or thin layers (see Chapter 5) can be used to effect the resolution of a class of alkoxy lipids according to both chain length and number of double bonds per molecule (*7,20,26,38,66*). However, gas–liquid chromatography is the method of choice for this type of separation. The latter technique yields the most efficient separations and is amenable to quantitative analysis. Samples that are too complex for complete resolution and unequivocal identification of constituents by gas–liquid chromatography should be prefractionated into simpler fractions by argentation chromatography (see, e.g., Fig. 4).

Experimental conditions worked out for the analysis of triglycerides by gas–liquid chromatography (see Chapter 7) are applicable to trialkyl glycerol ethers, dialkyl glycerides, and alkyl diglycerides (*50,74*), as well as the corresponding diol lipids (*50,74*). Alkyl glycerol ethers and dialkyl glycerol ethers, and also alkyl ethers of diols, are chromatographed in the form of less-polar derivatives. The most reliable methods are described here in detail. The first and second of these procedures are applicable to the analysis of alkyl glycerol-(1) ethers only, whereas the other two can be used for the fractionation of mixtures containing both symmetrical and asymmetrical alkyl glycerol ethers.

A. Analysis of Alkyl Glycerol-(1) Ethers as "Isopropylidene Derivatives" [Alkyl 2,3-O-Isopropylidene Glycerol-(1) Ethers] (*38*)

1. PRINCIPLE OF THE METHOD

Alkyl glycerol-(1) ethers are ketalized to alkyl 2,3-*O*-isopropylidene glycerol-(1) ethers by reacting them at room temperature with acetone in the presence of minute amounts of perchloric acid, which serves as a catalyst:

$$
\begin{array}{l}
\text{H}_2\text{C—O—R} \\
\quad | \\
\text{HC—OH} \\
\quad | \\
\text{H}_2\text{C—OH}
\end{array}
\quad\xrightarrow{\text{CH}_3\text{COCH}_3/\text{H}^+}\quad
\begin{array}{l}
\text{H}_2\text{C—O—R} \\
\quad | \\
\text{HC—O} \qquad \text{CH}_3 \\
\qquad\quad \diagdown \; C \diagup \\
\text{H}_2\text{C—O} \qquad \text{CH}_3
\end{array}
$$

The isopropylidene derivatives are isolated from the reaction mixture by adsorption TLC and are analyzed by GLC.

2. PREPARATION OF ISOPROPYLIDENE DERIVATIVES (*9,38*)

Alkyl glycerol-(1) ethers (1–5 mg), 5 ml of absolute acetone, and 5–10 μl of perchloric acid (72%) are placed in a glass-stoppered flask. The mixture is shaken at room temperature for 30 min and then applied onto a chromatoplate coated with silica gel G. After developing with hexane–diethyl ether 90:10, v/v, the opalescent band containing the isopropylidene derivative ($R_f \sim 0.6$) is scraped off. The product is eluted with ether and analyzed by GLC.

3. GAS-LIQUID CHROMATOGRAPHY OF ISOPROPYLIDENE DERIVATIVES (9)

Column dimensions, 190×0.4 cm; stationary phase, 20% EGS on Gas-Chrom E, 80–100 mesh; column temperature, 210°C; carrier gas, helium, 2.5 kg/cm². For an example, see Fig. 3.

B. Analysis of Alkyl Glycerol-(1) Ethers as Alkoxy Acetaldehydes (74)

1. PRINCIPLE OF THE METHOD

Alkyl glycerol-(1) ethers are cleaved at room temperature with sodium metaperiodate in pyridine solution to yield alkoxy acetaldehydes:

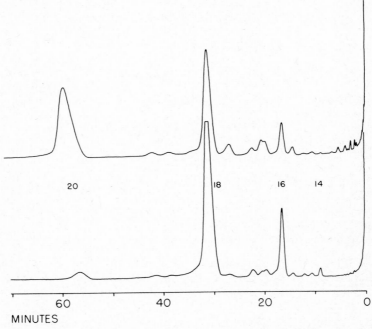

FIG. 3. Gas–liquid chromatography of alkyl glycerol ethers in the form of isopropylidene derivatives (*49*). (a) Alkyl glycerol ethers derived from the neutral plasmalogens of ratfish liver oil by catalytic hydrogenation and lithium aluminum hydride reduction; (b) alkyl glycerol ethers derived from the alkyl diglycerides of ratfish liver oil by hydrogenation and lithium aluminum hydride reduction. For experimental conditions, see text.

$$
\begin{array}{c}
\text{H}_2\text{C}\!-\!\text{O}\!-\!\text{R} \\
| \\
\text{HC}\!-\!\text{OH} \\
| \\
\text{H}_2\text{C}\!-\!\text{OH}
\end{array}
\xrightarrow[\text{pyridine}]{\text{NaIO}_4}
\begin{array}{c}
\text{H}_2\text{C}\!-\!\text{O}\!-\!\text{R} \\
|\;\text{H} \\
\text{C} \\
\text{O}
\end{array}
$$

The reaction mixture is subjected to GLC.

2. PREPARATION OF ALKOXY ACETALDEHYDES (74)

Alkyl glycerol-(1) ethers (10 mg), 100 mg of powdered sodium metaperiodate (No. 56072, Fluka A. G., Buchs SG, Switzerland), and 10 ml of dry pyridine are placed in a small reaction flask fitted with inlet and outlet tubes for dry nitrogen and magnetic stirrer. The mixture is stirred vigorously at room temperature for 36 hr, ether and air-free water are added, and the water layer is extracted twice with ether. The combined ether phases are washed consecutively with ice-cold water, 2 N sulfuric acid (until acidic), water, 1% potassium carbonate solution (until basic), and water and are dried over anhydrous sodium sulfate. The dry solution is analyzed by GLC.

3. GAS–LIQUID CHROMATOGRAPHY OF ALKOXY ACETALDEHYDES (74)

Column dimensions 150 × 0.4 cm; stationary phase, 20% EGS on Gas-Chrom P, 80–100 mesh; column temperature, 160°C; carrier gas, helium, 2.5 kg/cm².

C. Analysis of Alkyl Glycerol-(1) Ethers and Alkyl Glycerol-(2) Ethers as "TFA Derivatives" [Bitrifluoroacetyl Alkyl Glycerol Ethers] (47)

1. PRINCIPLE OF THE METHOD

Symmetrical and asymmetrical alkyl glycerol ethers are reacted at room temperature with anhydride of trifluoroacetic acid to yield bitrifluoroacetyl alkyl glycerol ethers:

$$
\begin{array}{c}
\text{H}_2\text{C}\!-\!\text{OH} \\
| \\
\text{HC}\!-\!\text{O}\!-\!\text{R} \\
| \\
\text{H}_2\text{C}\!-\!\text{OH}
\end{array}
\xrightarrow{(\text{CF}_3\text{CO})_2\text{O}}
\begin{array}{c}
\text{H}_2\text{C}\!-\!\text{O}\!-\!\overset{\displaystyle O}{\overset{\|}{\text{C}}}\!-\!\text{CF}_3 \\
| \\
\text{HC}\!-\!\text{O}\!-\!\text{R} \\
| \\
\text{H}_2\text{C}\!-\!\text{O}\!-\!\overset{\displaystyle O}{\overset{\|}{\text{C}}}\!-\!\text{CF}_3
\end{array}
$$

After evaporation of excess anhydride and the trifluoroacetic acid formed, the TFA derivatives are redissolved in trifluoroacetic anhydride and analyzed immediately by GLC. Samples containing polyunsaturated alkyl glycerol ethers turn brown on standing. Considerable losses of polyunsaturated compounds are encountered in the analysis of such solutions.

2. PREPARATION OF TFA DERIVATIVES (47)

Alkyl glycerol ethers (1 mg) and 1 ml of trifluoroacetic anhydride are placed in a glass-stoppered reaction tube. The mixture is shaken and then allowed to stand at room temperature for 15 min. The excess of trifluoroacetic anhydride and the trifluoroacetic acid formed are evaporated under a stream of dry nitrogen. The residue is diluted with trifluoroacetic anhydride, and this solution is analyzed by GLC.

3. GAS–LIQUID CHROMATOGRAPHY OF TFA DERIVATIVES (47)

Column dimensions, 160×0.4 cm; stationary phase, 15% EGSS-X on Gas-Chrom P, 100–120 mesh; column temperature, 170°C or temperature programming from 150° to 185°C; carrier gas, helium, 2.5 kg/cm^2. For an example, see Fig. 4.

D. Analysis of Alkyl Glycerol-(1) Ethers and Alkyl Glycerol-(2) Ethers as "TMS Derivatives" [Bitrimethylsilyl Alkyl Glycerol Ethers] (47,77)

1. PRINCIPLE OF THE METHOD

The reaction of symmetrical or asymmetrical alkyl glycerol ethers with trimethylchlorosilane in the presence of hexamethyldisilazane in pyridine leads to bitrimethylsilyl alkyl glycerol ethers:

$$
\begin{array}{ccc}
\text{H}_2\text{C—O—R} & & \text{H}_2\text{C—O—R} \\
| & \xrightarrow[\text{[(CH}_3)_3\text{Si]}_2\text{NH/pyridine}]{\text{(CH}_3)_3\text{SiCl}} & | \\
\text{HC—OH} & & \text{HC—O—Si(CH}_3)_3 \\
| & & | \\
\text{H}_2\text{C—OH} & & \text{H}_2\text{C—O—Si(CH}_3)_3
\end{array}
$$

The TMS derivatives are subjected to gas chromatographic analysis.

2. PREPARATION OF TMS DERIVATIVES (47,75)

Alkyl glycerol ethers (1 mg), 1 ml of pyridine, 0.2 ml of hexamethyldisilazane and 0.1 ml of trimethylchlorosilane are placed in a glass-stoppered reaction tube. The mixture is shaken for half a minute and then allowed to stand at room temperature for 5 min. Hexane (5 ml) and 5 ml of water are added; the water layer is extracted; and the phases are separated. After treating the water phase two times more with hexane, the combined organic layers are dried over anhydrous calcium sulfate. The solvent is evaporated to dryness under a stream of dry nitrogen. The product is dissolved in hexane; the solvent is evaporated until the odor of pyridine is not detectable any more; and the residue is analyzed by GLC.

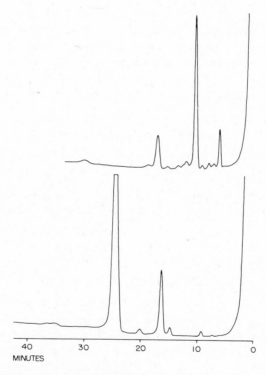

FIG. 4. Gas–liquid chromatography of alkyl glycerol ethers of dogfish liver oil in the form of TFA derivatives (77). Saturated and unsaturated ethers were isolated by argentation chromatography. The constituents of each of the two groups were then converted to TFA derivatives and analyzed by gas–liquid chromatography. (a) Saturated compounds; (b) unsaturated compounds. For experimental conditions, see text.

3. GAS–LIQUID CHROMATOGRAPHY OF TMS DERIVATIVES (*47*)

Column dimensions, 200 × 0.4 cm; stationary phase, 5% Apiezon L on Chromosorb W, 60–80 mesh; column temperature, 250°C; carrier gas, helium, 2.5 kg/cm².

These four procedures have been applied to the analysis of alkyl glycerol ethers that were isolated from dogfish liver oils or from the alkyl diglyceride fraction of such oils. Table 1 shows the results of these determinations. Also listed are values that were obtained by two other methods, namely, gas–liquid chromatography of the dimethyl ethers of alkyl glycerol ethers (*25,32*) and of the alkyl iodides derived from alkyl glycerol ethers (*37,76*). The results of these six analyses are in good agreement if consideration is given to the fact that they were obtained with different samples of dogfish liver oil.

Gas–liquid chromatography can be used for fractionating dialkyl glycerol ethers as such (*67*) or as acetates (*50*). However, chromatography of TFA derivatives and TMS derivatives yields better separations. The procedures used for the preparation of these compounds are the same as those described above for the formation of the corresponding derivatives of alkyl glycerol ethers; the experimental conditions for gas–liquid chromatography are similar to those specified above; and the derivatives of alkyl and dialkyl glycerol ethers can be analyzed simultaneously (*50*).

Alkyl ethers of ethanediol and propanediols are also chromatographed, in the vapor phase, as TFA derivatives or as TMS derivatives. It is possible to determine these derivatives simultaneously with the corresponding glycerol-derived ethers (*50,74*).

VI. Identification of Alkoxy Lipids

In addition to chromatographic analysis, the following methods are of value for establishing purity and identity of alkoxy lipids: IR spectroscopy (*23,57,78,80*), NMR spectroscopy (*23,57,73*), determination of optical activity (*15,46,49,67*), determinations of melting point and CST values (*46,62,66,79*), and mass spectrometry (*32,57*).

VII. Applications

Gas–liquid chromatography has been used for the analysis of alkyl glycerol ethers that were isolated from the neutral alkoxy lipids of human perinephric fat (*9*) and subcutaneous fat (*10*), beef heart (*48*),

TABLE 1

Alkyl Glycerol Ethers from the Liver Oil of Dogfish
(*Squalus acanthias*)

Chain length and number of double bonds	Methyl ethers[a] (32), %	Iodides[a] (37), %	Isopropylidene derivatives[b] (44), %	Trifluoro acetates[a] (77), %	Trimethylsilyl ethers[a] (77), %	Alkoxy acetaldehydes[b] (74), %
14:0	5.7	3.1	2.6	2.6	3.7	4.3
14:1				0.6	0.3	
15:br			0.6	0.1	0.3	
15:0	1.9	1.0	0.7	0.2	0.5	
16:br			0.7	0.3	0.2	
16:0	13.2	12.7	13.4	14.0	17.3	16.9
16:1	10.6	12.8	10.2	11.8	11.3	12.9
17:br			0.9	0.4	0.5	
17:0	3.0	0.5	0.3	0.3	0.4	0.9
17:1			1.1	0.9	0.9	1.1
18:br			0.7	0.3	0.4	
18:0	3.4	2.1	3.9	4.3	4.7	2.4
18:1	47.8	65.4	61.4	59.9	53.1	56.1
18:2	2.4		Trace	Trace	Trace	
19[c]	1.2		0.8	Trace	0.4	
20:0			0.2	0.5	2.0	
20:1	8.0	2.6	2.3	0.8	1.8	5.4
22:0						
22:1	2.7					

[a] Glycerol ethers from unsaponifiable fraction.
[b] Glycerol ethers from *O*-alkyl diglycerides.
[c] Total C_{19} alkyl glycerol ethers.

various shark liver oils (*28,44,49,74*), and molluscan tissues (*41*). The alkyl glycerol ethers obtained by hydrolysis of phosphatides of bovine erythrocytes (*38*) and bone marrow (*39*), terrestrial slugs (*40,42*), and molluscan tissues (*41*) have also been analyzed. The major alkyl glycerol ethers of all tissues investigated were saturated and monounsaturated compounds (see Tables 1, 2, and 3).

The composition of the alkyl glycerol ethers derived from neutral lipids has been compared with that of ethers obtained from phosphatides of the same tissue (*41*). As can be seen in Table 2, at least in molluscan tissues, the two lipids are very similar in regard to their constituent alkyl glycerol ethers.

Comparisons between the alkyl glycerol ethers from alkyl diglycerides and the aldehydes from neutral plasmalogens, as well as the fatty acids of these two lipid classes, have been made (*9,10,49*). As

TABLE 2

ALKYL GLYCEROL ETHERS IN NEUTRAL LIPIDS AND PHOSPHATIDES
OF MOLLUSCAN TISSUES (*41*)

Alkyl glycerol ether[a]	Clam (*Protothaca staminea*)		Hepatopancreas of octopus (*Octopus dofleini*)	
	Neutral lipids, %	Phosphatides, %	Neutral lipids, %	Phosphatides, %
14:0	3	5		1
15:br	1	3		2
15:0	1	3		1
16:br		1		
16:0	23	35	34	30
16:1	5	6		
17:br	8	11	5	6
17:0	3	3	4	6
18:br(?)	4	3	4	4
18:0	36	21	34	41
18:1	6	4	9	6
19:br				4
19:2(?)			7	
20:0	7			
20:1		4		
22:br		5		

[a] Analyzed by GLC of isopropylidene derivatives.

an example, Table 3, both the alkyl glycerol ethers and the aldehydes from neutral alkoxy lipids of human perinephric fat are mainly saturated and monounsaturated.

The alkyl glycerol ethers and aldehydes in phosphatides also have been compared, and it has been found that the composition of the two groups are very similar in regard to the chain lengths and number of double bonds in these compounds (*42*).

The dialkyl glycerol ethers derived from the phosphatides of halophilic bacteria have been thoroughly analyzed. They were found to

TABLE 3

ALDEHYDES IN NEUTRAL PLASMALOGENS AND ALKYL GLYCEROL ETHERS IN
O-ALKYL DIGLYCERIDES OF HUMAN PERINEPHRIC FAT (9)

Chain length and number of double bonds	Aldehydes, %, in neutral plasmalogens		Alkyl glycerol ethers, %, in O-alkyl diglycerides as isopropylidene derivatives
	As aldehydes	As dimethyl acetals	
16:0	56.6	58.0	37.6
18:0	26.9	25.8	39.1
18:1	12.0	11.5	19.3
Rest	4.5	4.7	4.0

consist, almost entirely, of a single compound, 1,2-bidihydrophytyl glycerol ether (56,57).

Considering the ubiquitous occurrence of alkoxy lipids, knowledge of their composition in various species, tissues, and cells is scarce. Additional information should aid in elucidating the metabolic inter-relationship that may exist between long-chain alkyl and alkenyl ethers and esters.

REFERENCES

1. E. Hardegger, L. Ruzicka, and E. Tagmann, *Helv. Chim. Acta,* **26,** 2205 (1943).
2. L. Svennerholm and H. Thorin, *Biochim. Biophys. Acta,* **41,** 371 (1960).
3. B. Hallgren and S. Larsson, *J. Lipid Res.,* **3,** 39 (1962).
4. O. Renkonen, *Biochim. Biophys. Acta,* **59,** 497 (1962).
5. J. R. Gilbertson and M. L. Karnovsky, *J. Biol. Chem.,* **238,** 893 (1963).
6. N. Tuna and H. K. Mangold, in *Evolution of the Atherosclerotic Plaque* (R. J. Jones, ed.), Univ. Chicago Press, Chicago, 1963, p. 85.
7. D. Todd and G. P. Rizzi, *Proc. Soc. Exptl. Biol. Med.,* **115,** 218 (1964).
8. B. Miller, C. E. Anderson, and C. Piantadosi, *J. Gerontol.,* **19,** 430 (1964).
9. H. H. O. Schmid and H. K. Mangold, *Biochem. Z.,* **346,** 13 (1966).
10. H. H. O. Schmid, N. Tuna, and H. K. Mangold, in preparation.
11. M. Tsutimoto and Y. Toyama, *Chem. Umschau Gebiete Fette, Öle Wachse, Harze,* **29,** 27, 35, 43 (1922).
12. Y. Toyama, *Chem. Umschau Gebiete Fette, Öle Wachse, Harze,* **29,** 245 (1922).
13. Y. Toyama, *Chem. Umschau Gebiete Fette, Öle Wachse, Harze,* **31,** 13, 61, 193 (1924).
14. E. André and A. Bloch, *Bull. Soc. Chim. France,* **2**(5), 789 (1935).
15. E. Baer and H. O. L. Fischer, *J. Biol. Chem.,* **140,** 397 (1941).

16. H. N. Holmes, R. E. Corbet, W. B. Geiger, N. Kornblum, and W. Alexander, *J. Am. Chem. Soc.,* **63,** 2607 (1941).

17. V. Prelog, L. Ruzicka, and P. Stein, *Helv. Chim. Acta,* **26,** 2222 (1943).

18. M. L. Karnovsky and W. S. Rapson, *J. Soc. Chem. Ind. (London),* **65,** 138 (1946).

19. M. L. Karnovsky, W. S. Rapson, and M. Black, *J. Soc. Chem. Ind. (London),* **65,** 425 (1946).

20. A. Emmerie, *Rec. Trav. Chim.,* **72,** 893 (1953).

21. M. L. Karnovsky and A. F. Brumm, *J. Biol. Chem.,* **216,** 689 (1955).

22. M. L. Karnovsky, S. S. Jeffrey, M. S. Thompson, and H. W. Deane, *J. Biophys. Biochem. Cytol.,* **1,** 173 (1955).

23. H. E. Carter, D. B. Smith, and D. N. Jones, *J. Biol. Chem.,* **232,** 681 (1958).

24. R. Blomstrand and J. Gürtler, *Acta Chem. Scand.,* **13,** 1466 (1959).

25. B. Hallgren and S. O. Larsson, *Acta Chem. Scand.,* **13,** 2147 (1959).

26. H. K. Mangold and D. C. Malins, *J. Am. Oil Chemists' Soc.,* **37,** 383 (1960).

27. D. C. Malins and H. K. Mangold, *J. Am. Oil Chemists' Soc.,* **37,** 576 (1960).

28. D. C. Malins, *Chem. Ind. (London),* **1960** 1359.

29. J. C. M. Schogt, P. Haverkamp Begemann, and J. Koster, *J. Lipid Res.,* **1,** 446 (1960).

30. J. Eichberg, J. R. Gilbertson, and M. L. Karnovsky, *J. Biol. Chem.,* **236,** PC 15 (1961).

31. D. J. Hanahan and R. Watts, *J. Biol. Chem.,* **236,** PC59 (1961).

32. B. Hallgren and S. Larsson, *J. Lipid Res.,* **3,** 31 (1962).

33. R. Pietruszko, *Biochim. Biophys. Acta,* **64,** 562 (1962).

34. R. Pietruszko and G. M. Gray, *Biochim. Biophys. Acta,* **56,** 232 (1962).

35. S. Nakagawa and J. M. McKibbin, *Proc. Soc. Exptl. Biol. Med.,* **111,** 634 (1962).

36. W. T. Norton and M. Brotz, *Biochem. Biophys. Res. Commun.,* **12,** 198 (1963).

37. K. E. Guyer, W. A. Hoffman, L. A. Horrocks, and D. G. Cornwell, *J. Lipid Res.,* **4,** 385 (1963).

38. D. J. Hanahan, J. Ekholm, and C. M. Jackson, *Biochemistry,* **2,** 630 (1963).

39. G. A. Thompson, Jr., and D. J. Hanahan, *Biochemistry,* **2,** 641 (1963).

40. G. A. Thompson, Jr., and D. J. Hanahan, *J. Biol. Chem.,* **238,** 2628 (1963).

41. G. A. Thompson, Jr., and P. Lee, *Biochim. Biophys. Acta,* **98,** 151 (1965).

42. G. A. Thompson, Jr., *J. Biol. Chem.,* **240,** 1912 (1965).

43. G. A. Thompson, Jr., *Biochemistry,* **5,** 1290 (1966).

44. D. C. Malins, J. C. Wekell, and C. R. Houle, *J. Lipid Res.,* **6,** 100 (1965).

45. H. H. O. Schmid and H. K. Mangold, *Biochim. Biophys. Acta,* **125,** 182 (1966).

46. W. J. Baumann, V. Mahadevan, and H. K. Mangold, *Z. Physiol. Chem.,* **347,** 52 (1966).

47. R. Wood and F. Snyder, *Lipids,* **1,** 62 (1966).

48. H. H. O. Schmid, L. L. Jones, and H. K. Mangold, *J. Lipid Res.,* in press.

49. H. H. O. Schmid, W. J. Baumann, and H. K. Mangold, *Biochim. Biophys. Acta,* in press.

50. R. Wood and F. Snyder, unpublished, 1967.

51. G. S. Harrison and F. Hawke, *J. S. African Chem. Inst.*, **5**, 13 (1952).
52. M. Popović, *Z. Physiol. Chem.*, **340**, 18 (1965).
53. G. V. Marinetti, J. Erbland, and E. Stotz, *J. Am. Chem. Soc.*, **81**, 861 (1959).
54. M. Faure, J. Maréchal, and J. Troestler, *Compt. Rend.*, **257**, 2187 (1963).
55. S. N. Sehgal, M. Kates, and N. E. Gibbons, *Can. J. Biochem. Physiol.*, **40**, 69 (1962).
56. M. Kates, P. S. Sastry, and L. S. Yengoyan, *Biochim. Biophys. Acta.*, **70**, 705 (1963).
57. M. Kates, L. S. Yengoyan, and P. S. Sastry, *Biochim. Biophys. Acta*, **98**, 252 (1965).
58. O. Renkonen, *J. Am. Oil Chemists' Soc.*, **42**, 298 (1965).
59. O. Renkonen, *Lipids*, **1**, 160 (1966).
60. L. J. Nutter and O. S. Privett, *Lipids,* **1**, 234 (1966).
61. L. D. Bergelson, V. A. Vaver, N. V. Prokazova, A. N. Ushakov, and G. A. Popkova, *Biochim. Biophys. Acta,* **116**, 511 (1966).
62. W. J. Baumann, H. H. O. Schmid, H. W. Ulshöfer, and H. K. Mangold, *Biochim. Biophys. Acta,* in press.
63. M. M. Rapport and W. T. Norton, *Ann. Rev. Biochem.*, **31**, 103 (1962).
64. D. J. Hanahan and G. A. Thompson, Jr., *Ann. Rev. Biochem.*, **32**, 215 (1963).
65. S. C. Gupta and F. A. Kummerow, *J. Org. Chem.*, **24**, 409 (1959).
66. W. J. Baumann and H. K. Mangold, *J. Org. Chem.*, **29**, 3055 (1964).
67. M. Kates, T. H. Chan, and N. Z. Stanacev, *Biochemistry,* **2**, 394 (1963).
68. W. J. Baumann and H. K. Mangold, *J. Org. Chem.*, **31**, 498 (1966).
69. W. J. Baumann and H. K. Mangold, *Biochim. Biophys. Acta,* **116**, 570 (1966).
70. H. E. Carter, P. Johnson, D. W. Teets, and R. K. Yu, *Biochem. Biophys. Res. Commun.*, **13**, 156 (1963).
71. K. Owens, *Biochem. J.*, **100**, 354 (1966).
72. W. G. Brown, *Org. Reactions*, **6**, 469 (1951).
73. H. Serdarevich and K. K. Carroll, *Can. J. Biochem. Physiol.*, **44**, 743 (1966).
74. W. J. Baumann, H. H. O. Schmid, and H. K. Mangold, unpublished, 1967.
75. R. D. Wood, P. K. Raju, and R. Reiser, *J. Am. Oil Chemists' Soc.*, **42**, 161 (1965).
76. D. J. Hanahan, *J. Lipid Res.*, **6**, 350 (1965).
77. R. D. Wood, private communication, 1967.
78. W. J. Baumann and H. W. Ulshöfer, unpublished, 1967.
79. H. H. O. Schmid, H. K. Mangold, W. O. Lundberg, and W. J. Baumann, *Microchem. J.*, **11**, 306 (1966).
80. H. R. Warner and W. E. M. Lands, *J. Am. Chem. Soc.*, **85**, 60 (1963).

GAS-LIQUID CHROMATOGRAPHY OF FATTY ACIDS AND DERIVATIVES

9

Robert A. Stein, Vida Slawson, and James F. Mead

DEPARTMENT OF BIOLOGICAL CHEMISTRY
UCLA SCHOOL OF MEDICINE
LOS ANGELES, CALIFORNIA

I. Introduction

Since formulation of the concept of a gaseous mobile phase in chromatography in 1941 (*1*) and its practical development somewhat later, there has been an almost explosive adoption of gas–liquid chromatography (GLC) technique in many areas of science. The broad spectrum of use, however, requires sufficient variation of operating parameters to make successful procedures in different fields functionally dissimilar. The biochemical and economic importance of lipid materials account for widespread interest in their chromatography, but the permutations that nature and man have wrought on the chains of fatty acids present such a wide array of separation and identification problems that it would appear presumptive to consider all acids as individuals.

The number of different phases used in gas chromatography seems to be limited only by the imagination. Many of these have specific properties for a unique separation, but in all cases they fit into the continuum that exists between miscibility and insolubility of the sample in the liquid phase. The use of "polar" and "nonpolar" for classification of GLC liquid phases is more convenient than rigorous, but this use does convey useful information. In GLC these terms suggest how strongly groups containing unshared electron pairs or hydrogen-bonding groups are bound in comparison with —CH$_2$— groups.

The availability of specific GLC data from abstracting services and review articles would make extensive repetition of these data redundant. The principal aim of this chapter is to point to some general relationships that exist between GLC behavior and the structural features found in fatty acids and to discuss some of the problems connected with the chromatography of fatty acids and their derivatives.

II. Normal Saturated Esters

GLC data for the homologous series of methyl esters of straight-chain fatty acids demonstrate a useful correlation between the skeletal length of the fatty acid and its retention time on the column. The retention time or other elution characteristic of a compound from the column is measured by the time the sample remains on the column or the volume of the carrier gas required to elute the compound. It has been found that the equation

$$\text{fatty acid chain length} = k \log \text{time}$$

holds true for homologous series over a wide range of chain length. The points that fit such an equation are easily plotted on semilog graph paper. A graph of this nature is of great value in predicting retentions of compounds that are not available for comparison. For example, with a plot of the type shown in Fig. 1, the retentions of the odd-numbered homologs can be predicted from the retentions of methyl laurate, myristate, palmitate, and stearate.

The semilog relationship is valid as long as the addition of the homologous unit, in this case, —CH$_2$— group, does not cause a signifi-

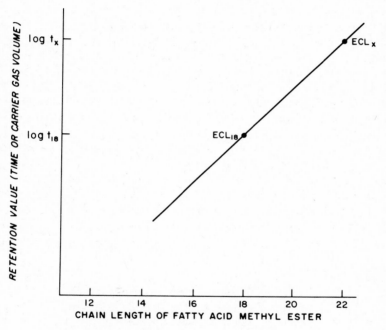

FIG. 1. Relationship of retention to chain length.

cant change in the polar character of the molecule. The relationship is linear over a wide range, but departure from linearity has been observed at the short end of the series (2). Adding a methylene group to a fatty acid molecule containing three carbons and one carboxyl is not the same as adding the same group to a molecule containing twenty carbons and one carboxyl. In a sense, for gas chromatographic purposes, a prorated amount of carboxyl group would have to be added to obtain a linear relationship through the complete range of fatty acids. Corrections of this type could be incorporated in the mechanics of plotting the data, but this is not necessary in the range of fatty acids usually encountered. When the fatty acid molecule becomes long enough, the change in contribution of the carboxyl group compared with the rest of the molecule becomes negligible. The molecular size necessary for a linear curve is a function of the liquid phase used. The more polar the phase, the longer the fatty acid chain must be before the curve becomes linear.

Plotting chromatographic data on semilog paper makes it possible to observe the consistency of homologous series, but the potential hazard of extrapolating too far from available data should always be borne in mind. Predictions of retentions or identifications of unknowns are more wisely obtained by interpolation.

Qualitatively, at least, deviations from linearity of the semilog plot of the GLC data of short-chain compounds can be predicted from any of the physical properties of alkyl compounds that imply some interaction of charge-separated species. For example, dipole moments of alkyl substituents change rapidly in the series methyl, ethyl, propyl and then become relatively constant.

III. Substituted Fatty Esters

A. Branched-Chain Esters

The GLC retention time of a branched-chain ester is always smaller than that of its straight-chain isomer. Type, number, and location of the substitution play significant roles in determining a compound's characteristic elution.

In a series of isomeric fatty esters containing a methyl branch, beginning at the methyl end of the chain, the retention time decreases in a gradual but slightly irregular manner until the branch occurs near the middle of the molecule (*3*). As it approaches the carboxyl end, a slight increase in retention times is observed until the 2 position is reached. At this point the retention is strikingly decreased. The cause of this latter is an apparent steric inhibition to solvation or to adsorption of the carboxyl group by the liquid phase.

A semilog plot of a homologous series of branched-chain esters will give a line that is parallel to the line for the normal isomers. The data for the homologous series of iso and anteiso fatty esters, for example, will form characteristic lines with slopes and intercepts dependent upon the liquid phase and operating conditions. In general, the polar columns will differentiate between the positions where branching occurs better than the nonpolar ones. This is noted particularly with branches that occur at either end of the molecule. On a nonpolar column, Apiezon M, the iso and anteiso compounds are eluted together, whereas on a polar column, polymeric ethylene glycol adipate, the iso form precedes the anteiso (4).

Multiple methyl branches decrease the retention further than single branches. Two methyl groups substituted on the penultimate carbon of a fatty acid (the so-called neo isomer) decrease the relative retention time below that for the iso or anteiso isomers on both polar and nonpolar columns (4). A 2,4,6-trimethyl-2-alkenoate was eluted earlier than the isomeric 10-methyl or normal carbon skeletons (5).

There is sufficient evidence for the parallel relation of the semilog plots of homologous series of fatty esters to indicate that this correspondence is usual. It is thus practical to predict the behavior of a series from the retention of one isomer and the semilog plot of the normal methyl esters run under the same conditions. However, this relationship does not invariably apply, even to the restricted category of saturated long-chain methyl esters. This can be demonstrated by the analysis of lipids from the lugworm (6) and the tubercule bacillus (5). In the tubercule bacillus, 10-methylheptadecanoic, 10-methyloctadecanoic, and 10-methylnonadecanoic acids have been found. The postulate of linear and parallel behavior of the gas chromatographic data would allow an extrapolation to shorter chain length, for example, to methyl 10-methyldodecanoate.

In the lugworm, a series of iso and some anteiso compounds were found in the C_{15} and higher range. Extrapolation of the anteiso line could be made to anteisotridecanoate. This also falls on the 10-methyl series line. One is forced into the position of concluding that either the 10-methyl and anteiso series are superimposed, that they are not parallel but happen to intersect at C_{13}, or that they are not linear. None of these possibilities contributes to firm confidence in extrapolation of data of this type. The example chosen is not an isolated one. Each methyl-branched fatty ester will fall on two series, in one of which the methyl end is constant and in the other of which the carboxyl end is constant. Limited experimental evidence shows that all of these series are not superimposable. Thus they are not straight lines or do not form straight lines parallel to the normal isomers.

Similar arguments could be developed for the polysubstituted acids, but one should be aware that the more complex the fatty acid, the more correlative evidence is needed to support ambiguous gas chromatographic extrapolations.

Chain branches larger than methyl offer an opportunity to explore the correlation of structural change and gas chromatographic behavior. There is a considerable amount of data available concerning the steric effects of various types of substituents (7). Empirical constants are obtained from chemical reaction data, and a correlation between chromatographic behavior of compounds containing these groups and the substituent constants would be valuable. A question one might ask is whether rearranging a part of a normal fatty acid into a branch on the chain has any relationship with the E_s (steric substituent constant) for the specific branch that was made. For example, can the chromatographic behavior of methyl, ethyl, isopropyl, etc., substituted fatty acids be related to the E_s values for methyl, ethyl, and isopropyl, etc., groups?

For a tentative answer to this question, a comparison was made between the methyl stearates substituted at the 12 position by methyl, ethyl, isopropyl, n-butyl, and isobutyl and the methyl esters of the isomeric normal fatty acids (8). The relative retention times (compared to the normal isomers) obtained from a silicone column were substituted into the Taft equation, $\log (k/k_0) = \rho E_s$, where k and k_0 are relative retention times of 12-alkyl and 12-methyl, respectively, ρ is a reaction constant, and E_s is the substituent constant. The linear correlation between the gas chromatographic data and E_s values is shown in Fig. 2.

This correlation covers a small range of E_s values and only one type of substituent. Future work will undoubtedly show the limitations of this type of relationship, but nonetheless, this approach presents an interesting bridge between gas chromatography and chemical kinetic data. In this regard, stearic acid esterified with alcohols containing the same alkyl groups as the substituents used in the previous correlation (e.g., methyl, ethyl, isopropyl stearates) did not show this linear relationship. Since these groups create electronic effects in the neighboring environment, it seems likely that, in the case of the esters, one is observing a combination of steric and polar effects.

B. Hydroxy Esters and Derivatives

Molecules containing hydroxyl groups, which are capable of participating in hydrogen bonds with the liquid phase, are retained on GLC columns considerably longer than estimated from molecular

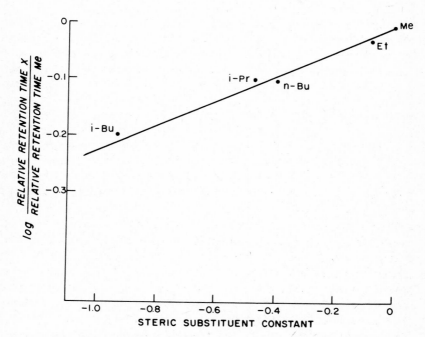

FIG. 2. Correlation between alkyl group substituent constants E_s and log [(relative retention time methyl 12-alkylstearate)/(relative retention time methyl 12-methylstearate)]. [From R. A. Stein (*8*).]

weight alone. A hydrophobic liquid phase will allow the alcohols to be eluted relatively more rapidly than will a hydrophilic phase. Fatty esters with hydroxyl groups on their chains are not exceptional in this regard.

The GLC behavior of the simple monohydroxyl fatty esters has been surveyed. All the positional isomers of methyl hydroxypalmitate (*9*) and hydroxystearate (*10*) have been chromatographed. A comparison of the relative retention time with position of the hydroxyl group shows the similarity of behavior of the isomers with the hydroxyl group located in the middle portion of the molecule. The ends of the positional isomeric series (positions 2, 3, 17, 18) are sufficiently different that they can be distinguished from one another. Hydroxyl groups at positions 2 or 3 impart retention times shorter than those of the other isomers (10). This effect has been attributed to an intramolecular hydrogen bonding that decreases the bonding to the liquid phase of the column (10). The 4- and 5-hydroxy acids are also rather special cases because of the possibility of lactone formation. Significantly, 4-hydroxystearate had a retention time identical with that of γ-stearolactone (10).

Since the hydroxy, oxo, and acetoxy acids can be interconverted, GLC comparison of all three derivatives is not difficult and gives considerably more information for identifying positional isomerism than any individual type. Although acetoxy derivatives have been chromatographed successfully, the possibility of a Chugaev elimination reaction occurring with this type of compound should be kept in mind. The oxo compounds showed progressively longer retention times on both polar and nonpolar columns as the position of the ketone was moved toward the methyl end of the molecule. 3-Oxo esters have been shown to decarboxylate to the corresponding methyl ketone (10).

Retentions for a series of long-chain alcohol derivatives, including pentafluoropropionyl and heptafluorobutyryl esters have been reported (11). This gives an indication of the characteristics such groups might impart to a fatty ester. Because of the high halogen content, these may offer greater sensitivity with an electron-capture detector than is now attainable for methyl esters in an argon or flame ionization detector.

An interesting correlation between the GLC behavior of $HOCH_2$
$\overset{|}{CH_2}$—, CH_3CHOH, and $MeOOCCH_2$— substituents has been reported (12). On a SE-30 column, $HOCH_2CH_2$— and $MeOOCCH_2$— were equivalent (identical retention times of methyl 10-hydroxydecanoate and dimethyl sebacate), and on a diethylene glycol succinate column $\overset{|}{CH_3CHOH}$ and $MeOOCCH_2$— were equivalent (identical retention times of methyl 9-hydroxydecanoate and dimethyl sebacate).

Although there are various reports of successful elution of hydroxy fatty esters from a wide range of liquid phases, other columns are not successful for this purpose (*12–14*). For example, methyl 2-hydroxystearate decomposed on an "old" silicone rubber column but could be successfully chromatographed on a new one (*15*).

In addition to the use of acetyl derivatives to circumvent the chromatography of the free alcohols, trifluoroacetates (*16*), methyl ethers (*13,17*), and trimethylsilyl ethers have been used (*18*).

The variability found in the chromatography of hydroxy-substituted esters indicates that the elution of such compounds from a GLC column is probably not an all or none affair. The appearance of a peak may not represent complete elution from the column, and quantitative calibration of response with one positional isomer is not necessarily valid for another. A pertinent observation was made that the oleyl alcohol to methyl myristate peak area ratio was inversely proportional to the temperature of a polyester GLC column (*19*). Further, the loss of alcohol was proportional to its retention time and to the age of the column. This loss can be explained on the basis of free carboxylic acid or anhydride groups in the liquid phase esterifying the hydroxyl groups of the sample or by an irreversible interaction with solid support, dehydration, or desorption occurring at such a slow rate that the total sample did not show up in the peak. The reaction of long-chain alcohols with silanol groups, such as might be found in diatomaceous supports, was demonstrated by chemical changes in silica heated with alcohol (*20*). Carbon contents as great as 25% were found after heating silica with alcohols. Consequently, if the accuracy of quantitative results is to be assured, the response of the system to a given type of compound must be assessed.

The reported successful GLC of 3-hydroxy fatty acid esters (*9,21, 22*) implies a heat stability for this type of compound that is not always observed. For further information in this regard, the products obtained from the Reformatsky reaction should be considered (*23*).

C. Halogen-Containing Fatty Esters

Halogenated derivatives of fatty esters have not been chromatographed extensively. This is due in part to the rarity of their natural occurrence and in part to the fact that transformation of another functional group to halogen does not usually impart advantageous

GLC characteristics. Changes in GLC behavior after treatment with bromine have been used for the presumptive identification of unsaturated methyl esters (4). Introduction of two bromine atoms into the middle of a fatty ester molecule has approximately the same effect as an equivalent increase in molecular weight, indicating the minor importance of polar effect of bromine on a polyester column (24). Although retentions of threo- and erythrodibromides derived from methyl oleate and methyl elaidate were different, separation of a mixture was not observed. In connection with the generalization that unsaturated compounds elute later than their saturated parents from polar columns, a contrary observation is that methyl 9(10)-bromo-9-octadecenoate will precede methyl 9-bromostearate on such a column (8). The carbon numbers for ω-fluoro methyl esters from decanoic to octadecanoic acids indicated a polar effect of this halogen on polar GLC columns (25). On Apiezon, the increased retention time due to the fluorine was not much different from that of an additional methylene group, but on polyethylene glycol adipate the carbon numbers due to fluorine increased by 3.2 units. This retardation is not inconsistent with the formation of hydrogen bonds to the halogen substituent (26,27) and the necessary orientation for such halogen–hydrogen bonding (28). Similar large retardations on phosphoric-acid-containing liquid phases were seen in the chromatography of short acids with a fluorine or chlorine atom in the 2 position (29,30).

Although bromides are relatively stable compounds, the possibility of decomposition of this derivative of a fatty ester (31) should be considered when assessing the purity of such materials by GLC. Evidence for debromination during GLC of methyl-9,10-dibromostearate to an olefin was convincing (32).

Because the differences in polarity of the cis and trans vinyl bromides derived from the methyl erythro- and threo-9,10-dibromostearates are greater than the polarity differences between the methyl elaidate and methyl oleate precursors of the dibromides, it is possible to make more accurate quantitative determinations of smaller amounts of methyl elaidate in mixtures of the geometric isomers than can be achieved with infrared analysis (33). Good resolution was found with polar packed ethylene glycol succinate columns. Reaction conditions appropriate to analysis of natural oils were described, and good quantitative results on mixtures of elaidic and oleic acids have been obtained (34).

IV. The Effect of Unsaturation

The presence of unsaturation in a fatty acid ester molecule will cause great variability in GLC behavior. A single double bond causes a fatty ester to be eluted before its saturated homolog from a relatively nonpolar phase, such as Apiezon or a silicone. This effect stems from the loss of the two C—H bonds from each double bond present. The decrease in dispersion forces resulting from this loss is not completely compensated for by an induced dipolar attraction between the π electrons of the olefin and the liquid phase. Accompanying these major effects, there will also be influences due to the geometry of the substituents on the double bond that can influence the approach of the double bond to a surface of the liquid phase. On a polar column, the olefin-containing molecule is retarded by increased dipolar interaction between the sample and the liquid phase. Thus an unknown compound may be tentatively characterized as unsaturated by its changing behavior on different liquid phases. The greater relative contribution of double bond in methyl 12:1 as compared with methyl 22:1 makes the separation between the shorter olefin ester and the corresponding saturated parent greater than in the longer-chain pair. This mirrors the decreasing contribution of polar effects to total retention as chain length increases.

Packed columns are relatively insensitive to the subtle differences among positional isomers, and such separations are usually made on capillary columns. In general, the capillary columns have HETP values similar to those of packed columns, but they can provide superior separations because of their greater length and the exceedingly small samples chromatographed (*35*). Positional isomeric monoene methyl esters have been separated on a 150-ft. Carbowax capillary column. The separation of the 7- and 9-ene isomers of methyl tetradecenoate was superior to the separation of the 7- and 9-ene isomers of methyl hexadecenoate on the capillary column, whereas they could not be distinguished on a packed column (*36*).

An ingenious method for distinguishing unsaturated straight-chain esters from simple branched isomers has been proposed (*37*). The effect of temperature change on the two types of ester was different for the two types of structures. An increase in operating temperature caused an increase in the separation factor for an unsaturated straight-chain ester, relative to the preceding saturated straight-chain ester, whereas the increase in temperature caused the separation factor to decrease for a saturated ester, relative to the preceding saturated component.

Geometric isomerism does not cause sufficiently large differences in GLC behavior to permit easy differentiation between *cis* and *trans* isomers of the long-chain esters with packed columns. Small differences in the retentions of methyl oleate and methyl elaidate have been observed on nonpolar columns (*38*), from which the *cis* form eluted more rapidly than the *trans*, and both preceded the saturated. Use of capillary columns resulted in a separation of methyl elaidate and methyl oleate on Apiezon (*39*). This was apparently improved by increasing the pressure during the course of the chromatogram (*40*). The *trans* isomer of the pair eluted before the *cis* on a polyester (DEGS) coated capillary (*41*).

The effect of positional isomerization is magnified by the presence of more than one double bond in the molecule, and appreciable differences among polyene isomers can be seen on both polar and nonpolar phases (*38*). Separation of the $\Delta^{8,11}$ and $\Delta^{11,14}$ isomers of methyl octadecadienoate was achieved on a polar packed column (*42*). A polar capillary column was needed for complete separation of the positional forms intermediate to these. The separation is shown in Fig. 3.

The number of geometric isomers possible for any positional isomer of a polyunsaturated fatty acid is 2^n, where n represents the number of double bonds. Columns with excellent resolving power are needed to separate all of them. A partial separation of the all-*cis* methyl linoleate from the all-*trans* form was made on a packed Apiezon column (*43*). Complete separation of all-*cis* and all-*trans* linoleate was possible with either a polar or nonpolar capillary column, but for analysis of the four possible geometric isomers, both columns were needed. It is interesting to note that both phases were capable of separating the 9-*cis*-12-*trans* from the 9-*trans*-12-*cis*.

Conjugated double bonds in a fatty ester molecule increase the retention considerably over that of the unconjugated isomer on both polar and nonpolar liquid phases, but the changes imparted by conjugation are larger on a polar phase than on a nonpolar. This may be construed as evidence for actual polarity of the so-called "nonpolar" phases.

Cis and *trans* isomers of conjugated dienes have been cleanly separated from each other on a polar phase (*44*), and, curiously, the methyl *cis-trans*-octadecadienoates eluted earlier than the *trans-trans*.

The effect of acetylenic bonds is qualitatively similar to olefinic unsaturation. On a nonpolar phase, the retention of methyl octadecynoate or methyl dodecynoate fell between that of the saturated and monoenoic analogs (45). The same effect of the acetylenic bond is found in more complex molecules. The methyl ester of 8-hydroxy-

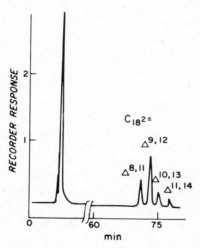

FIG. 3. Separation of positional isomers of methyl octadecadienoate on a capillary column coated with ethylene glycol glutarate. [From R. A. Landowne and S. R. Lipsky, 1961 (42).]

ximenynic acid (8-hydroxy-octadec-*trans*-11-en-9-yoic) had an ECL of 19.9 on Apiezon L (46), or 0.4 less than the methyl ester of 9-hydroxy-*trans,trans*-10,12-octadecadienoic acid (47). Contrary to these findings, the positional isomers of methyl nonynoates were found to elute after methyl nonanoate on an Apiezon L capillary (48). On a polyester phase, the retention due to the acetylenic bond in the 18 or 22 carbon esters is somewhat greater than that imparted by the two double bonds in methyl linoleate (48). No differentiation was seen between the methyl esters of 9-octadecynoic acid (stearolic) and the naturally occurring 6-octadecynoic acid (tariric) on either Apiezon or Resoflex (47). However, a capillary column coated with diethylene glycol succinate separated all the methyl nonynoate positional isomers except those with unsaturation at positions 4 and 5 (48).

V. GLC of Fatty Acids

Chromatography of free fatty acids rather than methyl esters is particularly appropriate for the low-molecular-weight acids, the methyl esters of which have sufficient volatility to cause loss during usual concentration procedures. Addition of an acid, such as phosphoric or stearic, to a silicone liquid phase has been shown to be effective in reducing the pronounced tailing of acids from formic to dodecanoic (49). Many variants of this procedure have been proposed. The addition of 2% phosphoric acid to a polar liquid phase permitted the separation of C_4–C_{22} saturated acids as well-defined, symmetrical peaks (50), with partial resolution of oleic acid from stearic acid. Phosphoric acid alteration of another polyester proved valuable in separation of a mixture of acetic, valeric, chloroacetic, decanoic, and dichloroacetic acids (30). Two disadvantages were inherent in this system. "Repeat peaks," small peaks representing components of an original mixture, appeared after injection of dichloroacetic acid. Second, thermistor, ^{90}Sr, and [226]Ra detectors all gave different responses to components of the mixture.

Further characterization of phosphoric acid–polyester liquid phases showed other disadvantages (51). The polyester selected was important, and low detector response was obtained if columns had not been carefully conditioned. For quantitative determinations, calibration factors were needed for each column as well as for the detector. The general observation of lower response for longer retention time and repeating of peaks or "ghosting" may be attributed to esterification with alcohol groups on the polyester. Either a phosphoric acid or a Tween 80 treated column permitted temperature-programmed separation of the normal carboxylic acids from C_1 to C_{14} and resolution of isobutyric and isovaleric acids (52). Failure of the branched acids to separate on a similarly altered LAC-3R-728 column was noted.

Other mineral acids were generally effective in making columns amenable to free carboxylic acid GLC (53), although the usefulness of an acid-treated column varied with the mineral acid (54). Other systems for the chromatography of free acids included the use of acid-washed quartz as the solid support (55), acid-washed glass beads (56a), polyethylene glycol 600 (57), and addition of formic acid to the carrier gas (58).

The use of porous polymer beads for chromatography is an innovation with particular value for the chromatography of free acids (*59*). The process is not gas–liquid chromatography in a rigorous sense, because the polymers are solids. It is functionally GLC because an important factor in separations is "solubility" in the polymer, which "assumes the partition properties of a highly extended liquid surface." Peaks, even for free acids, are quite symmetrical, and there are no large adsorptive losses (*8*). However, long retention times limit the molecular weight of the solute.†

VI. The Interrelationship of GLC Retention Data

Identification of a fatty acid derivative by a retention time on a given column is tentative. It is possible for two or more compounds to have the same retention time on any given GLC column. The probability of different compounds having identical retention times on different columns is inversely proportional to the number of columns used and to the number of changes made in the operating conditions of a single column. The identity of an unknown substance with a known compound can be reasonably shown by GLC on two different columns (*60*). Direct comparison is satisfactory when one has a sufficient number of standards. When this approach is impractical because of the unavailability of suitable reference compounds, it is necessary to use data obtained by other investigators or find support for an identification by other analytical techniques.

Because GLC retention values are sensitive to such variables as the liquid phase and its concentration, composition, particle size, and pretreatment of the solid support, and operating parameters such as temperature and carrier gas velocity, it is difficult to describe the elution characteristics of a molecule or specific functional group in absolute terms. For simplicity, the migration of a compound on a column is usually related to a standard compound or to a homologous series of compounds with elution characteristics that tend to define the system. In the GLC of fatty acid derivatives, this comparison is usually made with methyl palmitate, methyl stearate, or the homologous series of methyl esters of normal saturated fatty acids. Common ways of graphing empirical data are shown in Fig. 1. An unknown compound

† Porapak Q, 6 mm × 32 in., 225°C, 30 psig N₂, octanoic acid, 18.3 min.

with a retention time t_x has a relative retention time of t_x/t_{18} with respect to methyl stearate. For a given relative retention time, arbitrary selection of a value for either log t_x or log t_{18} makes it possible to calculate the other. A relative retention time gives both coordinates for one point on the line and only the ordinate value for the second point. Thus the relative retention by itself provides insufficient data for determining either the slope line of the homologous series to which it belongs, or the slope line of the normal saturated series with which it is being compared.

An alternative system for describing retention values of methyl esters has been termed "equivalent chain length" (ECL) (47) or "carbon number" (61). This system is more or less confined to the fatty acid field, because assignments are made by relating the retention of the compound under consideration to the retentions of the methyl esters of the normal saturated fatty acids. This is most easily done by establishing the semilog relation of the normal esters and reading the linear coordinate of a given retention time. In Fig. 1 a compound of retention time t_x would have an ECL or carbon number of 22. The ECL may also be derived arithmetically (62). The information transmitted by relative retention values is essentially identical with that by carbon numbers. Since the mental manipulation of numbers seems to be somewhat easier on a linear basis than on a logarithmic one, the use of the carbon-number system is gaining favor for conveying GLC order of elution without an actual plot of the data. The choice of either system is a subjective one, and should be made for clarity of a specific point of view.

A more rigorous expression of retention data is in terms of the "Kovats retention index" (63–66). Retention values are related to those of two hydrocarbons with similar retentions. Because of its great flexibility, the system is being used increasingly in varied fields, and numerous compilations of data exist. Although these values have not been used to any great extent in the area of the GLC of fatty esters, a recent investigation of the influence of polarity of liquid phase on the chromatography of numerous aliphatic and cyclic α-branched carboxylic acid methyl esters utilized the Kovats system (67).

VII. Radioactive Samples

The use of radioactive materials to investigate biochemical pathways makes it necessary to measure activity as well as identity and relative amount of the fatty acids. A detailed review of continuous and discontinuous methods of detecting radioactive substances in GLC has appeared (68). The activity of labeled material may be assayed after chromatography by collection and counting of individual fractions (see section on collection). This procedure is mandatory for low-activity substances. However, obtaining GLC and radioactivity measurements in a single operation is convenient and eliminates losses in transfer and, moreover, provides for the detection of activity in unexpected compounds or in traces of high-activity material for which no mass peak is seen (69). Continuous analysis is obtained by passing the effluent gas into a counter or incorporating a radioactivity detector into the GLC system.

For representative methods of the GLC of radioactive fatty esters (69–81), for data concerning recoveries (82–88), and for chromatographic ramifications (89–95,217) in the handling of minute quantities of labeled materials the reader is directed to the list of references.

VIII. Reactions Occurring on GLC Columns

Although most fatty acid methyl esters have sufficient stability to survive GLC without degradation, introduction of certain structural features may increase their lability. It is generally conceded that monoenes and polyenes containing the methylene-interrupted pattern survive GLC unchanged. Conjugated trienes have been found to undergo *cis-trans* isomerization, although conjugated dienes did not (44). In addition to geometric isomerization, conjugated trienes undergo extensive bond migrations (96).

Hydroxyl groups in a fatty acid can be stable or unstable depending on their location in the molecule, nature of the column, and operating conditions. When the hydroxyl is α to a double bond (44,97), or to a carbonyl group, as in α-hydroxy esters (15), there may be a tendency to decompose. Although saturated hydroxy esters in general are not susceptible to structural alteration (44,98), it is probable that they react with certain column packings. The failure of methyl 18-hydroxy stearate to elute from a polyester phase from which other positional isomers did elute was attributed to the greater reactivity of the primary hydroxyl group in esterification reactions with the liquid phase (10). The argument for transesterification is augmented by the necessity for adequate conditioning of ethylene glycol adipate columns for chromatography of long-chain alcohols (99), the demonstration of the importance of support acidity (10), and the nature of the polyester preparation (99).

Cyclopropenoid (97,100), epoxy (101,102), and bromine (31,32) groups in fatty acid methyl esters are sufficiently labile to decompose during GLC. GLC columns are also used as sites for planned reactions. Microhydrogenolyses of sulfur-containing (103) and nitrogen-containing (104) compounds and organometallic (105) derivatives of hydrocarbons have been performed on GLC columns to obtain carbon skeletons that could be used for identification. This method of hydrogenolysis could be effectively applied to many types of compounds, including fatty acids and esters (106–108). Hydrogenation of unsaturated methyl esters on GLC columns gave results comparable with those obtained by the conventional methods (109).

Tetramethyl ammonium salts of aliphatic, hydroxy, and dicarboxylic acids have been converted to the methyl esters in good yields in the preheater of a gas chromatograph at 220–270°C (110). The first section of a column consisting of Chromosorb W coated with 20% phosphoric acid effected conversion of the potassium salts of fatty acids to free acids, but the quantitative aspect was not studied (111). A "flash exchange" procedure involving potassium ethyl sulfate and the sodium and potassium salts of fatty acids formed ethyl esters on the column (112,113). On a phosphoric-acid-altered column (51), α-sulfonated fatty acids were desulfurated and amides were dehydrated to nitriles.

Direct injection of ozonized fatty esters with thermal decomposition to aldehyde was proposed as a convenient method for analysis of micro quantities (114). Chromatography on a column packed with Carbowax 6000 and KOD caused the exchange of enolizable H atoms of certain ketones with deuterium, but results for esters and aldehydes were less satisfactory (115).

IX. Methods

A. Preparation of Methyl Esters for GLC

The GLC of fatty acids is most commonly performed with methyl esters. The possible loss of volatile esters in a mixture can be circumvented by the use of less volatile derivatives (2,116,117). However, it is not always appreciated that the vapor pressure of methyl laurate and methyl myristate is sufficient to cause appreciable losses during concentration (*118*). Methods for obtaining methyl esters of fatty acids depend considerably on the nature and stability of the starting material. Since standard references cover many procedures that have been successful for preparative and industrial use, the methods reviewed here will be specific applications found successful in the conversion of biologically encountered fatty acids in phosphatides, triglycerides, cholesterol esters, or sphingolipids to simple esters.

Although in common usage "esterification," "transesterification," and "alcoholysis" are frequently interchanged, these terms properly refer to formation of an ester from an alcohol and an acid, exchange of acyl groups among different esters, and exchange of an alkoxy group by another alcohol, respectively.

TABLE 1

RANGE OF EXPERIMENTAL CONDITIONS FOR CONVERTING LIPIDS TO FATTY ACID METHYL ESTERS

Acid concentration,[a] w/v	Temp., °C	Time	Ref.
2.5–5% HCl	60–100	1–17 hr	*122–129*
1–10% H_2SO_4	65–100	1–16 hr	*71, 126, 130–134*
1.2–10% HCl + 2,2-dimethoxypropane	22–65	2–8 hr	*135–137*
5% H_2SO_4 + 2,2-dimethoxypropane	70	1–4 hr	*138*
3.5–14% BF_3	68–100	2 min–16 hr	*58, 133, 139–141*

[a] Concentration in methanol, with addition of benzene when necessary to solubilize the lipid.

Differences in esterification rates of fatty acids are small. Attention has been called to the erroneous tabulation of reaction rates in International Critical Tables (*119*). Although both acid and base are effective catalysts for esterification reactions, there has been a pronounced tendency for lipid chemists to favor acid catalysis. This attitude may have been fostered by the possible isomerization of methylene-interrupted double bonds in the presence of excessive base (*4,120,121*).

Most of the acid-catalyzed esterifications represent modifications of the Fischer method, involving dry HCl in excess alcohol. In Table 1 is shown the range of acid concentration and reaction conditions that have been applied to lipids. The table indicates that the concentration of acid for effective catalysis is not critical. Time and temperature for

completion are a function of lipid class, with sphingolipids and cerebrosides requiring the more rigorous conditions (124,125,130,140).
Strong Lewis acids can be as effective as mineral acid catalysts, and
their use has achieved some popularity (133,139–145). 2,2-Dimethoxy-
propane (acetone dimethyl ketal) is sometimes included in acid-
catalyzed esterification mixtures to react with water formed during
the reaction and drive the esterification to completion (136,137,146).
Its influence in transesterifications, where water is not formed, is
probably that of a simple drying agent for the solvent.

Basic catalysis has also been used for methanolysis reactions, (147–
150). The efficacy of sodium methoxide can be estimated from the
observation that methanolysis of lecithin at 0°C was completed in 24
min (151). Potassium methoxide was shown to be somewhat more
effective than sodium methoxide in methanol at reflux temperatures
(152). The loss of short-chain methyl esters because of incomplete
extraction and volatilization during concentration can be prevented by
injection of the reaction mixture directly onto the gas column (153).
In methanol basic hydrolysis followed by BF_3 catalysis without isolation of the soaps gives good yields (154) of methyl esters.

Direct esterification of fatty acids may be obtained with diazomethane. Diazomethane is suitable for esterification of sterically
hindered acids (155), and ^{14}C-diazomethane offers a method of forming labeled esters (156). The reaction of diazomethane with carboxylic
acids is rapid, relative to its reaction with other functional groups, but
the nature of the solvent has a considerable effect. A detailed investigation of reaction rates of diazomethane showed that in dry ethyl ether
at 20°C (156) esterification of stearic acid was incomplete after 1 hr.
Addition of methanol (2.67 M) or water to the ether accelerated the
reaction so that it went to completion within 5 min. Transesterification
or the formation of pyrazoline or cyclopropane derivatives by addition
to olefins did not occur under these conditions, but discretion is necessary when using diazomethane with acids containing other functional
groups (157,158).

Commonly employed precursors for diazomethane are nitroso-
methylurea (159), easily prepared but unstable at room temperature;
N-methyl-N-nitroso-N'-nitroguanidine (160), reasonably stable but
a possible cause of dermatitis; N-methyl-N-nitroso-p-toluenesulfona-
mide (161), stable and nonirritating; and bis(N-methyl-N-nitroso)
terephthalamide (162). The latter two are commercially available.

Methods of ester-forming reactions involving acid catalysis and diazomethane (*123,163,164*) give similar results. Three base-catalyzed esterifications applied to several natural fats gave substantially identical results (*165*). However, anomalous results have been reported for all esterification methods. Undistilled diazomethane may give extraneous peaks (*166–168*). Diazomethane may leave a colorless oil with the GLC characteristics of C_{10}–C_{20} esters, and whose infrared spectrum and elemental analysis indicated more complex material than polymethylene (*169*).

Methanol and HCl react with each other. A 5% solution of HCl in methanol is converted in 50% yield to methyl chloride within 6 weeks at room temperature (*125,170*). Other products increased with storage of the reagent (*171*). In the presence of cholesterol esters, acid-catalyzed transmethylation may give rise to unsuspected cholesterol derivatives (*133,138,172*). Significant quantities of extraneous material were produced by the transesterification of bread dough or flour lipid with 12.5% BF_3 in methanol (*173*). These compounds were identified as methoxylated esters produced in the reaction, like the methyl methoxystearate found when oleic acid was esterified with 51% BF_3 in methanol (*143*). Traces of a yellow material in dimethoxypropane–acid–methanol mixtures, presumably polymer, could be removed by passage through Florisil (*137*). Polymers formed during acid-catalyzed esterifications in the presence of dimethoxypropane eluted in the region of C_4–C_{12} esters from GLC columns with polar phases (*136*) and Ucon HB 2000 (*174*). The formation of polymers was inhibited by the addition of dimethyl sulfoxide (*174*).

To protect unsaturated fatty acids from autoxidation, common practice is to esterify under nitrogen or to add antioxidant (*130,175*). Lipid fractions separated by thin-layer chromatography may be scraped off the plate and methylated in the presence of silicic acid (*130,131,176*). Methanolysis in the presence of adsorbent required a higher temperature for comparable results (*132*).

B. Deactivation of Supports

Detrimental adsorptive effects of GLC supports are particularly evident when the amount of liquid phase is small (*177*), but they can also be significant at higher levels (*178,179*). The practical importance of the solid phase in successful GLC is attested to by the variety of supports proposed for GLC use.

Initial consideration of the support as a holder of liquid phase has been supplanted by realization of the participation of the solid support during the chromatographic process. This participation may be physical or chemical. Although steel, glass, polymers, and salts have been used for special applications, the most universally used support is diatomaceous. Early investigators, noting the reduction in polarity achieved by silane treatment of the silicic acid supports for reverse-phase chromatography (180,181) used similar treatments of diatomaceous earth with dichlorodimethylsilane for inactivation before coating with liquid phase. Peak symmetry was improved, especially for polar substances, and silanization of silica-containing supports is now common. For best results, extremely fine particles must be removed from the support before silanization (182), and the support should be dried (180–185). Current practice is to siliconize for 15 or 20 min with a 1–5% solution of dichlorodimethylsilane (181,186,187), or to pass gas saturated with the reagent through the support (56b, 180,185). The treated material should be washed to remove excess silane and HCl, then dried at about 100°C. HCl liberated in the reaction may alter characteristics of the support in the same manner as mild acid washing. Since silyl ethers undergo slow hydrolysis in moist air (188), siliconized supports should be protected against moisture during storage and used with dry carrier gas.

A satisfactory treatment or reaction time for one diatomaceous support is not necessarily adequate for another (189–192a). Chromatography of hydroxyl-containing steroids was used for evaluation of supports for low-loaded columns because of the susceptibility of these compounds to adsorption (193). The variability of Gas Chrom P subjected to different treatments is shown in Fig. 4. Acid-washed Gas Chrom P was superior to the acid-washed trimethylchlorosilane-treated support, although acid-washed dichlorodimethylsilane-treated support was superior to either.

Hexamethyldisilazane is less toxic and has a lower volatility than dimethyldichlorosilane (192a). It also has the theoretical advantage of reacting with a single hydroxyl group on the surface of the silicate to form a trimethylsilyl ether. Dichlorodimethylsilane, on the other hand, may react with a single —OH group or with two neighboring groups. Reaction with a single group leaves a reactive Si—Cl group, which can be hydrolyzed to yield a hydroxylated surface resembling the original or esterified by alcohol washing to leave an alkoxy site.

With two appropriately situated hydroxyl groups, the following reaction may occur (*179*).

$$-Si-O-Si- + SiCl_2Me_2 \rightarrow -Si-\!\!-\!\!-O-\!\!-\!\!-Si- + 2HCl$$

Suitable reaction conditions are refluxing 25 g of support in 60–80°C petroleum ether containing 7 ml of hexamethyldisilazane (*192a*) for 6 (Celite) to 10 hr (firebrick), washing with *n*-propanol and petroleum

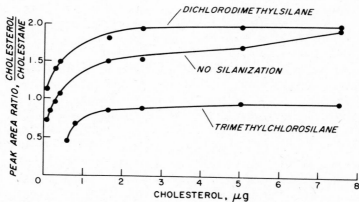

FIG. 4. Effect of support pretreatment on recovery of a polar compound. Liquid phase, 1% F-60. [From E. C. Horning, K. C. Maddock, K. V. Anthony, and W. J. A. Vanden Heuvel, 1963 (*193*).]

ether, and drying in vacuo. Supports prepared in this manner are stable to at least 200°C (*192a*). Hexamethyldisilazane has also been used for in situ deactivation by injection onto an Apiezon-coated column at operating temperature (*194*). The procedure has the functional advantage of eliminating the fragmentation and resultant exposure of new reactive surfaces frequently experienced in the handling of diatomite supports. This in situ method is not applicable in the presence of polyester liquid phases or materials that react with hexamethyldisilazane.

Comparison of different silane treatments was made by chromatographing dimethyl adipate, benzyl alcohol, and p-anisaldehyde on a 20% Apiezon column (195a). Acid-washed Chromosorb W was pretreated by refluxing for 2 hr with trimethylchlorosilane (TMCS), dimethyldichlorosilane (DMCS), or hexamethyldisilazane (HMDS). These differences are shown in Table 2.

TABLE 2

EFFECT OF DIFFERENT SILANIZING AGENTS ON CHROMOSORB W[a]

Support treatment	Theoretical plates	A_s, asymmetry[b]	Surface area, m²/g
TMCS	1020	1.51, 1.68	0.58, 0.54
DMCS	1070	0.95, 1.01	0.46, 0.49
HMDS	1020	1.15, 1.11	0.58, 0.50

[a] From J. J. Kirkland, 1963 (195a).
[b] A measure of asymmetry; A_s is equal to 1 for a symmetrical peak.

On this evidence, dimethyldichlorosilane-treated support seems somewhat more effective than the other silanes for the chromatography of esters under these conditions. A similar use of peak asymmetry of normal hydrocarbons was made to evaluate support deactivation (196). Other methods for reducing tailing of peaks include heating (197), metal plating (198), and addition of small amounts of polar organic substances such as polyvinylpyrrolidone (199) or polyethylene glycols (179,200).

The reactivity of unaltered siliceous material may be deduced from the mildness of the conditions that sufficed for the reaction of the surface of amorphous silica with normal and branched-chain alcohols (20). The reaction with alcohols was increased with partially dehydrated silica (20). Reaction with the silanol groups of the surface of porous glass was suggested as a possible explanation for the deactivation caused by alcohol soaking (201) and decrease in the number of surface silanol groups after treatment with alcohol vapor (202). Similarly, improved performance of a glass capillary after silanization indicated the high initial activity of a material sometimes categorized as unreactive (203).

Although silanization of GLC supports to improve their performance has become routine in the lipid field, it is not always beneficial. Criteria for improvement are usually surface-area reduction and decrease in retention volume of polar compounds. These means of evaluation may not be sufficient for characterizing GLC properties of supports. For the liquid phase to cover the support with the thin, stable film necessary for high efficiency, wetting of the surface must occur. Diatomaceous supports usually have a sufficiently high surface energy to be wetted by most organic liquid phases (*195b*). Modification by the addition of organic groups, as by silanization, may reduce the surface energy to the point where wetting by the liquid phase is impaired. Poor wetting characteristics can account for the inferiority of Fluoropak as a support for Carbowax 400 (*204*) and the increased tailing found with supports inactivated with more than 0.1% dichlorodimethylsilane (*183*) and may account for some of the failures experienced in the preparation of polyester-coated capillary columns.

C. Collection of GLC Samples

Collection of esters eluting from GLC columns is necessary in preparative work and in separation of individual components for identification. Methyl esters emerging from a GLC column tend to form fogs or solid aerosols on cooling of the vapor phase, making quantitative recovery difficult (*93,192b,205*). The probability of fog formation is increased when the vapor is cooled rapidly (*206,207*). With rapid cooling, high-molecular-weight solutes may not diffuse to the walls of the cooling vessel and supersaturation occurs.

The yields of recovered samples are quite variable. Collections of compounds from the effluent in cooled U tubes or on solvent-dampened plugs of cotton and glass wool give fatty ester recoveries of 40–90% (*84,87,94*). Efficiency of collection can be improved by trapping the solute on Celite coated with polyester (*208*) or uncoated (*71,209*). An easily made trap incorporating sintered-glass Drechsel bottles affords 90% recoveries of polyenoic esters from a preparative column (*210*). Radioactive fatty esters have been collected on anthracene crystals (*95*) or on anthracene coated with various trapping agents. Of the several different liquid coatings used on the anthracene, silicone proved most satisfactory (*70*). Small samples collected on uncoated glass beads (*74*) were not easily recovered. Precoating the beads with mineral oil facilitated elution from the beads but contaminated the sample.

In connection with the quantitative recovery of very small quantities of lipid from glass surfaces, it should be borne in mind that the bonding forces between adsorbed molecules and the surface can be strong enough to resist removal by lipid solvents (211). A superior removal of 1-^{14}C-stearic acid from frosted glass by the more polar solvents than by the nonpolar ones is demonstrated in Fig. 5. Prolonged washing in carbon tetrachloride was manifestly inadequate for complete re-

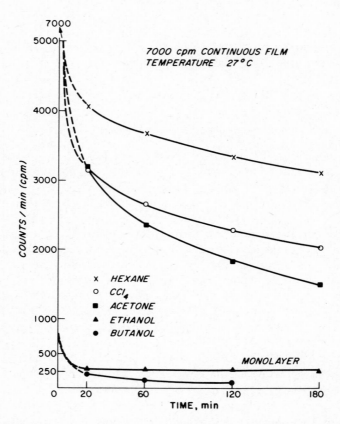

FIG. 5. Solvent efficiency in removal of ^{14}C-stearic acid from glass. [From J. C. Harris and J. Satanek, 1961 (211).]

moval. This adsorption might be due to the metal or metal oxide content of glass, since adsorption of long-chain acids and esters on such surfaces has been demonstrated (212). The adsorption of fatty acids on active surfaces can result in formation of fatty acid metal salts (213–216).

Potential errors arising from several sources of gas chromatographic procedure were evaluated by collection of 1-^{14}C-labeled methyl esters from Apiezon and polyester phases (*94,95,217*). Trapping of fogs on fine-pore (0.8 μ) filters was more efficient than collection on solvent-wetted cotton (*93*). Recoveries of methyl 1-^{14}C-stearate, -oleate, and -lignocerate were better than 97%, with the discrepancy attributable to the chromatographic process (on SE-30) rather than deficiencies in collection.

Other techniques may be used. The ester fog escaping condensation may be reheated, then recondensed (*218*). Sudden cooling of high-boiling-point compounds, rather than gradual gradient cooling, is said to produce a cloud of charged particles that condense on chilling and re-form on warming (*219*). Electrostatic precipitation greatly increases collection efficiency of these compounds (*219,220*). This technique has been successfully applied to the collection of methyl esters (*221*). Alternating current is more effective than direct, but its use may lead to decomposition of fatty esters (*221*). When carbon dioxide is used as the carrier gas, it can be condensed in a liquid-nitrogen trap, entraining the solute (*205*). Sublimation of the carbon dioxide leaves the residual sample. Similarly, a solvent may be injected into a noncondensible carrier gas effluent and subsequently frozen to trap volatiles emerging from the column (*222*). Apparatus for introduction of carbon tetrachloride into a GLC system in this fashion is commercially available.

D. Peak-Area and Retention-Time Measurements

The mechanical aspects of GLC peak-area measurement have not received as much attention as other aspects of GLC. Errors in peak-area measurement can cause appreciable inaccuracies in the values for mixtures. The most common methods of assessing peak area are by multiplication of the peak height by its width at half height and by triangulation (evaluating the area enclosed by tangents to the curve). Values for Gaussian peaks obtained by these methods represent less than the true area (*157b,223,224*). Measurement of the peak width at two points is said to compensate for peak asymmetry and give true area (*224*):

$$\text{area} = \frac{\text{height} \times (\text{width at 0.15 height} + \text{width at 0.85 height})}{2}$$

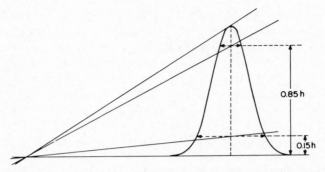

FIG. 6. Template for rapid location of 0.15*h* and 0.85*h* positions on the altitude of a peak. [From L. Condal-Bosche, 1964 (*224*).]

Use of the template shown in Fig. 6 makes measurement of these widths relatively rapid. Errors in measurement are variable. The errors that may be associated with different methods of area estimation (*225*) are listed in Table 3. Improvements in accuracy have been

TABLE 3

PRECISION OF MEASUREMENT BY DIFFERENT METHODS[a]

Method	Mean error[b] for following peak areas, mm^2		
	0–50, av. 25	51–100, av. 75	101–300, av. 200
Planimetry			
Mean error, mm^2	10.0	6.84	10.0
Mean error, %	40	9.1	5.0
Fivefold planimetry			
Mean error, mm^2	5.75	3.84	4.61
Mean error, %	23.0	5.1	2.3
Weighing			
Mean error, mm^2	4.37	7.53	5.50
Mean error, %	17.5	10.0	2.8
Triangulation			
Mean error, mm^2	4.18	7.14	6.8
Mean error, %	16.7	9.5	3.4
Height × width at half height			
Mean error, mm^2	4.80	9.38	8.52
Mean error, %	19.2	12.5	4.3

[a] From J. Janak, 1960 (*225*).
[b] $\sqrt{\Sigma\Delta^2/(N-1)}$.

claimed for measurement of peak width at 45.4% of height multiplied by peak height (*226*) and for counting the number of squares of chart paper circumscribed by the peak (*227*). A method for estimating the areas of unsymmetrical peaks has also been described (*228*).

An ideal chromatogram consists of a discrete, symmetrical peak for each component superimposed on a constant base line. In practice, it is not unusual for a chromatogram to be deficient in one or more of these respects. Although overlapping peaks may be esthetically distressing, "resolution must be very poor indeed to interfere with quantitative results" (*229*). For increased accuracy, overlapping peaks may be corrected by application of a construction method (*229*) or experi-

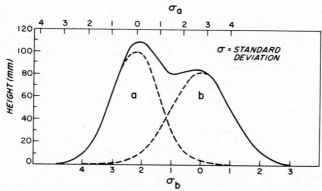

FIG. 7. Overlapping normal probability curves. [From D. M. Smith and J. C. Bartlet, 1961 (*231*).]

mental correction factors (*230*). In the case of overlapping peaks with discernible peak maxima, the positions of the maxima are altered slightly from those of the pure compounds. The mathematical aspect of the change for two additive normal probability curves is shown in Fig. 7 (*231*). It demonstrates the contribution of one curve to the apparent height of the other, as well as horizontal displacement of the maxima.

Another cause for variation of retention times is an actual alteration of the liquid phase by the presence of the sample. That is, the passage of a compound *A* through the column can, by its solubility in the stationary phase, change the partition coefficient for a compound *B* following it. It is not improbable that this alteration of the liquid phase could be responsible for the effects of concentration on the relative retention time of a minor component of a mixture (*232*). Although the shifts obtained were small, they were considered sufficient for misidentification. In connection with the accurate measurement of retention time, notice should be taken of the effect of sample size (*233*). Retention time can be either decreased or increased by an increase in sample size (*234,235*). The general increase in standard deviation of R_f with increase in retention time that was noted (*233*), but not explained, could arise from such sample-size effects.

Asymmetric peaks present a problem in assessing both retention time and area. A method for estimating the relative retention of a tailing peak was described (*236*). Mathematical correction for contribution of an asymmetrical peak to the height of an overlapping peak can be made (*237*). In connection with area calculation for peaks on sloping base lines, a mathematical justification for use of the vertical height as an approximation of the peak height has been presented (*238*).

The convenience of obtaining quantitative and qualitative information simultaneously contributes to the utility of gas chromatographic techniques. Quantitation, however, can be a vexing aspect of fatty acid ester separations. Detector response represents only one source of error in a system. Transesterification with liquid phases, reaction with the solid support, nonlinear splitting of the gas stream, and lability of specific substituents may contribute further error, or even compensating error. For rigorously accurate quantitative results it is imperative to standardize the complete system used. A special report detailing calibration procedures specific for the GLC of higher fatty esters has appeared (*166*).

Flame ionization detectors, in general, give a response that is correlated with carbon content, but substituents and spatial arrangement may alter this somewhat (*11,239,240*). Response of the methyl esters of the long-chain fatty acids is reasonably close to weight per cent, with little alteration by olefinic unsaturation (*133,241–243*). However, there are indications that ignoring oxygen-bonded carbons may give improved accuracy (*227,244–246*).

Argon ionization detectors represent a family of instruments the operating characteristics of which are exceedingly dependent upon design (*247*). The response is proportional to mass for compounds of a given class above a molecular weight of about 150. Compounds with a large cross-sectional area give a greater response, whereas halogenated and nitro compounds capable of capturing free electrons give a smaller response (*248,249*). Contradictory reports, however, emphasize the individuality of the response of the argon detector to higher esters and acids (*17,71,134,164,165,243,248,250–252*).

The simplicity and precision of the thermal conductivity cell recommend it, although the low sensitivity of older models is disadvantageous when samples are limited. The cell responds to a difference in thermal conductivity of the carrier gas and the eluted compound. Since thermal conductivity is a function of the square root of the molecular weight, changes in the thermal conductivity of the higher members of a homologous series are small, and response for high-molecular-weight compounds approximates weight composition (*206b*). Particularly when the molecular-weight range in the sample is small, calculation of weight per cent composition from area per cent composition can give accurate results for long-chain fatty esters (*253,254*). General linearity of detector response has been demonstrated (*150, 255–257*), although accuracy may be improved by the application of correction factors (*109,166,258*).

REFERENCES†

1. A. J. P. Martin and R. L. M. Synge, *Biochem. J.*, **35**, 1358 (1941).
2. K. Oette and E. H. Ahrens, Jr., *Anal. Chem.*, **33**, 1847 (1961).
3. S. Abrahamsson, S. Stallberg-Stenhagen, and E. Stenhagen, in *Progress in the Chemistry of Fats and Other Lipids* Vol. 7 (R. T. Holman and T. Malkin, eds.), Macmillan, New York, 1963, pp. 100–104.
4. J. W. Farquhar, W. Insull, Jr., P. Rosen, W. Stoffel, and E. H. Ahrens, Jr., *Nutrition Rev. Suppl.,* **17**, 1 (1959).
5. J. Cason, G. L. Lange, and H. R. Urscheler, *Tetrahedron*, **20**, 1955 (1964).
6. W. Cocker, T. Dahl, and T. B. H. McMurry, *J. Chem. Soc.*, **1963**, 1654.
7. R. W. Taft, Jr., in *Steric Effects in Organic Chemistry* (M. S. Newman, ed.), Wiley, New York, 1956, pp. 556–675.
8. R. A. Stein, unpublished work.
9. J. S. O'Brien and G. Rouser, *Anal. Biochem.*, **7**, 288 (1964).
10. A. P. Tulloch, *J. Am. Oil Chemists' Soc.*, **41**, 833 (1964).
11. W. J. A. Vanden Heuvel, W. L. Gardinier, and E. C. Horning, *J. Chromatog.*, **19**, 263 (1965).
12. N. Weaver, J. H. Law, and N. C. Johnston, *Biochim. Biophys. Acta*, **84**, 305 (1964).

† Literature containing information applicable to the chromatography of radioactive lipid material is designated with an asterisk.

13. Y. Kishimoto and N. S. Radin, *J. Lipid Res.*, **1**, 72 (1960).
14. H. P. Tulloch, J. F. T. Spencer, and P. A. J. Gorin, *Can. J. Chem.*, **40**, 1326 (1962).
15. I. Kitagawa, M. Sugai, and F. A. Kummerow, *J. Am. Oil Chemists' Soc.*, **39**, 217 (1962).
16. R. Wood, E. L. Bever, and F. Snyder, *J. Am. Oil Chemists' Soc.*, **43**, 139A (1966).
17. K. K. Carroll, *J. Lipid Res.*, **3**, 263 (1962).
18. R. D. Wood, P. K. Raju, and R. Reiser, *J. Am. Oil Chemists' Soc.*, **42**, 81 (1965).
19. R. G. Ackman and R. D. Burgher, *J. Chromatog.*, **6**, 541 (1961).
20. C. C. Ballard, E. C. Broge, R. K. Iler, D. S. St. John, and J. R. McWhorter, *J. Phys. Chem.*, **65**, 20 (1961).
21. M. van Ammers, M. H. Deinema, C. A. Landheer, and M. H. M. van Rooyen, *Rec. Trav. Chim.*, **83**, 708 (1964).
22. A. P. Tulloch and J. F. T. Spencer, *Can. J. Chem.*, **42**, 830 (1964).
23. R. L. Shriner, *Org. Reactions,* **1**, 1 (1942).
24. R. A. Stein, *J. Chromatog.*, **6**, 118 (1961).
25. P. F. V. Ward and R. J. Hall, *Nature*, **201**, 611 (1964).
26. D. E. Martire and L. E. Pollara, in *Advances in Chromatography*, Vol. 1 (J. C. Giddings and R. A. Keller, eds.), Dekker, New York, 1965, pp. 335–363.
27. G. C. Pimentel and A. L. McClellan, *The Hydrogen Bond,* Freeman, San Francisco, pp. 195–201.
28. A. B. Littlewood, *Anal. Chem.*, **36**, 1441 (1964).
29. H. Gershon and J. A. A. Renwick, *J. Chromatog.*, **20**, 134 (1965).
30. E. D. Smith and A. B. Gosnell, *Anal. Chem.*, **34**, 438 (1962).
31. R. A. Landowne and S. R. Lipsky, *Nature*, **182**, 1731 (1958).
32. W. R. Koehler, J. L. Solan, and H. T. Hammond, *Anal. Biochem.*, **8**, 353 (1964).
33. R. A. Stein, *J. Am. Oil Chemists' Soc.*, **38**, 636 (1961).
34. G. P. Cartoni, A. Liberti, U. Pallotta, and R. Palombari, *Riv. Ital. Sostanze Grasse,* **40**, 653 (1963).
35. K. Fontell, R. T. Holman, and G. Lambertsen, *J. Lipid Res.*, **1**, 391 (1960).
36. C. Panos, *J. Gas Chromatog.*, **3**, 278 (1965).
37. R. A. Landowne and S. R. Lipsky, *Biochim. Biophys. Acta*, **47**, 589 (1961).
38. A. T. James, *Analyst,* **88**, 572 (1963).
39. S. R. Lipsky, J. E. Lovelock, and R. A. Landowne, *J. Am. Chem. Soc.*, **81**, 1010 (1959).
40. S. R. Lipsky, R. A. Landowne, and J. E. Lovelock, *Anal. Chem.*, **31**, 852 (1959).
41. C. Litchfield, R. Reiser, and A. F. Isbell, *J. Am. Oil Chemists' Soc.*, **40**, 302 (1963).
42. R. A. Landowne and S. R. Lipsky, *Biochim. Biophys. Acta,* **46**, 1 (1961).
43. C. Litchfield, A. F. Isbell, and R. Reiser, *J. Am. Oil Chemists' Soc.*, **40**, 330 (1962).
44. L. J. Morris, R. T. Holman, and K. Fontell, *J. Lipid Res.*, **1**, 412 (1960).

45. I. Zeman, *J. Gas Chromatog.*, **3**, 18 (1965).
46. F. D. Gunstone and A. J. Sealy, *J. Chem. Soc.*, **1963**, 5772.
47. T. K. Miwa, K. L. Mikolajczak, F. R. Earle, and I. A. Wolff, *Anal. Chem.*, **32**, 1739 (1960).
48. H. Rakoff and R. E. Anderson, *J. Am. Oil Chemists' Soc.*, **42**, 459A (1965).
49. A. T. James and A. J. P. Martin, *Biochem. J.*, **50**, 679 (1952).
50. L. D. Metcalfe, *Nature*, **188**, 142 (1960).
51. L. D. Metcalfe, *J. Gas Chromatog.*, **1**, 7 (1963).
52. E. M. Emery and W. E. Koerner, *Anal. Chem.*, **34**, 1196 (1962).
53. P. Jowett and B. J. Horrocks, *Nature*, **192**, 966 (1961).
54. J. G. Nikelly, *Anal. Chem.*, **36**, 2244 (1964).
55. H. Yunoki and C. Tsutamura, *Hiroshima J. Med. Sci.*, **13**, 323 (1964).
56. A. Prevot, in *Manuel Pratique de Chromatographie en Phase Gazeuse* (J. Tranchant, ed.), Masson, Paris, 1964, (a) p. 56, (b) p. 55.
57. W. R. Mayberry and G. J. Prochazha, *J. Gas Chromatog.*, **3**, 232 (1965).
58. R. G. Ackman, R. D. Burgher, and J. C. Sipos, *Nature*, **200**, 777, (1963).
59. O. L. Hollis, *Anal. Chem.*, **38**, 309 (1966).
60. A. T. James, *J. Chromatog.*, **2**, 552 (1959).
61. F. P. Woodford and C. M. van Gent, *J. Lipid Res.*, **1**, 188 (1960).
62. T. K. Miwa, *J. Am. Oil Chemists' Soc.*, **40**, 309 (1963).
63. E. Kovats, *Helv. Chim. Acta*, **41**, 1915 (1958).
64. E. Kovats and H. Strickler, *J. Gas Chromatog.*, **3**, 244 (1965).
65. E. Kovats, in *Advances in Chromatography*, Vol. 1 (J. C. Giddings and R. A. Keller, eds.), Dekker, New York, 1965, pp. 229–247.
66. L. S. Ettre, *Anal. Chem.*, **36**, 31A (1964).
67. G. Schomburg, *J. Chromatog.*, **14**, 157 (1964).
*68. J. Adloff, *Chromatog. Rev.*, **4**, 19 (1962).
*69. C. Hitchcock and A. T. James, *J. Lipid Res.*, **5**, 593 (1964).
*70. A. Karmen and H. R. Tritch, *Nature*, **186**, 150 (1960).
*71. F. Davidoff and E. D. Korn, *J. Biol. Chem.*, **238**, 3199 (1963).
*72. H. Goldfine and K. Bloch, *J. Biol. Chem.*, **236**, 2596 (1961).
*73. J. L. Gellerman and H. Schlenk, *J. Protozool.*, **12**, 178 (1965).
74. L. S. Grey, L. D. Metcalfe, and S. W. Leslie, in *Gas Chromatography* (L. Fowler, ed.), Academic Press, New York, 1963, pp. 203–212.
*75. M. G. Horning, D. B. Martin, A. Karmen, and P. R. Vagelos, *J. Biol. Chem.*, **236**, 669 (1961).
*76. A. T. James and E. A. Piper, *J. Chromatog.*, **5**, 265, (1961).
*77. A. T. James and E. A. Piper, *Anal. Chem.*, **35**, 515 (1963).
*78. W. J. Lannarz, R. J. Light, and K. Bloch, *Proc. Natl. Acad. Sci. U.S.*, **48**, 840 (1962).
*79. L. H. Mason, H. J. Dutton, and L. R. Bair, *J. Chromatog.*, **2**, 322 (1959).
*80. H. Mohrhauer and R. T. Holman, *J. Am. Oil Chemists' Soc.*, **42**, 639 (1965).
*81. G. Popjak, A. E. Lowe, D. Moore, L. Brown, and F. A. Smith, *J. Lipid Res.*, **1**, 29 (1960).
*82. R. Blomstrand and J. Gurtler, *Acta Chem. Scand.*, **19**, 249 (1965).

*83. S. C. Brooks, V. C. Godefroi, and W. L. Simpson, *J. Lipid Res.*, **7**, 95 (1966).

*84. W. R. Harlan and S. J. Wakil, *J. Biol. Chem.*, **238**, 3216 (1963).

*85. J. G. Hildebrand, *Biochemistry*, **3**, 1304 (1964).

*86. A. Karmen, M. Whyte, and D. S. Goodman, *J. Lipid Res.*, **4**, 312 (1963).

*87. J. H. Law, H. Zalkin, and T. Kaneshiro, *Biochim. Biophys. Acta*, **70**, 143 (1963).

*88. O. K. Reiss, J. G. Warren, and J. K. Newman, *Lipids*, **1**, 230 (1966).

*89. A. Karmen, L. Guiffrida, and R. L. Bowman, *J. Lipid Res.*, **3**, 44 (1962).

*90. A. Karmen, I. McCaffrey, J. W. Winkelman, and R. L. Bowman, *Anal. Chem.*, **35**, 537 (1963).

*91. M. A. Kirschner and M. B. Lipsett, *J. Lipid Res.*, **6**, 7 (1965).

*92. A. Murray and D. L. Williams, *Organic Syntheses with Isotopes*, Wiley (Interscience), New York, 1958, pp. 1661–1663.

*93. A. K. Hajra and N. S. Radin, *J. Lipid Res.*, **3**, 131 (1962).

*94. H. Meinertz and V. P. Dole, *J. Lipid Res.*, **3**, 140 (1962).

*95. M. Pascaud, *J. Chromatog.*, **10**, 125 (1963).

96. K. L. Mikolajczak and M. O. Bagby, *J. Am. Oil Chemists' Soc.*, **41**, 391 (1964).

97. I. A. Wolff and T. K. Miwa, *J. Am. Oil Chemists' Soc.*, **42**, 208 (1965).

98. L. J. Morris, R. T. Holman, and K. Fontell, *J. Am. Oil Chemists' Soc.*, **37**, 323 (1960).

99. J. W. Farquhar, *J. Lipid Res.*, **3**, 21 (1962).

100. J. A. Cornelius and G. Shone, *Chem. Ind. (London)*, **1963**, 1246.

101. A. P. Tulloch, B. M. Craig, and G. A. Ledingham, *Can. J. Microbiol.*, **5**, 485 (1959).

102. S. F. Herb, P. Magidman, and R. A. Barford, *J. Am. Oil Chemists' Soc.*, **41**, 222 (1964).

103. C. J. Thompson, H. J. Coleman, C. C. Ward, and H. T. Rall, *Anal. Chem.*, **32**, 424 (1960).

104. C. J. Thompson, H. J. Coleman, C. C. Ward, and H. T. Rall, *Anal. Chem.*, **34**, 151 (1962).

105. C. J. Thompson, H. J. Coleman, R. L. Hopkins, and H. T. Rall, *Anal. Chem.*, **37**, 1042 (1965).

106. M. Beroza and R. Sarmiento, *Anal. Chem.*, **35**, 1353 (1963).

107. M. Beroza and R. Sarmiento, *Anal. Chem.*, **36**, 1744 (1964).

108. M. Beroza and R. Sarmiento, *Anal. Chem.*, **37**, 1040 (1965).

109. T. L. Mounts and H. J. Dutton, *Anal. Chem.*, **37**, 641 (1965).

110. E. W. Robb, in *Gas Chromatography 1962* (M. van Swaay, ed.), Butterworth, London, 1962, p. 354.

111. G. F. Thompson and K. Smith, *Anal. Chem.*, **37**, 1591 (1965).

112. J. W. Ralls, *J. Agr. Food Chem.*, **8**, 141 (1960).

113. I. J. Hunter, *J. Chromatog.*, **7**, 288 (1962).

114. V. L. Davison and H. J. Dutton, *J. Am. Oil Chemists' Soc.*, **42**, 461A (1965).

*115. M. Senn, W. J. Richter, and A. L. Burlingame, *J. Am. Chem. Soc.*, **87**, 680 (1965).

116. G. Clement and J. Bezard, *Compt. Rend.,* **253,** 564 (1961).

117. B. M. Craig, A. P. Tulloch, and N. L. Murty, *J. Am. Oil Chemists' Soc.,* **40,** 61 (1963).

118. R. G. Ackman and R. D. Burgher, *J. Lipid Res.,* **5,** 130 (1964).

119. L. Hartman, *J. Am. Oil Chemists' Soc.,* **42,** 664 (1965).

120. N. W. R. Daniels and J. W. Richmond, *Nature* **187,** 55 (1960).

121. H. T. Ast, *Anal. Chem.* **35,** 1539 (1963).

122. J. L. Foote, R. J. Allen, and B. W. Agranoff, *J. Lipid Res.,* **6,** 518 (1965).

123. T. J. Heath, E. P. Adams, and B. Morris, *Biochem. J.,* **92,** 511 (1964).

124. M. Kates, *J. Lipid Res.,* **5,** 132 (1964).

125. Y. Kishimoto and N. S. Radin, *J. Lipid Res.,* **6,** 435 (1965).

126. G. J. Nelson and N. K. Freeman, *J. Biol. Chem.,* **235,** 578 (1960).

127. W. A. Spencer and R. Schaffrin, *Can. J. Biochem.,* **42,** 1659 (1964).

128. W. Stoffel, F. Chu, and E. H. Ahrens, Jr., *Anal. Chem.,* **31,** 307 (1959).

129. J. S. O'Brien, D. L. Fillerup, and J. F. Mead, *J. Lipid Res.,* **5,** 329 (1964).

130. D. E. Bowyer, W. M. F. Leat, A. N. Howard, and G. A. Gresham, *Biochim. Biophys. Acta,* **70,** 423 (1963).

131. A. M. Chalvardjian, *Biochem. J.,* **90,** 518 (1964).

132. G. L. Feldman and G. Rouser, *J. Am. Oil Chemists' Soc.,* **42,** 290 (1965).

133. W. R. Morrison and L. M. Smith, *J. Lipid Res.,* **5,** 600 (1964).

134. P. Ways, C. F. Reed, and D. J. Hanahan, *J. Clin. Invest.,* **42,** 1248 (1963).

135. J. F. Berry, W. H. Cevallos, and R. R. Wade, Jr., *J. Am. Oil Chemists' Soc.,* **42,** 492 (1965).

136. M. E. Mason and G. R. Waller, *Anal. Chem.,* **36,** 583 (1964).

137. N. S. Radin, A. K. Hajra, and Y. Akahori, *J. Lipid Res.,* **1,** 250 (1960).

138. I. L. Shapiro and D. Kritchevsky, *J. Chromatog.,* **18,** 599 (1965).

139. O. S. Duron and A. Nowotny, *Anal. Chem.,* **35,** 370 (1963).

140. S. A. Hyun, G. V. Vahouny, and C. R. Treadwell, *Anal. Biochem.,* **10,** 193 (1965).

141. L. D. Metcalfe and A. A. Schmitz, *Anal. Chem.,* **33,** 363 (1961).

142. K. Abel, J. I. Peterson, and H. de Schmertzing, *J. Bacteriol.,* **85,** 1039 (1963).

143. A. K. Lough, *Biochem. J.,* **90,** 4c (1964).

144. J. I. Peterson, H. de Schmertzing, and K. Abel, *J. Gas Chromatog.,* **3,** 126 (1965).

145. W. R. Supina, in *Biomedical Applications of Gas Chromatography* (H. A. Szymanski, ed.), Plenum Press, New York, 1964 (a) p. 276, (b) pp. 271–305.

146. N. B. Lorette and J. H. Brown, Jr., *J. Org. Chem.,* **24,** 261 (1959).

147. B. M. Craig and N. L. Murty, *J. Am. Oil Chemists' Soc.,* **36,** 549 (1959).

148. G. G. Esposito and M. H. Swann, *Anal. Chem.,* **34,** 1048 (1962).

149. E. J. Gauglitz, Jr. and L. W. Lehman, *J. Am. Oil Chemists' Soc.,* **40,** 197 (1960).

150. M. E. Mason, M. E. Eager, and G. R. Waller, *Anal. Chem.,* **36,** 587 (1964).

151. G. V. Marinetti, *Biochemistry,* **1,** 350 (1962).

152. F. E. Luddy, R. A. Barford, and R. W. Riemenschneider, *J. Am. Oil Chemists' Soc.,* **37,** 447 (1960).

153. J. M. deMan, *J. Dairy Sci.,* **47,** 546 (1964).

154. L. D. Metcalfe, A. A. Schmitz, and J. R. Pelka, *Anal. Chem.,* **38,** 514 (1966).

155. K. S. Markley, ed., in *Fatty Acids. Their Chemistry, Properties, Production, and Uses,* 2nd ed., Wiley (Interscience), New York, 1961, p. 760.
156. H. Schlenk and J. L. Gellerman, *Anal. Chem.,* **32,** 1412 (1960).
157. H. P. Burchfield and E. E. Storrs, *Biochemical Applications of Gas Chromatography,* Academic, New York, 1962 (*a*) p. 589, (*b*) p. 122, (*c*) pp. 267–279, 527–558.
158. G. G. McKeown and S. I. Read, *Anal. Chem.,* **37,** 1781 (1965).
159. E. D. Amstutz and R. R. Meyers, in *Organic Syntheses,* Collective Vol. 2 (A. H. Blatt, ed.), Wiley, New York, 1943, p. 462.
160. A. F. McKay, *J. Am. Chem. Soc.,* **70,** 1974 (1948).
161. T. J. deBoer and J. H. Backer, *Rec. Trav. Chim.,* **73,** 229 (1954).
162. L. F. Fieser and M. Fieser, *Advanced Organic Chemistry,* Reinhold, New York, 1961, pp. 377–378.
163. H. P. Kaufman and G. Mankel, *Fette Seifen Anstrichmittel,* **65,** 179 (1963); *Anal. Abstr.,* **11,** 1117 (1964).
164. M. L. Vorbeck, L. R. Mattick, F. A. Lee, and C. S. Pederson, *Anal. Chem.,* **33,** 1512 (1961).
165. G. R. Jamieson and E. H. Reid, *J. Chromatog.,* **17,** 230 (1965).
166. E. D. Horning, E. H. Ahrens, Jr., S. R. Lipsky, F. H. Mattson, J. F. Mead, D. A. Turner, and W. H. Goldwater, *J. Lipid Res.,* **5,** 20 (1964).
167. A. T. James, in *Methods of Biochemical Analysis* Vol. 8 (D. Glick, ed.), Wiley (Interscience), New York, 1960, (*a*) p. 19, (*b*) pp. 1–59.
168. A. T. James and J. P. W. Webb, *Biochem. J.,* **66,** 515 (1957).
169. W. R. Morrison, T. D. V. Lawrie, and J. Blades, *Chem. Ind. (London),* **1961,** 1534.
170. S. R. Carter and J. A. V. Butler, *J. Chem. Soc.,* **125,** 963 (1924).
171. P. V. Johnston and B. I. Roots, *J. Lipid Res.,* **5,** 477 (1964).
172. L. F. Eng, Y. L. Lee, and B. Gerstl, *J. Lipid Res.,* **5,** 128 (1964).
173. J. B. M. Coppock, N. W. R. Daniels, and P. W. R. Eggitt, *J. Am. Oil Chemists' Soc.,* **42,** 652 (1965).
174. P. G. Simmonds and A. Zlatkis, *Anal. Chem.,* **37,** 302 (1965).
175. C. J. F. Bottcher, F. P. Woodford, E. Boelsma-van Houte, and C. M. van Gent, *Rec. Trav. Chim.,* **78,** 794 (1959).
176. A. Rosenberg and N. Stern, *J. Lipid Res.,* **7,** 122 (1966).
177. D. T. Sawyer and J. K. Barr, *Anal. Chem.,* **34,** 1052 (1962).
178. J. J. Kirkland, *Anal. Chem.,* **35,** 2003 (1963).
179. J. Bohemen, S. H. Langer, R. H. Perrett, and J. H. Purnell, *J. Chem. Soc.,* **1960,** 2444.
180. G. A. Howard and A. J. P. Martin, *Biochem. J.,* **46,** 532 (1950).
181. F. A. Vandenheuvel and D. R. Vatcher, *Anal. Chem.,* **28,** 838 (1956).
182. W. R. Supina, R. S. Henly, and R. F. Kruppa, *J. Am. Oil Chemists' Soc.,* **43,** 202 A (1966).
183. W. L. Holmes and E. Stack, *Biochim. Biophys. Acta,* **56,** 163 (1962).
184. P. Urone and J. F. Parcher, *J. Gas Chromatog.,* **3,** 35 (1965).
185. E. C. Horning, E. A. Moscatelli, and C. C. Sweeley, *Chem. Ind. (London),* **1959,** 751.
186. J. E. Arnold and H. M. Fales, *J. Gas Chromatog.,* **3,** 131 (1965).

187. J. Sjovall, C. R. Meloni, and D. A. Turner, *J. Lipid Res.,* **2,** 317 (1961).

188. E. J. Hedgley and W. G. Overend, *Chem. Ind. (London),* **1960,** 378.

189. H. Buhring, *J. Chromatog.,* **11,** 452 (1963).

190. I. Hornstein and P. F. Crowe, *Anal. Chem.,* **33,** 310 (1961).

191. R. H. Perrett and J. H. Purnell, *J. Chromatog.,* **7,** 455 (1962).

192. H. Purnell, *Gas Chromatography,* Wiley, New York, 1962, (*a*) pp. 236–238, (*b*) p. 261.

193. E. C. Horning, K. C. Maddock, K. V. Anthony, and W. J. A. Vanden Heuvel, *Anal. Chem.,* **35,** 526 (1963).

194. E. P. Atkinson and G. A. P. Tuey, *Nature,* **199,** 482 (1963).

195. J. J. Kirkland, in *Gas Chromatography* (L. Fowler, ed.), Academic Press, New York, 1963, (*a*) pp. 88–91, (*b*) pp. 91–94.

196. S. Dal Nogare and J. Chiu, *Anal. Chem.,* **34,** 890 (1962).

197. J. F. K. Huber and A. I. M. Keulemans, *J. Gas Chromatog.,* **1,** 30 (1963).

198. E. C. Ormerod and R. P. W. Scott, *J. Chromatog.,* **2,** 65 (1959).

199. W. J. A. Vanden Heuvel, W. L. Gardiner, and E. C. Horning, *Anal. Chem.,* **35,** 1745 (1963).

200. T. Johns, in *Gas Chromatography* (V. J. Coates, H. J. Noebels, and I. S. Fagerson, eds.), Academic Press, New York, 1958, pp. 31–39.

201. H. L. MacDonell, J. M. Noonan, and J. P. Williams, *Anal. Chem.,* **35,** 1253 (1963).

202. M. Folman, *Trans. Faraday Soc.,* **57,** 2000 (1961).

203. A. V. Kiselev, in *Gas Chromatography 1962* (M. van Swaay, ed.), Butterworth, London, 1962, pp. xxxiv–lii.

204. D. T. Sawyer and J. K. Barr, *Anal. Chem.,* **34,** 1518 (1962).

205. I. Hornstein and P. Crowe, *Anal. Chem.,* **37,** 170 (1965).

206. S. Dal Nogare and R. S. Juvet, Jr., *Gas-Liquid Chromatography, Theory and Practice,* Wiley (Interscience), New York, 1962, (*a*) p. 253, (*b*) pp. 196–197.

207. A. Klinkenberg, in *Vapour Phase Chromatography* (D. H. Desty, ed.), Butterworth, London, 1957, p. 211.

208. W. Kemp and O. Rogne, *Chem. Ind. (London),* **1965,** 418.

209. D. A. Shearer, B. C. Stone, and W. A. McGugan, *Analyst,* **88,** 147 (1963).

210. W. R. Hardy, *J. Chromatog.,* **17,** 177 (1965).

211. J. C. Harris and J. Satanek, *J. Am. Oil Chemists' Soc.,* **38,** 169 (1961).

212. F. A. Matsen, A. C. Makrides, and N. Hackerman, *J. Chem. Phys.,* **22,** 1800 (1954).

213. A. W. Adamson, *Physical Chemistry of Surfaces,* Wiley (Interscience), New York, 1960, p. 581.

214. E. L. Cook and N. Hackerman, *J. Phys. Colloid Chem.,* **55,** 549 (1951).

215. E. B. Greenhill, *Trans. Faraday Soc.,* **45,** 625 (1949).

216. V. L. Burton, *J. Am. Chem. Soc.,* **71,** 4117 (1949).

*217. J. Bezard, P. Boucrot, and G. Clement, *J. Chromatog.,* **14,** 386 (1964).

218. H. Schlenk and D. M. Sand, *Anal. Chem.,* **34,** 1676 (1962).

219. P. Kratz, M. Jacobs, and B. M. Mitzner, *Analyst,* **84,** 671 (1959).

220. A. E. Thompson, *J. Chromatog.,* **6,** 454 (1961).

221. L. Borka and O. S. Privett, *Lipids,* **1,** 104 (1966).

222. J. H. Jones and C. D. Ritchie, *J. Assoc. Offic. Agr. Chem.,* **41**, 753 (1958).

223. A. T. James and V. R. Wheatley, *Biochem. J.,* **63**, 269 (1956).

224. L. Condal-Bosche, *J. Chem. Educ.,* **41**, A 235 (1964).

225. J. Janak, *J. Chromatog.,* **3**, 308 (1960).

226. P. Jaulmes and R. Mestres, *Compt. Rend.,* **248**, 2752 (1959).

227. J. L. Moore, T. Richardson, and C. H. Amundson, *J. Gas Chromatog.,* **2**, 318 (1964).

228. B. C. Cox and B. Ellis, *J. Chromatog.,* **20**, 600 (1965).

229. R. O. Brace, in *Principles and Practice of Gas Chromatography,* (R. L. Pecsok, ed.), Wiley, New York, 1959, p. 149.

230. K. K. Carroll, *Nature,* **191**, 377 (1961).

231. D. M. Smith and J. C. Bartlet, *Nature,* **191**, 688 (1961).

232. R. G. Ackman, *J. Gas Chromatog.,* **3**, 15 (1965).

233. M. J. E. Golay, *Nature,* **202**, 489 (1964).

234. J. E. Funk and G. Houghton, *J. Chromatog.,* **6**, 193 (1961).

235. A. B. Littlewood, *Gas Chromatography. Principles, Techniques, and Applications,* Academic Press, New York and London, 1962, pp. 39–41.

236. T. Gerson, *J. Chromatog.,* **6**, 178 (1961).

237. J. C. Bartlet and D. M. Smith, *Can. J. Chem.,* **38**, 2057 (1960).

238. S. J. Hawkes and C. P. Russell, *J. Gas Chromatog.,* **3**, 72 (1965).

239. R. A. Dewar, *J. Chromatog.,* **6**, 312 (1961).

240. J. C. Sternberg, in *Gas Chromatography* (L. Fowler, ed.), Academic Press, New York, 1963, pp. 161–191.

241. A. Carisano and L. Gariboldi, *J. Sci. Food Agr.,* **15**, 619 (1964).

242. L. S. Ettre and F. J. Kabot, *J. Chromatog.,* **11**, 114 (1963).

243. P. Ways and D. J. Hanahan, *J. Lipid Res.,* **5**, 318 (1964).

244. R. G. Ackman and J. C. Sipos, *J. Am. Oil Chemists' Soc.,* **41**, 377 (1965).

245. R. Kaiser, *Gas Phase Chromatography,* Vol. III (P. H. Scott, transl.), Butterworth, London, 1963, p. 101.

246. J. C. Sternberg, W. S. Gallaway, and D. T. L. Jones, in *Gas Chromatography* (N. Brenner, J. E. Callen, and M. D. Weiss, eds.), Academic Press, New York, 1962, p. 265.

247. J. E. Lovelock, in *Gas Chromatography 1960* (R. P. W. Scott, ed.), Butterworth, London, 1960, pp. 16–29.

248. C. J. F. Bottcher, G. F. G. Clemens, and C. M. van Gent, *J. Chromatog.,* **3**, 582 (1960).

249. J. E. Lovelock, A. T. James, and E. A. Piper, *Ann. N.Y. Acad. Sci.,* **72**, 720 (1959).

250. G. L. Feldman and C. K. Grantham, *J. Gas Chromatog.,* **2**, 12 (1964).

251. G. V. Novitskaya, *J. Chromatog.,* **18**, 20 (1965).

252. R. K. Tandy, F. T. Lindgren, W. H. Martin, and R. D. Wills, *Anal. Chem.,* **33**, 665 (1961).

253. B. M. Craig and N. L. Murty, *J. Am. Oil Chemists' Soc.,* **36**, 549 (1959).

254. S. F. Herb, P. Magidman, and R. W. Riemenschneider, *J. Am. Oil Chemists' Soc.,* **37**, 127 (1960).

255. L. A. Horrocks, D. G. Cornwell, and J. B. Brown, *J. Lipid Res.,* **2**, 92 (1961).

256. G. R. Jamieson, *J. Chromatog.*, **3**, 464 (1960).
257. J. V. Killheffer, Jr., and E. Jungerman, *J. Am. Oil Chemists' Soc.*, **37**, 456 (1960).
258. W. A. Pons, Jr., and V. L. Frampton, *J. Am. Oil Chemists' Soc.*, **42**, 786 (1965).

BIBLIOGRAPHY

Book Sections Pertinent to GLC of Lipids

H. P. Burchfield and E. E. Storrs, *Biochemical Applications of Gas Chromatography*, Academic Press, 1962, New York, pp. 267–279, 527–558.
E. C. Horning and W. J. A. Vanden Heuvel, *Ann. Rev. Biochem.*, **32**, 709 (1963).
E. C. Horning, A. Karmen, and G. C. Sweeley, in *Progress in the Chemistry of Fats and Other Lipids*, Vol. 7 (R. T. Holman and T. Malkin, eds.), Pergamon, New York, 1964, pp. 167–247.
A. T. James, in *Methods of Biochemical Analysis* Vol. 8 (D. Glick, ed.), Wiley (Interscience), New York, 1960, pp. 1–59.
R. Kaiser, *Gas Phase Chromatography* (P. H. Scott, transl.), Butterworth, London, 1963, Chap. 7.
L. J. Morris, in *Chromatography* (E. Heftmann, ed.), Reinhold, New York, 1961, pp. 428–458.
W. R. Supina, in *Biomedical Applications of Gas Chromatography*, based on Canisius 1963 Lectures (H. A. Szymanski, ed.), Plenum, New York, 1964, pp. 271–305.
F. P. Woodford, in *Fatty Acids. Their Chemistry, Properties, Production, and Uses*, 2nd ed. (K. S. Markley, ed.), Wiley (Interscience) New York, 1964, pp. 2249–2282.

Review Articles and Retention Data
Pertinent to GLC of Lipids

D. T. Downing, *Rev. Pure Appl. Chem.*, **11**, 196 (1961) (Hydroxy acids).
J. W. Farquhar, W. Insull, Jr., P. Rosen, W. Stoffel, and E. H. Ahrens, Jr., *Nutr. Rev., Suppl.*, **17**, 1 (1959) (General GLC of fatty acids, retention data).
K. Fontell, R. T. Holman, and G. Lambertsen, *J. Lipid Res.*, **1**, 391 (1960) (Methods for fatty acid separation, including GLC).
R. S. Henly, *J. Am. Oil Chemists' Soc.*, **42**, 673 (1965) (Preparative GLC of Lipids).
E. C. Horning and W. J. A. Vanden Heuvel, *J. Am. Oil Chemists' Soc.*, **41**, 707 (1964) (Quantitative and qualitative GLC of lipids).
A. T. James, *Analyst*, **88**, 572 (1963) (General GLC of fatty acids).
T. K. Miwa, *J. Am. Oil Chemists' Soc.*, **40**, 309 (1963) (ECL of common esters, not a review).
T. K. Miwa, K. L. Mikolajczak, F. R. Earle, and I. A. Wolff, *Anal. Chem.*, **32**, 1739 (1960) (ECL of common esters).

L. J. Morris, R. T. Holman, and K. Fontell, *J. Lipid Res.*, **1**, 412 (1960) (Alteration on GLC, not a review).

N. S. Radin, *J. Am. Oil Chemists' Soc.*, **42**, 569 (1965) (Hydroxy acids).

A. G. Vereshchagin, *Russ. Chem. Rev. (English Transl.)*, (11), **33**, 578 (1964) (General GLC of fatty acids).

I. A. Wolff and T. K. Miwa, *J. Am. Oil Chemists' Soc.*, **42**, 208 (1965) (Unusual acids and deviations from normal GLC).

10

GAS CHROMATOGRAPHY OF THE LONG-CHAIN ALDEHYDES

G. M. Gray

LISTER INSTITUTE OF PREVENTIVE MEDICINE
LONDON, ENGLAND

I. Introduction†

Unlike the long-chain fatty acids, which are components of a very wide range of materials of biological origin, the corresponding long-chain aldehydes are far less widely distributed and do not usually

† Gas chromatography terms and conventions used in this chapter are as defined by the special group (Analytical Chemistry Division) of the International Union of Pure and Applied Chemistry. See *Pure Appl. Chem., 8, 553* (1964).

occur in their free state. Carbonyl compounds are partially responsible for the pleasant flavors and odors of many fruits and flowers, and gas chromatography has shown the wide range of aromatic and lower aliphatic aldehydes that are present in many essential oils from these sources (*1,2*). However, even in these "aldehyde-rich" oils only very small traces of the longer-chain or "fatty" aldehydes with ten or more carbon atoms have been found (*2*).

Interest in the long-chain aldehydes was aroused indirectly from studies that were being carried out in the phosphatide field during the 1950s. The phosphatides had been known (*3–5*) for many years as important lipid constituents of mammalian and plant tissues, but because of the complexities of the lipid mixtures and the technical difficulties involved in analyzing and separating individual members, little progress had been made since the classic studies of the Macleans (*4*) prior to 1930. However, the wide interest in the use of chromatographic techniques following the pioneering studies of Martin and Synge (*6*) resulted in the development of several methods that were admirably suited to the separation and isolation of lipids. The effective use of these methods has been in part responsible for the very rapid expansion and the healthy, dynamic state of the lipid field at the present time. Because of their involvement in mammalian cell structure and function the phosphatides have commanded special interest, and most of the known compounds have been isolated from tissue extracts by chromatographic techniques (*7*). Of direct relevance to this chapter is a family of phosphatides known as plasmalogens. These compounds were originally detected and named by Feulgen et al. (*8*), but it was the work of Klenk and Debuch (*9,10*) and Rapport and co-workers (*11,12*) that provided the first proof of their correct structure (**1**).

$$
\begin{array}{ccc}
\text{CH}_2\text{—O—CH=CHR} & & \text{CH}_2\text{—OH} \\
| & \xrightarrow[\text{acid}]{\text{mild}} & | \\
\text{CH—O—CO·R}' & & \text{CH—O—CO·R}' \\
| & & | \\
\text{CH}_2\text{—O—P—O—X} & & \text{CH}_2\text{—O—P—O—X} + \text{RCH}_2\text{CHO} \\
\quad\swarrow\;\diagdown & & \quad\swarrow\;\diagdown \\
\text{O}\qquad\text{OH} & & \text{O}\qquad\text{OH} \\
\textbf{(1)} & & \textbf{(2)} \qquad\qquad \textbf{(3)}
\end{array}
$$

where X = choline, ethanolamine, or serine

Treatment of (**1**) with mild acid disrupts the α,β-unsaturated ether linkage and yields two products, a lysophosphatide (**2**) and a long-chain aldehyde (**3**).

In the course of detailed studies on the plasmalogen composition of different tissues (*13,14*) and on the distribution of fatty acids in the plasmalogens and in the corresponding diester phosphatides it became relevant to study the aldehydogenic chain distribution as well. The success of gas chromatography for the analysis of fatty acids made it the obvious technique for the analysis of the corresponding long-chain aldehydes, and the results that were obtained (*15*) showed very clearly the potential of the method. It appears to the author that, with the exception of the field of essential oils, those wishing to study the long-chain aldehydes at the moment will almost certainly be studying problems in the lipid field related to the plasmalogens or to the sphingolipids (see Weiss, Chapter 11), and it is therefore of some practical interest to discuss in detail the isolation of the long-chain aldehydes from lipid extracts of biological materials (see page 407) as well as their analysis by gas chromatography.

II. Volatile Derivatives of the Long-Chain Aldehydes

Aldehydes can be analyzed and separated by gas chromatography (*16*) without the necessity of prior conversion to another derivative, and the separations obtained are satisfactory. A major disadvantage, however, is the instability of aldehydes over a period of time, since it is well known that the lower aldehydes undergo condensation and polymerization with relative ease to form aldol- and polymer products. The long-chain aldehydes, and by "long-chain" is meant those compounds with ten or more carbon atoms, also polymerize on standing to dimer and trimer forms that are relatively nonvolatile (*17*). The reaction is accelerated by alkaline conditions. As it is not always convenient to chromatograph the aldehydes immediately, they are isolated from a particular source; and since any inherent instability would make quantitative results suspect, it is better practice to convert them to stable volatile derivatives. There are several that can be prepared quantitatively, and the relevant ones will be described.

A. Conversion of Aldehyde to a Dimethyl Acetal (DMA)

$$R\cdot CHO + 2MeOH \xrightarrow{H^+} R\cdot C\underset{\substack{|\\H}}{\overset{\substack{OMe\\ \diagup}}{\diagdown}}_{OMe} + H_2O$$

The dimethyl acetals of the long-chain aldehydes are excellent derivatives, being easily formed in quantitative yield, readily volatile, and very stable at neutral and alkaline pH. They are made under the following conditions (18). The aldehyde is refluxed for 2 hr with 2% anhydrous methanolic hydrogen chloride (solute/solvent 1:20, w/v). The reaction mixture is cooled, and the hydrogen chloride is neutralized with a slight excess of anhydrous sodium carbonate. Equal volumes of petroleum ether (b.p. 40–60°C) and water are added to the methanolic solution and shaken. The petroleum ether containing the dimethyl acetal is removed, and the aqueous methanol is extracted with a further volume of petroleum ether. As a precaution the combined petroleum ether extracts are vigorously shaken for several minutes with saturated sodium metabisulfite solution to remove any traces of unreacted aldehyde, washed twice with water, and finally dried over anhydrous sodium carbonate. The solvent is removed by evaporation under reduced pressure. If the aldehyde sample is contaminated with a fatty acid, which would now be present as a methyl ester (ME), the acetal is refluxed with 0.5 N methanolic sodium hydroxide for 2 hr. This procedure converts the methyl ester to the sodium salt of the fatty acid. The solution is cooled and diluted with water, and the acetal is extracted with petroleum ether. The petroleum ether is washed with water and dried; the acetal is recovered as before. The author has found that yields of acetals have been quantitative under these conditions and in practice the metabisulfite washing stage can be safely left out when microquantities of aldehydes are being converted to their dimethyl acetals.

Farquhar (19) obtained quantitative conversion of the aldehydogenic chains of plasmalogens to their dimethyl acetals by heating the lipid with 5% anhydrous methanolic hydrogen chloride at 90°C in a sealed tube for 2 hr, and recently Morrison and Smith (20) prepared dimethyl acetals from lipids with 14% boron trifluoride–methanol complex, heating for 30 min in a closed tube at 100°C. All three methods are suitable for standard use on a macro- or microscale. The methods mentioned require anhydrous solvents, but it appears that strictly anhydrous conditions are not always necessary. Sloane-Stanley (21) has investigated the conversion of aldehyde to dimethyl acetal in some detail and found that quantitative conversion takes place in chloroform–methanol 4:1, v/v, containing 2.5 ml conc. HCl/100 ml of solvent after refluxing for 20 min. It is worth mentioning that the necessity of preparing anhydrous methanolic hydrogen chloride is avoided by using this procedure or that of Morrison and Smith (*20*). In the author's opinion the dimethyl acetals are the most suitable derivatives for the analysis of the long-chain aldehydes by gas chromatography based on the following criteria: ease of formation, not only from pure aldehydes but from aldehydes and aldehydogenic moieties present in crude lipid extracts from biological sources; quantitative yields on macro- or microscale; simple purification procedures; and excellent stability.

B. Conversion of Aldehyde to Alcohol and Alcohol Acetate

Information obtained from the gas chromatographic behavior of a compound and its derivatives as compared with compounds of known structures can help in determining its correct chemical structure. With this in mind Farquhar (*19*) prepared alcohol and alcohol acetate derivatives for gas chromatography by reducing the long-chain aldehydes with a 3% anhydrous ether solution of lithium aluminum hydride at −20°C for 2 hr. The average recovery was 82%. The alcohol acetates were prepared by treating the alcohols with acetic anhydride–pyridine 3:6, v/v, at 37°C for 15 min. After the addition of water the acetates were extracted with petroleum ether.

C. Conversion of Aldehyde to Acid

The oxidation of the long-chain aldehydes to the corresponding acids was used as a means of structural identification by the author (*22*), by Schogt et al. (*23*), and by Farguhar (*19*). If the aldehydes are initially isolated as their dimethyl acetals, it is necessary to recover them prior to oxidizing them to acids. The aldehyde dimethyl acetals (100 mg) are dissolved in 4.0 ml 90% acetic acid; 0.25 ml 5 N methanolic HCl is added; and the solution is heated for 1 hr at 90–100°C. The solution is diluted with one volume of water and extracted (4×) with 1.5 ml petroleum ether (b.p. 40–60°C). The petroleum ether is removed under vacuum, and the residue is dissolved in 2.5 ml glacial acetic acid. The solution of aldehydes is warmed slightly (approx. 40°C), and small amounts of chromium trioxide (*24*) in glacial acetic acid are added until excess is present. The solution is then poured into water (2 vol.) and extracted with benzene; the benzene solution is washed with water and evaporated to dryness and the residue is dissolved in dry ether. The fatty acids are precipitated as their sodium salts by the addition of sodium methoxide, and the precipitate is centrifuged, washed with a little petroleum ether, and dissolved in water. The acids are liberated with HCl, extracted with ether, and converted to their methyl esters with anhydrous methanolic HCl. Gas chromatography of the methyl esters is carried out in the usual way (see Chapter 8). An alternative oxidation procedure using alkaline silver oxide has been described by Farquhar (19). The usual procedures of oxidation unfortunately cause destruction of unsaturated compounds, and so the above procedure is only useful for studying saturated aldehydes.

Identification of unsaturated components can be obtained from the results of gas chromatography before and after hydrogenation, and aldehyde dimethyl acetals are conveniently hydrogenated in ethanol (solvent/solute 100:1, v/v) using a platinum oxide catalyst. It is sometimes necessary to add up to 20% of chloroform to keep the acetals dissolved throughout the reaction period.

III. Synthesis of Reference Aldehydes

Reference to standard compounds of known structure is an essential requirement in the analysis and identification of compounds by gas chromatography, and since aldehydes can be oxidized to acids, a series of fatty acid methyl esters could be used as reference compounds. However, problems would arise with unsaturated aldehydes, and it has been pointed out that the dimethyl acetals are excellent derivatives and in practice a range of long-chain aldehyde dimethyl acetals of known structure are the ideal reference compounds. They should represent, if possible, members of the normal straight-chain series, both saturated and unsaturated, and of the branched-chain iso and anteiso series.

Long-chain aldehydes are not easy to obtain commercially, and it is usually better to synthesize the required compounds in the laboratory from pure materials. The author used the procedure of Wegand et al. (25) and obtained satisfactory yields of several different aldehydes. Hexadecanal (palmitaldehyde) and octadecanal (stearaldehyde) were obtained in 80% yield, oleylaldehyde in 70% yield, and linoleylaldehyde and linolenylaldehyde in 50 and 10% yields, respectively. The starting product of the synthesis is the acid chloride of the corresponding long-chain fatty acid. This is converted to an N-methylanilide by a Schotten–Baumann reaction, and the anilide is dissolved in anhydrous tetrahydrofuran or ether and slowly added, at 0°C, to a tetrahydrofuran (or ether) solution of lithium aluminium hydride. After 4 hr at 0°C the organometal complex is hydrolyzed with dilute hydrochloric acid and the aldehyde formed is extracted with ether and converted to its dimethyl acetal.

Mangold (*26*) successfully synthesized oleyl-, linoleyl-, and linolenylaldehydes by a modified Grundmann synthesis using lithium aluminum hydride instead of aluminum isopropylate. The yields were reasonable, and because of the high purity of the product the method is very suitable for the synthesis of microquantities. Recently Mahadevan (*27*) reported the conversion of some long-chain alcohol tosylates to their corresponding aldehydes by oxidation with dimethyl sulfoxide. This method appears generally applicable to unsaturated long-chain compounds, and the yields obtained by Mahadevan were very satisfactory (60–70%). Spectral examination of the products revealed that no *cis-trans* isomerization had occurred under the conditions of synthesis. It is common practice to isolate aldehydes from reaction mixtures with 2,4-dinitrophenylhydrazine; and if this is done, the hydrazone derivatives can be directly converted to the aldehyde dimethyl acetals for gas chromatographic studies by the method described by Mahadevan et al. (*28*).

IV. Isolation of Aldehydes from Biological Sources

At the present time the main purpose of the gas chromatography of the long-chain aldehydes is to analyze and identify the different components in mixtures that are isolated from natural sources. For the purposes of accurate analysis it is desirable that the mixture to be chromatographed contain only aldehydes (as a suitable derivative) and not other unknown components as well; therefore, an important preliminary to the analysis is the preparation of a pure sample of aldehydes.

The long-chain aldehydes, in the form of plasmalogens, have been detected throughout the animal kingdom (*29*) and also recently in some bacteria (30,31). Comparative studies of the plasmalogens might involve a comparison of the total aldehydes from different tissues or of the aldehydes present in the different plasmalogens of a particular tissue (32). In the former case the total lipid, or a sample of it, extracted from the tissue by established methods (33,34) is refluxed with 0.5 N anhydrous methanolic hydrogen chloride† for 2 hr. Transesterification takes place, and methyl esters of all the fatty acids in the lipids are formed and the aldehydogenic moieties of the plasmalogens are converted to aldehyde dimethyl acetals. The acetals are recovered, and the fatty acid methyl esters are removed as described previously page (404). In the latter case the whole lipid extract is separated into neutral lipids and phosphatides (14) and the phosphatides are then fractionated on silicic acid (13). This procedure separates choline plasmalogen from ethanolamine plasmalogen (plus serine plasmalogen). The ethanolamine and serine plasmalogens are separated by the method of Rouser et al. (35) using a silicic acid–ammonium silicate solumn.

The aldehyde dimethyl acetals are prepared from each fraction. Each of these three plasmalogen fractions will also contain the corresponding diester phosphatide, and if it is required to examine the fatty acids as well as the aldehydogenic chains in the plasmalogens, the preparation of the aldehyde dimethyl acetals is modified thus: the fraction is treated with 90% acetic acid (solvent/solute 4:1, v/w) for 18 hr at 37°C. The acetic acid is removed (as an azeotrope) by evaporating with excess carbon tetrachloride under reduced pressures at 40–45°C. The residue is redissolved in chloroform and chromatographed on silicic acid. The free aldehydes are eluted with the solvent front (chloroform) and are quickly converted into their dimethyl acetals. The other hydrolysis product of the plasmalogen is the lysophosphatide (2, Section I), which is easily separated from the diester phosphatide by silicic acid chromatography using different concentrations of methanol in chloroform (32).

Aldehyde dimethyl acetals that are prepared from lipid extracts of biological origin are likely to be contaminated, not only with methyl esters but also with unsaponifiable lipids such as cholesterol and glycerol ethers, and the alkali purification method will only remove the methyl esters. Molecular distillation (36) has been used but is not

† The other methods mentioned on page 404 are equally suitable.

suitable for micro quantities. Thin-layer chromatography on silicic acid is ideally suited to the separation of milligram quantities of compounds, and good separations of methyl esters and dimethyl acetals are obtained in benzene (*20*), toluene, and xylene (*37*). The dimethyl acetals can readily be separated from cholesterol using hexane–diethyl ether solvent mixtures. According to Eng et al. (*37*), the average recoveries of dimethyl acetals from thin-layer plates were slightly on the low side (85%), but they found the method especially useful for removing contaminants from dimethyl acetal fractions prepared from brain lipids.

V. Practical Conditions for Gas Chromatography

Since the original publications describing the gas chromatography of the long-chain aldehydes (*15,19*) some 6 years ago, advances on the technical side of the method have simplified a number of problems. There is now available a wide choice of excellent commercial apparatus with facilities for all methods of gas chromatography. Many commercial detectors have been thoroughly tested and evaluated (*38*), and pure, standardized liquid phases and supports are readily available from a number of supply houses. The "art" of the early days of the technique has virtually disappeared except perhaps for special separations requiring "modified" liquid phases and column supports.

A. Column Supports

The first separations of the aldehyde dimethyl acetals were carried out with liquid phases supported on Celite 545 of a standard mesh size (80–100 and 100–120 mesh) that had been treated in the laboratory with acid and alkali according to James and Martin (*39*) as a means of decreasing the number of active adsorption sites in the material. Heavy loadings of liquid phase (20–30% by weight) also minimized the adsorption effects of the support, but this often resulted in "slow" columns with large retention volumes. In the intervening years a range of "treated" diatomaceous earths have been produced commercially, specially for gas chromatography,† the object in all

---

† For example, Gas-chrom, Anakrom, Chromosorb.

cases being to produce a standard, reproducible product with minimum adsorption characteristics. Silanized supports are advisable for packed columns containing less than 5% (by weight) liquid phase (*40*), but the author has found that with columns of 10–15% (by weight) of liquid phase, commercial acid- and alkali-washed products (80–100 mesh) are completely satisfactory.

B. Liquid Phases

One of the established methods of analyzing homologous series of organic compounds is to examine their chromatographic behavior on columns containing liquid phases having different chemical characteristics and therefore different selective properties toward different functional groups in the solutes. A broad classification of the liquid phases is into polar or "selective" phases and nonpolar or "nonselective" phases.† The use of different phases proved invaluable in the analysis of long-chain fatty acid methyl esters (*42*) and was used with equal success for the analysis of the aldehyde dimethyl acetals (*15,19*).

The most versatile nonpolar phases are the Apiezon greases and the silicone oils, but for the analysis of aldehyde dimethyl acetals and fatty acid methyl esters the Apiezon greases are superior. From a range of these greases Apiezon L is the usual choice, though Apiezon M gives very similar results (*19*). There is a very wide choice of polar phases for gas chromatography, but experience has shown that the "polyester" phase synthesized from a glycol and a dibasic acid is the most suitable. The first analyses of aldehyde dimethyl acetals were carried out on Reoplex 400‡ (*15*), but subsequently ethylene glycol adipate (EGA) was used. Diethylene glycol succinate (DEGS) (*43*) and butanediol succinate (*44*) have also been used, but for good all-round properties with regard to separation of individual components, thermostability, and column life ethylene glycol adipate is probably the most suitable choice. It is easy to synthesize in the laboratory (*42*), but excellent products are available commercially (for example, from Applied Science Laboratories, State College, Pennsylvania).

Recently a series of new polyester phases were introduced (by Applied Science Laboratories) in which ethylene glycol succinate was chemically modified to varying degrees with a silicone (e.g., methyl

---

† For a stricter definition see Brown (*41*).

‡ A poly(oxyalkalene adipate); The Geigy Chemical Company.

silicone). Available data (*45*) indicated that separations of methyl esters were very good, especially oleate from stearate, and it seemed likely that the methyl silicone might confer a greater thermostability on the phase and give columns of longer life. The separation of the aldehyde dimethyl acetals on one of these siloxane copolymers, EGSSX as compared with EGA, is described later (page 420).

C. Coating Procedures

Though long-chain aldehyde dimethyl acetals are stable derivatives at neutral and alkaline pH, acetals, as a group, are very acid-labile and slightly acid conditions, especially at elevated temperatures, will rapidly destroy the acetal structure. Because of this, the support requires treatment before it is coated with Apiezon L or M, as these phases, on untreated supports, have caused partial breakdown of the dimethyl acetals. Treatment of the support is as follows: 10 g in a beaker is gently stirred with 100 ml 0.5 N methanolic sodium hydroxide. The solution is allowed to stand for 15 min, and the excess solution is filtered off on a Büchner funnel. The support is washed once with 50 ml methanol and air-dried by keeping the flask at reduced pressure. 2–3 ml of water shaken with about 0.5 g of the support should have a pH value of 7.5–8. The support is then coated with the required percentage (10–15%) of Apiezon L by adding a solution of the Apiezon L in petroleum ether (b.p. 40–60°C) to it in a round-bottomed flask. This is attached to a rotary evaporator, and the solvent is removed under reduced pressure, leaving the support uniformly coated with liquid phase.

Breakdown of aldehyde dimethyl acetals on columns containing some of the polyester phases did occur in some of the early studies, and this was probably due to the presence of small amounts of an acid catalyst used in the synthesis of the ester. The problem was solved with EGA by making it without a catalyst and separating the high-molecular-weight polymers by solvent fractionation (*46*).

Columns packed with Apiezon L are "conditioned" at 220–230°C for 12 hr before use at temperatures between 170° and 200°C, but columns containing a polyester phase require much longer conditioning periods in order to "bleed out" the more volatile polymers. At least 48–72 hr at 190–200°C is required for columns used at temperatures between 150° and 180°C. Nitrogen (oxygen-free) at an inlet pressure of 5–10 lb/in.² is used throughout the conditioning period. Analyses in the author's laboratory were carried out on straight glass columns (120 cm long, 4 mm diam.) and on coiled glass columns (150 cm long, 4 mm diam.). Preparative work can be carried out satisfactorily at a milligram level on similar columns by increasing the amount of liquid phase on the support, and 25–30% (by weight) of Apiezon L has given good separations. A 150-cm-long 12-mm-diam. glass column of 25% Apiezon L on celite 545 (60–80 mesh) gave good separations of 20- to 50-mg quantities of a mixture of branched-chain aldehyde dimethylacetals (22).

Most commercial gas chromatography instruments offer a wide choice of detection systems, but the two most sensitive and most popular are the β-ray argon ionization detector and the hydrogen flame ionization detector. The author's experience has been limited to these two types, and both have been very satisfactory with regard to sensitivity and linearity. More care is required in using the β-ray argon ionization detector for quantitative work, as it has a much narrower linear range (38) than the hydrogen flame ionization detector, and it is essential to check the linearity initially at a given voltage with standard compounds. Because of their very high sensitivity both detectors need postcolumn stream splitters for preparative work to prevent detector overload and, in the case of the hydrogen flame detector, to prevent all the sample being destroyed in the flame.

VI. The Gas Chromatographic Behavior of the Long-Chain Aldehyde Dimethyl Acetals

The chromatographic properties of the long-chain aldehyde dimethyl acetals on Apiezon L and Reoplex 400 have been reported previously (15). They were also compared with the properties of the corresponding fatty acid methyl esters chromatographed under similar conditions. Farquhar (19) extended these studies by examining the properties of the aldehyde dimethyl acetals on Apiezon M and EGA and compared them with the corresponding fatty acid esters, acetates, alcohols, and free aldehydes under similar chromatographic conditions.

The introduction of standardized supports (Gas-Chrom, Chromosorb, Anakrom, etc.) and standard pure stationary phases over the last few years has been mentioned, and some of these materials have been used in a reevaluation of the earlier work. It is standard practice to plot the carbon numbers of individual members of a homologous series against the logarithms of their retention volumes relative to an internal standard that is usually one of the series. Straight-line plots are obtained; from them it is possible to identify the carbon numbers of unknown compounds. As shown by James (*42*), more information with regard to structure can be obtained by plotting the logarithms of the retention volumes on two different stationary phases (usually a polar and a nonpolar one) against each other. This method was most useful in helping to identify some of the branched-chain aldehydes (*22*). A modification of the retention-volume versus carbon-number plot introduced by Woodford and van Gent (*47*) more closely relates the "carbon number" of a compound with chemical constitution. The carbon-number parameter has a distinct advantage of not being so temperature-sensitive as the relative retention volume and remains in practice almost constant over a relatively wide range of temperature (e.g., 150–180°C). A recent study group† has recommended that all retention volumes of compounds be related to an internal standard, and for preference this standard should be one of the *n*-alkanes. However, when studying a particular series of compounds, it is still very useful in practice to know the retention volumes relative to a common member of the series. For the naturally occurring series of long-chain aldehydes hexadecanal (palmitaldeyde) dimethyl acetal is a suitable standard for the dimethyl acetals. If a second independent standard is required, methyl hexadecanoate (methyl palmitate) is a convenient choice.

A. Identification of Long-Chain Aldehydes

The relative retention volumes and carbon numbers, on nonpolar and polar liquid phases, of the majority of the naturally occurring long-chain aldehydes are listed in Tables 1 and 2. The structures of the aldehydes were confirmed either by comparison with synthetic standards or, in the case of the branched-chain compounds, by oxidation to the corresponding acids, which were then compared with branched-chain acids of known structures (*22*).

† The Gas Chromatography Discussion Group of the Institute of Petroleum. See the Proceedings of the Fifth International Symposium of Gas Chromatography, 1964.

RELATIVE RETENTION VOLUMES r AND CARBON NUMBERS OF LONG-CHAIN
ALDEHYDE DIMETHYL ACETALS ON APIEZON L LIQUID PHASE[a]

Aldehyde	Designation	Relative retention volume r	Carbon number
Dodecanal	12:0	0.170	12.0
11-Methyldodecanal	iso-br 13:0	0.221	12.6
10-Methyldocecanal	anteiso-br 13:0	0.229	12.7
Tridecanal	13:0	0.267	13.0
12-Methyltridecanal	iso-br 14:0	0.345	13.6
11-Methyltridecanal	anteiso-br 14:0	0.363	13.7
Tetradecanal (myristaldehyde)	14:0	0.412	14.0
13-Methyltetradecanal	iso-br 15:0	0.541	14.6
12-Methyltetradecanal	anteiso-br 15:0	0.563	14.7
Pentadecanal	15:0	0.645	15.0
14-Methylpentadecanal	iso-br 16:0	0.840	15.6
13-Methylpentadecanal	anteiso-br 16:0	0.895	15.7
Hexadecanal (palmitaldehyde)	16:0	1.00	16.0
Hexadecenal (palmitoleic aldehyde)	16:1	0.840	15.6
15-Methylhexadecanal	iso-br 17:0	1.30	16.6
14-Methylhexadecanal	anteiso-br 17:0	1.36	16.7
Heptadecanal	17:0	1.55	17.0
Octadecanal (stearaldehyde)	18:0	2.39	18.0
cis-Octadeca-9-enal (oleylaldehyde)	cis 18:1[9]	2.02	17.6
Octadecenal	18:1	2.14	17.75
cis-cis-Octadeca-9,12-enal (linoleylaldehyde)	cis, cis 19:2[9,12]	1.89	17.45
Octadeca-9,12,15-enal (linolenylaldehyde)	18:3[9,12,15]	1.89	17.45
Methyl hexadecanoate (palmitate)	16:0	0.775	

[a] 12.5% Apiezon L on "alkaline" Gas-Chrom P (80–100 mesh). Temperature, 190°C; column, glass, 120 cm long, 4 mm diam, argon flow rate, 100 ml/min at 20 lb/in.[2]

On Apiezon L, Apiezon M, EGA, Reoplex 400, and EGSS-X the retention volumes of the aldehyde dimethyl acetals relative to hexadecanal dimethyl acetal (C_{16}DMA) are similar to those of the corresponding fatty acid methyl esters relative to methyl hexadecanoate (C_{16}ME) under the same experimental conditions (15,42). Their actual retention volumes on polar and nonpolar phases differ considerably

TABLE 2

RELATIVE RETENTION VOLUMES AND CARBON NUMBERS OF LONG-CHAIN-ALDEHYDE DIMETHYL ACETALS ON EGA AND EGSSX LIQUID PHASES[a]

Aldehyde	Designation	Ethylene glycol adipate (EGA)		EGSS-X	
		Relative retention volume r	Carbon number	Relative retention volume r	Carbon number
Dodecanal	12:0	0.214	12.0	0.251	12.0
11-Methyldodecanal	iso-br 13:0	0.265	12.56	0.298	12.52
10-Methyldodecanal	anteiso-br 13:0	0.279	12.7	0.324	12.7
Tridecanal	13:0	0.314	13.0	0.353	13.0
12-Methyltridecanal	iso-br 14:0	0.392	13.55	0.417	13.5
11-Methyltridecanal	anteiso-br 14:0	0.412	13.7	0.450	13.7
Tetradecanal (myristaldehyde)	14:0	0.460	14.0	0.500	14.0
13-Methyltetradecanal	iso-br 15:0	0.570	14.56	0.593	14.5
12-Methyltetradecanal	anteiso-br 15:0	0.596	14.68	0.628	14.67
Pentadecanal	15:0	0.68	15.0	0.705	15.0
14-Methylpentadecanal	iso-br 16:0	0.846	15.56	0.84	15.5
13-Methylpentadecanal	anteiso-br 16:0	0.890	15.7	0.90	15.7
Hexadecanal (palmitaldehyde)	16:0	1.00	16.0	1.00	16.0
Hexadecenal (palmitoleic aldehyde)	16:1	1.13	16.35	1.15	16.4
15-Methylhexadecanal	iso-br 17:0	1.25	16.57	1.18	16.5
14-Methylhexadecanal	anteiso-br 17:0	1.32	16.7	1.28	16.7
Heptadecanal	17:0	1.47	17.0	1.43	17.0
Octadecanal (stearaldehyde)	18:0	2.16	18.0	2.01	18.0
cis-Octadeca-9-enal (oleylaldehyde)	cis 18:1^9	2.43	18.3	2.33	18.4
Octadecenal	18:1				
cis,cis-Octadeca-9,12-enal (linoleylaldehyde)	cis, cis 18:2^{9,12}	2.92	18.8	—	—
Octadeca-9,12,15-enal (linolenylaldehyde)	18:3^{9,12,15}	3.67	19.4	—	—
Methyl hexadecanoate (palmitate)	16:0	1.16		1.25	

[a] 10% EGA on Gas Chrom CLH (80–100 mesh), 15% EGSSX on Gas Chrom CLH (80–100 mesh). Columns, 150 cm × 4 mm glass coils. Temperature, 150°C; nitrogen flow rate, 50 ml/min.

from those of the corresponding methyl esters. The dimethyl acetals are less polar than the methyl esters and, as might be expected, have larger retention volumes than the methyl esters on the nonpolar phases and smaller retention volumes on the polar phases (Table 3). The

TABLE 3

A Comparison of the Behavior of Long-Chain Aldehyde, Aldehyde Dimethyl Acetal, and Fatty Acid Methyl Ester on Different Liquid Phases

	10% EGA on Gas Chrom CLH 80–100 mesh; temp., 150°C	15% EGSSX on Gas Chrom CLH 80–100 mesh; temp., 150°C	12.5% Apiezon on alkaline Gas Chrom P 80–100 mesh; temp., 190°C
Retention volume relative to methyl hexadecanoate ($C_{16:0}ME$)			
Hexadecanal dimethyl acetal ($C_{16:0}DMA$)	0.86	0.80	1.29
Hexadecanal ($C_{16:0}$ald.)	0.75	0.80	—[a]
Resolution			
$C_{16:0}ME/C_{16:0}DMA$	1.37	2.0	3.0
$C_{16:0}DMA/C_{16:0}$ald.	1.29	0	—
$C_{16:0}DMA$/iso-br $C_{16:0}DMA$	1.39	1.48	1.95
$C_{17:0}DMA$/anteiso-br $C_{17:0}DMA$	1.14	0.87	1.56
$C_{16:0}ME$/iso-br $C_{16:0}ME$	1.55	1.55	—
Column efficiency (theoretical plate no.)			
$C_{16:0}DMA$	1320	1260	2440
$C_{16:0}ME$	1520	1370	2330
$C_{16:0}$ald.	1580	1300	—
Adjusted retention times, min			
$C_{16:0}DMA$	14.1	15.6	23.9
$C_{16:0}ME$	16.0	19.2	18.6
$C_{16:0}$ald.	12.4	15.6	—

[a] Aldehyde not eluted on this column.

elution behavior of these and other series has been examined by Farquhar (19) in some detail (see Fig. 1).

Farquhar has also presented useful data that demonstrate the effect of temperature on the efficiency of Apiezon M (Fig. 2) and EGA (Fig. 3) liquid phases for dimethyl acetatals and other functional groups. He noted that aldehydes were not eluted from his nonpolar column at the temperatures between 173° and 204°C. There is no doubt that on an Apiezon column that had been alkali-treated free aldehydes would probably form relatively nonvolatile aldol condensation products or polymers, but the author has found that free aldehydes chromatograph satisfactorily on untreated Apiezon L columns, though under these conditions there is considerable breakdown of the dimethyl acetals.

Optimum conditions for the analysis of mixtures are dependent on the choice of liquid phase, amount of liquid phase, column tempera-

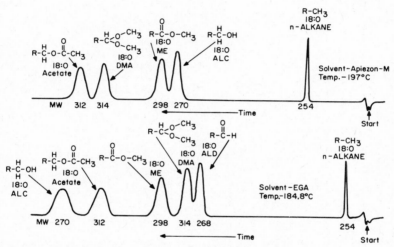

FIG. 1. Gas chromatographic behavior of long-chain aliphatic compounds containing different functional groups on polar and nonpolar liquid phases. [From Farquhar (*19*). Reproduced by kind permission of the author and the *Journal of Lipid Research.*]

ture, gas-flow rate, particle size of support, etc., and restrictions are also imposed by the possible instability of compounds at elevated temperatures and by a practical time period suitable for the analysis. Use of the less dense commercial standardized supports instead of ordinary celite as in the past (*15*) has greatly reduced retention volumes, and analyses are therefore much quicker and can be made at lower temperatures, a distinct advantage when using polyester-type liquid phases.

Considerable experience in the analysis of mixtures of long-chain aldehydes dimethyl acetals of varied composition by using Apiezon L liquid phases has indicated the most suitable conditions for their general analysis (Table 4, Fig. 4). The use of silanized supports, such as those treated with hexamethyldisilazane, for "alkaline" columns does not appear to offer any practical advantages with regard to peak

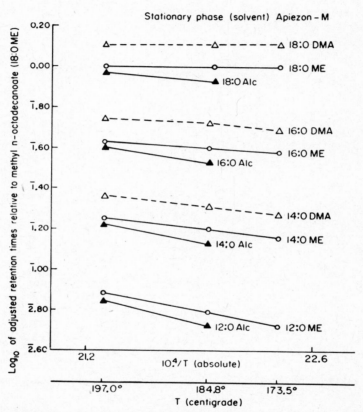

FIG. 2. Effect of temperature on the efficiency of the nonpolar liquid phase, Apiezon M, as a solvent for certain functional groups. [From Farquhar (*19*). Reproduced by kind permission of the author and the *Journal of Lipid Research.*]

symmetry, resolution, lower retention volumes, or column efficiency. As a point of interest, the author has found that such columns are not quite as efficient (about 80%) as those on nonsilanized support.

The long-chain aldehyde dimethyl acetals can be analyzed on a number of different polar liquid phases, but for efficiency of separation and long column life EGA and, more recently, EGSSX are the ones of choice and optimum conditions for general analysis are listed in Table 4. A comparison of these liquid phases (Table 3) shows some interesting differences with regard to the separation of fatty acid methyl esters, aldehyde dimethyl acetals, and free aldehydes. The

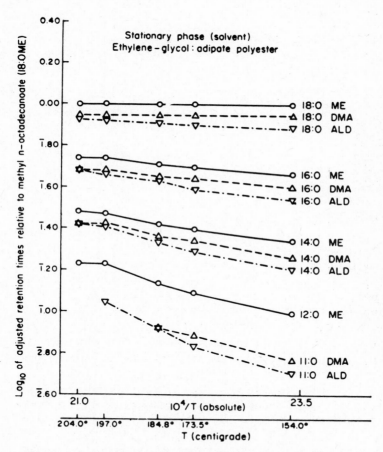

FIG. 3. Effect of temperature on the efficiency of the polar liquid phase ethylene glycol adipate (EGA) as a solvent for certain functional groups. [From Farquhar (*19*). Reproduced by kind permission of the author and the *Journal of Lipid Research*.]

TABLE 4

Experimental Conditions for Analyses of Long-Chain Aldehyde
Dimethyl Acetals on Nonpolar and Polar Liquid Phases

Liquid phase	Apiezon L	Ethylene glycol adipate (EGA) and ethylene glycol succinate–silicone copolymer (EGSSX)
Support	Gas Chrom P (or equivalent) alkali-treated,[a] pH 7.5–8, 80–100 mesh	Gas Chrom CLH (or equivalent), 80–100 mesh
% Liquid phase on support, by wt.	10–15	10–15
Column temp., °C	190	150
Carrier gas flow rate, ml/min	100	50
Average analysis time to $C_{18:0}$DMA, min	60[b]	30[c]

[a] See page 411.
[b] For column 120 cm long, 4 mm diam.
[c] For column 150 cm long, 4 mm diam.

resolution of C_{16} DMA and C_{16}ME is much better on EGSSX, but at
the temperatures used (150°, 160°, and 170°C) there is no resolution
of C_{16}DMA and C_{16} aldehyde, whereas a complete separation of these
compounds is obtained by using EGA. Farquhar found, however, that
at temperatures above 200°C resolution of C_{16}DMA and C_{16} aldehyde
(Fig. 3) is not obtained on EGA either. Under similar conditions the
efficiency of EGA is slightly better than EGSSX for C_{16}ME, C_{16}DMA,
and C_{16} aldehyde and EGA is able to accommodate much larger sample
loadings than EGSSX.

Both liquid phases give a better resolution of "close-running" fatty
acid methyl esters than of the corresponding aldehyde dimethyl acetals
(Fig. 5, Table 3). EGSSX gives a better resolution of iso from anteiso
branched-chain compounds (ME and DMA, see Fig. 5, Table 5) than
EGA, and the resolution of octadecanal (stearaldehyde) DMA and
octadecenal (oleylaldehyde) DMA is also superior. To sum up, good
separations of mixtures of the long-chain aldehyde dimethyl acetals are
obtained on both EGA and EGSSX, but in neither case is the resolu-

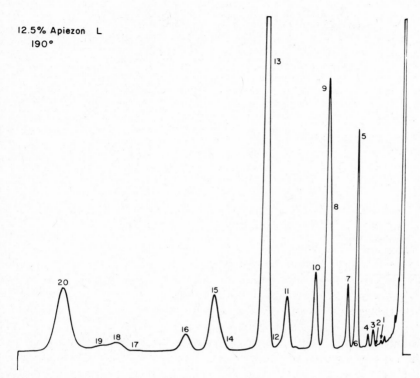

FIG. 4. Analysis of dimethyl acetals of aldehydes isolated from ox-spleen choline plasmalogen. Column, 12.5% Apiezon L on alkali-treated Gas Chrom P (80–100 mesh). Column size, 120 cm long, 4 mm diam.; temperature, 190°C; argon flow rate, 100 ml/min at 20 lb/in.^{2} Loading, approx. 20 μg mixture. Peak identification: 1, dodecanal; 2, 11-methyldodecanal; 3, 10-methyldodecanal; 4, tridecanal; 5, 12-methyltridecanal; 6, 11-methyltridecanal; 7, tetradecanal; 8, 13-methyltetradecanal; 9, 12-methyltetradecanal; 10, pentadecanal; 11, 14-methylpentadecanal; 12, 13-methylpentadecanal; 13, hexadecanal; 14, 15-methylhexadecanal; 15, 14-methylhexadecanal; 16, heptadecanal; 17, octadecadienal; 18, *cis*-octadeca-9-enal; 19, octadecenal; 20, octadecanal.

tion quite as good as with corresponding mixtures of the fatty acid methyl esters (see Fig. 5).

Recently Goldfine (*31*) isolated some aldehydes (as DMA) from the plasmalogens in *Clostridium butyricum* and identified in the mixture aldehydes that contained cyclopropane ring structures. These

TABLE 5

Nonintegral Carbon Numbers for a Series of Iso and Anteiso Branched-Chain Aldehyde Dimethyl Acetals and Fatty Acid Methyl Esters on EGA and EGSSX

	10% ethylene glycol adipate (EGA)[a]	15% ethylene glycol succinate–silicone copolymer (EGSSX)[a]
Iso branched chain		
Methyl esters	$n^b + 0.55$	$n + 0.46$
Dimethyl acetals	$n + 0.56$	$n + 0.50$
Anteiso branched chain		
Methyl esters	$n + 0.67$	$n + 0.67$
Dimethyl acetals	$n + 0.70$	$n + 0.70$

[a] Columns, glass, 150 cm long, 4 mm diam.; temp., 150°C, N_2 flow rate, 50 ml/min; support, Gas Chrom CLH (80-100 mesh).
[b] $n = 13, 14, 15, 16, 17$.

compounds appeared to be analogous to the cyclopropane fatty acids, like cis-9,10-methylene hexadecanoic acid, which have been isolated from several types of bacteria (48,49). From the gas chromatographic data reported by Goldfine the carbon numbers on Apiezon L and EGA are very similar to the monounsaturated long-chain aldehyde dimethyl acetals of similar chain length (Table 6), and obviously care must be exercised in identifying, purely on retention volumes, compounds containing cyclopropane ring structures. Analysis before and after hydrogenation of the dimethyl acetal derivatives would confirm the presence of unsaturated components.

VII. Biochemical Investigations

The technique of gas chromatography is of tremendous value in the chemical and biochemical studies of compounds that are present in biological materials. It has shown that the plasmalogens in many mammalian tissues contain a range of aldehydogenic chains from which can be isolated after mild acid hydrolysis up to twenty or so different long-chain aldehydes (15,19). Though the analysis of the total aldehyde composition of a particular tissue may provide useful data, as for instance in comparative biochemical studies of different

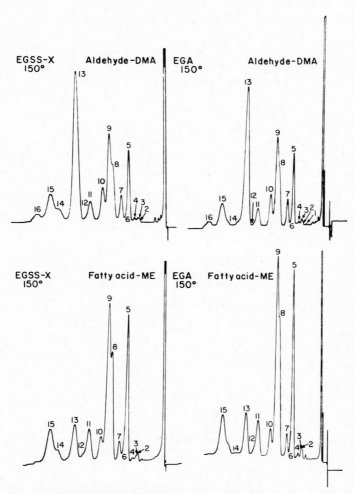

FIG. 5. A comparison of the separation of (a) a mixture of aldehyde dimethyl acetals and (b) the corresponding fatty acid methyl esters on ethylene glycol adipate (EGA) and ethylene glycol succinate–silicone copolymer (EGSSX) under similar experimental conditions (see Table 3). Peak identification as for Fig. 4. The peak numbers of the fatty acid methyl esters correspond to those for the aldehyde dimethyl acetals, e.g., peak 13 = (A) hexadecanal dimethyl acetal = (B) methyl hexadecanoate.

tissues and different animal species or in nutritional investigations, far more information is obtained by studying the aldehyde compositions of each of the different plasmalogens in the tissue. Some interesting results were obtained from a comparison of the aldehyde compositions of the choline plasmalogens from different tissues with those of the ethanolamine plasmalogens (some of which contained serine plasmalogen). The choline plasmalogens showed a marked preference for aldehydogenic chains of 16 carbon atoms (Fig. 6), whereas the ethanolamine plasmalogens, though on the average, containing a higher percentage of aldehydogenic chains of 18 carbon atoms than the choline plasmalogens showed no distinct preferance for either

TABLE 6

CARBON NUMBERS OF CYCLOPROPANE RING ALDEHYDE DIMETHYL ACETALS

	Designation	10% Apiezon M, 197°C	15% ethylene glycol adipate (EGA), 184.4°C
C_{17} cyclopropane[a] aldehyde	△17:0	16.75	17.37
C_{19} cyclopropane[a] aldehyde	△19:0	18.75	19.32
cis-Octadeca-9-enal[b] (oleylaldehyde)	18:1⁹	17.6	18.3
Octadecenal[b]	18:1 isomer	17.75	18.35

[a] From data of Goldfine (*31*).
[b] From data of Farquhar (*19*).

C_{16} or C_{18} aldehydogenic moieties. These results reflected in a less dramatic way the preferences shown by phosphatidyl choline and phosphatidyl ethanolamine for C_{16} and C_{18} saturated fatty acids (*32*). Hexadecanal and octadecanal were the major aldehyde components from the plasmalogens present in most of the tissues studied up to the present time. Exceptions to this distribution pattern were the aldehydes from ox liver choline plasmalogen and ox spleen choline and ethanolamine plasmalogen. The major components of these mixtures were the iso and anteiso branched-chain C_{15} aldehydes. These branched-chain compounds may originate from certain ruminant bacteria (*30*), which also contain branched-chain aldehydes. In this respect it is of interest that the spleen of a nonruminant animal, the pig, does not contain plasmalogens with branched aldehydogenic chains. Aldehydes from the plasmalogens of human brain white matter (*43*) also differ from the usual distribution pattern by having as the major component a monounsaturated aldehyde, oleylaldehyde.

Gas chromatography has indicated the nature and complexity of aldehyde mixtures from many natural sources, but the biosynthesis of

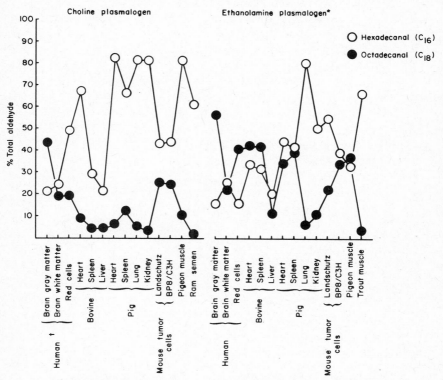

FIG. 6. Distribution of hexadecanal (palmitaldehyde) and octadecanal (stearaldehyde) in the plasmalogens of different tissues and species. *: some of the ethanolamine plasmalogen fractions contained serine plasmalogen. †: figures for brain taken from O'Brien, Fillerup, and Mead (43) and those for human red cells from Farquhar (52).

the α,β-unsaturated ether linkage in the plasmalogen molecule that is the source of the long-chain aldehydes is not yet known. Prudent choice of possible intermediates with specific radioactive labels may help to indicate the way in which the linkage is formed. If the intermediates, when incubated with the appropriate enzyme systems (a heart tissue homogenate?), produce a net synthesis of plasmalogen, a mild acid hydrolysis of this should yield radioactive aldehyde that can be identified by radio–gas chromatography (51). Some experiments carried out by Keenan (50) with a heart–lung preparation indicated that an activated long-chain alcohol may be a precursor. The function of plasmalogens in the mammalian cells is not known, but with other phosphatides they are situated in the membranes of the cell and cell organelles and as such are likely to be intimately involved with membrane function and stability. Studies of the composition and structure of lipids in isolated cell membrane preparations require ultra-microanalytical techniques, and identification of long-chain aldehydes by gas chromatography offers an ultrasensitive method for detecting a particular lipid type (i.e., the plasmalogens) in the very small amounts of lipid extracted from such preparations.

REFERENCES

1. R. A. Bernhard, *J. Chromatog.*, **3**, 471 (1960).
2. R. M. Ikeda, L. A. Rolle, S. H. Vannier, and W. L. Stanley, *J. Agr. Food Chem.*, **10**, 98 (1962).
3. J. L. W. Thudicum, *A Treatise on the Chemical Constitution of the Brain*, Tindall & Cox, London, 1884.
4. H. Maclean and I. S. Maclean, *Lecithin and Allied Substances*, Longmans, Green, London, 1927.
5. J. Folch, *J. Biol. Chem.*, **146**, 35 (1942).
6. A. J. P. Martin and R. L. M. Synge, *Biochem. J.*, **35**, 1358 (1941).
7. G. B. Ansell and J. N. Hawthorn, *Phospholipids*, Elsevier, Amsterdam, 1964.
8. R. Feulgen, K. Imhäuser, and M. Behrens, *Z. Physiol. Chem.*, **180**, 161 (1929).
9. E. Klenk and H. Debuch, *Z. Physiol. Chem.*, **296**, 179 (1954).
10. E. Klenk and H. Debuch, *Z. Physiol. Chem.*, **229**, 66 (1955).
11. M. M. Rapport and R. E. Franzl, *J. Biol. Chem.*, **225**, 851 (1957).
12. M. M. Rapport, B. Lerner, N. Alonzo, and R. E. Franzl, *J. Biol. Chem.*, **225**, 859 (1957).
13. G. M. Gray and M. G. Macfarlane, *Biochem. J.*, **70**, 409 (1958).
14. G. M. Gray, *Biochem. J.*, **77**, 82 (1960).
15. G. M. Gray, *J. Chromatog.*, **4**, 52 (1960).
16. C. C. Sweeley and E. A. Moscatelli, *J. Lipid Res.*, **1**, 40 (1960).
17. H. Stephen, *J. Chem. Soc.*, **1925**, 1874.
18. E. Klenk and E. Friedrichs, *Z. Physiol. Chem.*, **290**, 169 (1952).
19. J. W. Farquhar, *J. Lipid Res.*, **3**, 21 (1962).
20. W. R. Morrison and L. M. Smith, *J. Lipid Res.*, **5**, 600 (1964).
21. G. H. Sloane-Stanley, unpublished studies, 1962.
22. G. M. Gray, *J. Chromatog.*, **6**, 236 (1961).

23. J. C. M. Schogt, P. H. Begemann, and J. Koster, *J. Lipid Res.*, **1**, 446 (1960).

24. A. Pollard, A. C. Chibnall, and S. H. Piper, *Biochem. J.*, **25**, 2111 (1931).

25. F. Wegand, G. Eberhardt, H. Linden, F. Schafer, and I. Eigen, *Angew. Chem.*, **65**, 525 (1953).

26. H. K. Mangold, *J. Org. Chem.*, **24**, 405 (1959).

27. V. Mahadevan, *J. Am. Oil Chemists' Soc.*, **41**, 520 (1964).

28. V. Mahadevan, F. Phillips, and W. O. Lundberg, *J. Lipid Res.*, **6**, 434 (1965).

29. M. M. Rapport and W. T. Norton, *Ann. Rev. Biochem.*, **31**, 103 (1962).

30. M. J. Allison, M. P. Bryant, I. Katz, and M. Keeney, *J. Bacteriol.*, **83**, 1084 (1962).

31. H. Goldfine, *J. Biol. Chem.*, **239**, 2130 (1964).

32. G. M. Gray and M. G. Macfarlane, *Biochem. J.*, **81**, 480 (1961).

33. J. Folch, M. Lees, and G. H. Sloane-Stanley, *J. Biol. Chem.*, **226**, 497 (1957).

34. E. G. Bligh and W. J. Dyer, *Can. J. Biochem. Physiol.*, **37**, 911 (1959).

35. G. Rouser, J. S. O'Brien, and D. Heller, *J. Am. Oil Chemists' Soc.*, **38**, 14 (1961).

36. W. Stoffel, F. Chu, and E. H. Ahrens, *Anal. Chem.*, **31**, 307 (1959).

37. L. F. Eng, Y. L. Lee, R. B. Hayman, and B. Gerstl, *J. Lipid Res.*, **5**, 128 (1964).

38. I. A. Fowlis, R. J. Maggs, and R. P. W. Scott, *J. Chromatog.*, **15**, 471 (1964).

39. A. T. James and A. J. P. Martin, *Biochem. J.*, **50**, 679 (1952).

40. E. C. Horning, W. J. A. VandenHeuvel, and B. G. Creech, *Methods Biochem. Anal.*, **11**, 69 (1963).

41. I. Brown, *Nature*, **188**, 1021 (1961).

42. A. T. James, *J. Chromatog.*, **2**, 552 (1959).

43. J. S. O'Brien, D. L. Fillerup, and J. F. Mead, *J. Lipid Res.*, **5**, 329 (1964).

44. N. Stanacev and M. Kates, *Can. J. Biochem. Physiol.*, **41**, 1330 (1965).

45. E. C. Horning, K. C. Maddock, K. V. Anthony, and W. J. A. VandenHeuvel, *Anal. Chem.*, **35**, 526 (1963).

46. J. W. Farquhar, private communication, 1961.

47. F. P. Woodford and C. M. van Gent, *J. Lipid. Res.*, **1**, 188 (1960).

48. K. Hofmann, R. A. Lucas, and S. M. Sax, *J. Biol. Chem.*, **195**, 473 (1952).

49. T. Kaneshiro and A. G. Marr, *J. Biol. Chem.*, **236**, 2615 (1961).

50. R. W. Keenan, J. B. Brown, and B. H. Marks, *Biochem. Biophys. Acta*, **51**, 226 (1961).

51. A. T. James and E. A. Piper, *J. Chromatog.*, **5**, 265 (1961).

52. J. W. Farquhar, *Biochem. Biophys. Acta*, **60**, 80 (1962).

11

THIN-LAYER CHROMATOGRAPHY AND GAS CHROMATOGRAPHY OF SPHINGOSINE AND RELATED COMPOUNDS

Benjamin Weiss

DEPARTMENTS OF BIOCHEMISTRY
NEW YORK STATE PSYCHIATRIC INSTITUTE AND
COLLEGE OF PHYSICIANS AND SURGEONS
COLUMBIA UNIVERSITY
NEW YORK, NEW YORK

The interest in sphingolipids combined with the application of the new analytical techniques of thin-layer chromatography (TLC) and gas chromatography (GC) have resulted in many developments in our knowledge of sphingosine and related long-chain bases. In order to achieve a coherent exposition concerning the use of these methods, it was considered desirable to summarize briefly current research on the chemistry and biochemistry of these compounds. It was thought that such an approach would provide the necessary orientation for understanding the purpose behind the employment of these techniques.

I. Chemistry of Bases

The classical work by Carter and associates (*1–7*) established the structure and configuration of four natural bases [Fig. 1(a)–(d)] (a) sphingosine, D-*erythro*-1,3-dihydroxy-2-amino-4-*trans*-octadecene; (b) dihydrosphingosine, D-*erythro*-1,3-dihydroxy-2-aminooctadecane; (c) phytosphingosine, D-*ribo*-1,3,4-trihydroxy-2-aminooctadecane; and (d) dehydrophytosphingosine, D-*ribo*-1,3,4-trihydroxy-2-amino-8-*trans*-octadecene. For a comprehensive review of glycolipids containing these bases see Carter et al. (*8*).

As new bases and new sources for these bases have been reported, the nomenclature of these compounds has become both contradictory and confusing; this, however, is the almost inevitable result of the dynamism in an area actively open to investigation. In this review, the names of the parent long-chain bases are retained [Fig. 1(a)–(k)]. New compounds, differing only in the length of the aliphatic hydrocarbon chain, will be prefaced by a number indicating the total number of carbon atoms in the base. These compounds will be designated according to the parent base they resemble. It will be assumed in these instances that the positions of the functional groups bear the same relationship to those in the parent base and that the configuration at the carbon atoms bearing these groups is undetermined unless otherwise stated.

The presence of other long-chain bases in the acid hydrolysates of sphingolipids was shown by Carter et al. (*9*). They isolated two crystalline *O*-methyl ethers of sphingosine, which were named I and II without assignment of configuration. It was concluded from their studies of these compounds that an allylic rearrangement may have occurred and that during hydrolysis a carbonium ion was formed and gave rise to two diastereoisomers of opposite configuration at carbon atom 3. Later it was found that the *O*-methyl ethers were a mixture (*10*) that consisted of about 70% 1-hydroxy-2-amino-5-methoxy-3-*trans*-octadecene, formed by allylic rearrangement, and 30% of 3-*O*-methylsphingosine [Fig. 1(e)–(h)]. The configurations at carbon atoms 5 of the 5-methoxy isomer [Fig. 1(f) and (h)] in bases I and II were shown to be D and L, respectively. Similarly, Reindel et al. (*11*) found that anhydrophytosphingosine was formed during acid hy-

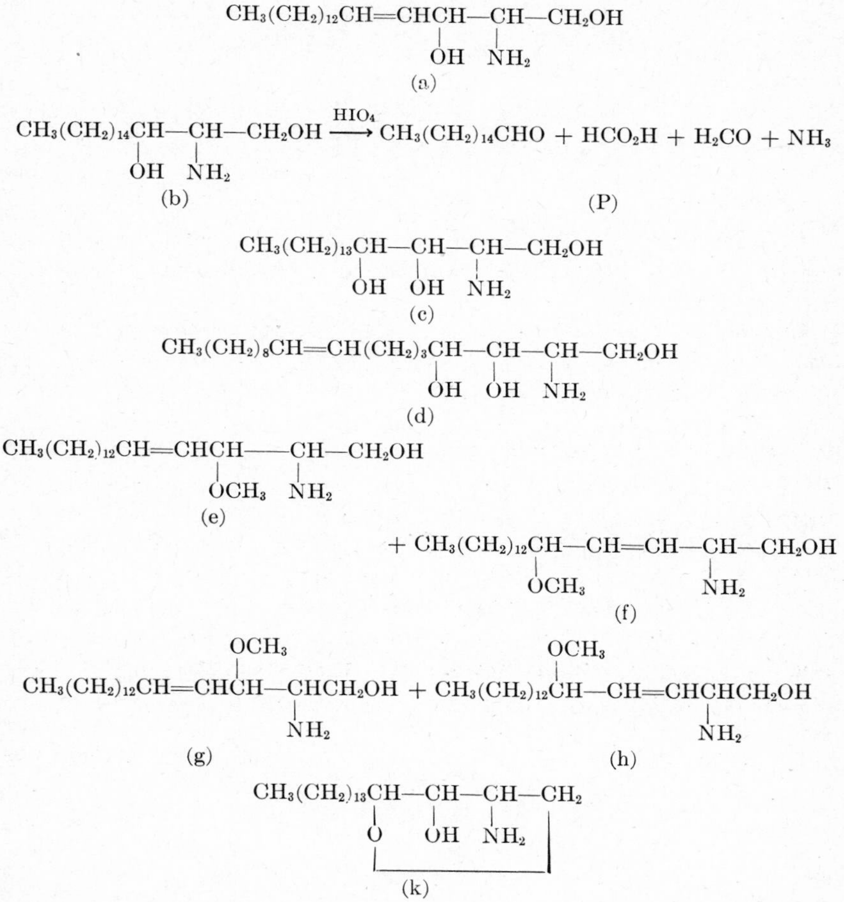

$$CH_3(CH_2)_{12}CH{=}CHCH{-}CH{-}CH_2OH$$
$$\underset{OH}{\mid}\ \underset{NH_2}{\mid}$$
(a)

$$CH_3(CH_2)_{14}CH{-}CH{-}CH_2OH \xrightarrow{HIO_4} CH_3(CH_2)_{14}CHO + HCO_2H + H_2CO + NH_3$$
$$\underset{OH}{\mid}\ \underset{NH_2}{\mid}$$
(b)   (P)

$$CH_3(CH_2)_{13}CH{-}CH{-}CH{-}CH_2OH$$
$$\underset{OH}{\mid}\ \underset{OH}{\mid}\ \underset{NH_2}{\mid}$$
(c)

$$CH_3(CH_2)_8CH{=}CH(CH_2)_3CH{-}CH{-}CH{-}CH_2OH$$
$$\underset{OH}{\mid}\ \underset{OH}{\mid}\ \underset{NH_2}{\mid}$$
(d)

$$CH_3(CH_2)_{12}CH{=}CHCH{-\!-\!-}CH{-}CH_2OH$$
$$\underset{OCH_3}{\mid}\ \underset{NH_2}{\mid}$$
(e)

$$+\ CH_3(CH_2)_{12}CH{-}CH{=}CH{-}CH{-}CH_2OH$$
$$\underset{OCH_3}{\mid}\ \ \ \ \ \ \ \ \ \ \underset{NH_2}{\mid}$$
(f)

$$\overset{OCH_3}{\mid}\qquad\qquad\qquad\qquad \overset{OCH_3}{\mid}$$
$$CH_3(CH_2)_{12}CH{=}CHCH{-}CHCH_2OH\ +\ CH_3(CH_2)_{12}CH{-}CH{=}CHCHCH_2OH$$
$$\underset{NH_2}{\mid}\qquad\qquad\qquad\qquad\qquad \underset{NH_2}{\mid}$$
(g)   (h)

$$CH_3(CH_2)_{13}CH{-}CH{-}CH{-}CH_2$$
$$\underset{O}{\mid}\ \ \underset{OH}{\mid}\ \underset{NH_2}{\mid}\ \ \Big|$$
$$\underline{\qquad\qquad\qquad\qquad}$$
(k)

FIG. 1. The compounds represented are (a) sphingosine; (b) dihydrosphingosine; (c) phytosphingosine; (d) dehydrophytosphingosine; (e) and (f) *O*-methyl ethers of sphingosine I; (g) and (h) *O*-methyl ethers of sphingosine II; (k) anhydro-phytosphingosine. The pairing of the 3-methoxy and 5-methoxy isomers is not known. P represents the oxidation products after treatment of dihydrosphingosine with periodate; similar results would be obtained upon oxidation of related bases. See text for details.

drolysis of phytosphingolipids. It was proved by Carter et al. (*12*) and O'Connell and Tsein (*13*) to have a tetrahydrofuran structure [Fig. 1(k)].

II. Biochemistry of Bases

The structural and configurational characterization of sphingosine and phytosphingosine made possible the study of the biosynthesis of these compounds. It was shown by in vivo experiments that carbon atoms 1 and 2, along with their hydrogen atoms, and nitrogen of sphingosine are derived from L-serine (*14,15*), whereas carbon atoms 3 to 18 originate from acetate (*16*); Carter et al. (*3*) had suggested earlier that serine might be a precursor of sphingosine. The preparation from rat brain tissue of an enzyme system was described; it catalyzed the formation of sphingosine from either palmityl coenzyme A and reduced nicotinamide adenine dinucleotide (*17*) or palmitaldehyde in the absence of reduced coenzyme (*18*) in the presence of L-serine, pyridoxal phosphate, and Mn^{2+}.

In a study of the biosynthesis of phytosphingosine, Greene et al. (*19*) recently showed that carbon atom 3 of serine became carbon atom 1 of the base and that palmitic acid provided the framework for the hydrocarbon chain. It appears that in the biosynthesis of both phytosphingosine and sphingosine, dihydrosphingosine is an intermediate.

III. Thin-Layer Chromatography

A. Methods

Although many adsorbents are available, silicic acid containing 5–15% calcium sulfate has been the adsorbent of choice for TLC of long-chain bases; unless otherwise stated, all the work reviewed was done with this material and with the ascending technique. For most separations, a layer 250 to 300 μ thick is satisfactory. The coated plate is activated by heating at 105–120°C for 30–120 min and stored in a desiccator. The amount of sample applied to a 20×20 cm plate with a microsyringe ranges usually from 10 γ, about the lowest level of visualization for a single component, to 10 mg for preparative work. In the latter case, the silicic acid may have to be washed prior to use, e.g., with chloroform–methanol 2:1, to remove contaminating oils, which, if present, may be seen at the solvent front as a yellow, ultraviolet-absorbing band. Generally, the solvents used for development have been mixtures of chloroform and methanol, alone or in combination with a third more polar component such as water, acetic acid, or NH_4OH. The solvents should be free of reactive contaminants such as phosgene and peroxides. No problems of demixing with secondary front formation have been reported with the multicomponent systems used. When samples are to be recovered, it is essential to locate the compounds by nondestructive agents, such as fluorescent dyes, ultraviolet absorption derivatives, or iodine vapor. If guide strips are used, or if recovery is not necessary, then the bases may be detected by destructive means, such as periodate–Schiff reagent, ninhydrin, or charring. Radioactive samples may be radioautographed, after which the radioactive zones may be removed for scintillation counting. For quantitative estimation of the long-chain bases, several colorimetric procedures may be used (*20–23*) after elution of the compounds from adsorbent with chloroform–methanol 1:1.

B. Separation of C_{18} Bases

The sharp resolution of *erythro*- and *threo*-sphingosines with chloroform–methanol 4:1 or chloroform acetone 2:1 enabled Fujino and Zabin (*24*) to study the isomeric composition of the bases formed by rat brain homogenates with palmityl coenzyme A and radioactive L-serine as substrates. Lipids were seen as blue spots on a light blue background after spraying with 0.04% bromphenol blue. In accord with earlier work on the configuration of sphingosine in cerebrosides by Carter and Fujino (*25*), it was concluded on the basis of the tribenzoyl derivative of the ^{14}C-labeled product that *erythro*-dihydrosphingosine, together with some *erythro*-sphingosine, was the main compound formed in this system; no *threo* base formation was noted. Kochetkov et al. (*26–28*) identified dihydrosphingosine by TLC, along with small quantities of sphingosine, in a new minor component from brain cerebrosides. The compound was named sphingoplasmalogen because of the vinyl ether linkage of the substituent on the secondary hydroxyl group of the base; the remaining functional groups were substituted in the same manner as in cerebrosides.

Carter and Hendrickson (*7*) separated the *N*-benzoyltriacetyl derivatives of phytosphingosine and dehydrophytosphingosine on siliconized silica gel G with acetonitrile–acetic acid–water 70:10:5. Unsaturated compounds were detected by exposure of the plate to iodine vapor. The saturated derivatives were observed as white spots on a purple background after spraying with a 1% solution of α cyclodextrin in 30% ethanol followed by exposure to iodine vapor.

The separation of a variety of long-chain bases with chloroform–methanol–2 *N* NH₄OH 40:10:1 (Fig. 2) was accomplished by Sambasivarao and McCluer (*29*). Similar resolution (*30*) was obtained with chloroform–methanol–water 49:49:2; the bases were observed by spraying with 0.2% ninhydrin in *n*-butanol–pyridine 95:5. The compounds separated in order of increasing R_f were phytosphingosine, *O*-methyldihydrosphingosine II, DL-*threo*-dihydrosphingosine, D-*erythro*-dihydrosphingosine, DL-*threo-trans*-sphingosine, *O*-methylsphingosine II, and D-*erythro-trans*-sphingosine (*29,30*). The *O*-methyl ethers I and II of sphingosine [Fig. 1 (e)–(h)] were not resolved into their components in these systems. Purified natural sphingosine was resolved into dihydrosphingosine, sphingosine, and a fast-moving component (*30*). The last component was generated by the acidity of the adsorbent; it consisted of a mixture of the cationic forms of dihydrosphingosine and sphingosine; other sphingosine derivatives gave rise to similar artifacts. Resolution of the bases was affected by the presence of various salts, such as $CoCl_2$, $CuCl_2$, $MnSO_4$, and $FeCl_3$, which reduced the pH of the solution; under these conditions the bases were converted to their cationic forms, which gave the same R_f values. Amides of sphingosine and dihydrosphingosine, made visible by spraying with *t*-butyl hypochlorite starch–KI reagent (*31*), were easily separated from the free bases but not from each other (*30*).

The enzymatic synthesis of *N*-palmitylsphingosine from sphingosine and palmitic acid was observed by Gatt (*32*) after development with chloroform–acetic acid–methanol 94:5:1. The components were seen after spraying with 2 *N* H_2SO_4 and charring at 180°C. The soluble enzyme preparation from rat brain also catalyzed the hydrolysis of various long-chain fatty acid amides of sphingosine and dihydrosphingosine; it did not effect the synthesis of cerebroside from galactosylsphingosine and palmitic acid. No adenosine triphosphate, coenzyme A, or metal cofactors were required.

In a study of the biosynthesis of phytosphingosine by *Hansenula*

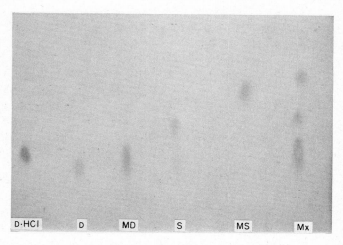

FIG. 2. Thin-layer chromatography of long-chain bases. A 20 × 20 cm glass plate was coated by means of a commercial spreader with a 250-μ-thick layer of Adsor-bosil-1, 10% CaSO₄ (Applied Science Laboratories, Inc., State College, Pennsylvania) which was prepared by mixing 15.0 g with 30 ml of water. After standing at room temperature for 30 min, the plate was activated by heating at 105°C for 2 hr. The bases, 75 γ in 2 λ of chloroform (chloroform–methanol 2:1 for di-hydrosphingosine), were applied with a microsyringe to the cooled plate, which was placed in a rectangular glass tank lined with Whatman No. 1 paper and containing 250 ml of chloroform–methanol–2 N NH₄OH 40:15:1.5 (*29*). When the solvent front had ascended within about 3 cm from the top, the plate was removed and promptly sprayed with 0.2% ninhydrin in isopropanol–collidine 97.5:2.5. The chromatogram was photographed approximately 18 hr after spraying at maximum color intensity. D·HCl, D-*erythro*-dihydrosphingosine·hydrochloride; D, D-*erythro*-dihydrosphingosine; MD, *O*-methyldihydrosphingosine II; S, D-*erythro-trans*-sphingosine; MS, *O*-methylsphingosine II; Mx, mixture (50 γ of each) D-*erythro*-dihydrosphingosine·hydrochloride; D-*erythro*-dihydrosphingosine; *O*-methyldihydrosphingosine II; D-*erythro-trans*-sphingosine; *O*-methylsphingosine II.

ciferri, Greene et al. (*19*) separated tetraacetylphytosphingosine from triacetyldihydrosphingosine and the *N*-acetyl from the polyacetyl bases with petroleum ether–chloroform–methanol 7:3:1. The triacetyl derivatives of sphingosine and dihydrosphingosine were not resolved in this system; also, the *N*-acetyl derivatives of phytosphingosine and sphingosine were not separated. Detection of the compounds was effected by exposure to iodine vapor followed by spraying with 1%

starch solution. Radioactive material was observed by radioautography.

Gaver and Sweeley (*33*) by means of TLC investigated the conditions necessary to minimize the formation of *O*-methyl ethers of sphingosine during hydrolysis of sphingolipids. They examined the base patterns after development with chloroform–methanol–water 100:42:6 and color development with iodine vapor. Although no solvent that completely eliminated the formation of the *O*-methyl ethers was found, the presence of water in the methanolic HCl during hydrolysis lowered the production of the ethers to a considerable extent with a concomitant increase in the yield of sphingosine.

C. Separation of Other Long-Chain Bases

A C_{20}-dihydrosphingosine was isolated from a mixture of dinitrophenylated (DNP) long-chain bases from gangliosides of human and bovine brain by Karlsson (*34*). The pure DNP C_{20}-dihydrosphingosine was identical on TLC with the compound obtained after hydrogenation of C_{20}-sphingosine. The ratio of C_{18}- to C_{20}-sphingosine was 1:2, and the ratio of saturated to unsaturated base was 1:10. In an analysis of the bases from gangliosides obtained from various sources, Sambasivarao and McCluer (*35*) identified a C_{20}-dihydrosphingosine by the presence of stearic acid after permanganate oxidation of the base mixture. Spots that corresponded to *erythro*- and *threo*-sphingosines and *O*-methylsphingosine were found. C_{20}-sphingosine was a component of the *erythro* and *threo* spots and was not resolved from sphingosine by chloroform–methanol–2 N NH_4OH 40:10:1. It was concluded that the threo isomer was produced during hydrolysis and that the C_{20}-sphingosine, on the basis of infrared spectra and TLC behavior, probably had the *erythro-trans*-configuration (*36*) similar to sphingosine. This finding extends previous observations of a C_{20}-sphingosine in horse and bovine brain lipids (*37,38*) and bovine gangliosides (*39,40*).

Karlsson (*41*) identified phytosphingosine by TLC as the DNP derivative along with sphingosine and dihydrosphingosine from kidney cerebrosides; phytosphingosine comprised about 50% of the total bases. Hair contained sphingosine, dihydrosphingosine, and phytosphingosine along with each of their C_{20} homologs; 5–20% of the total bases consisted of phytosphingosine and its C_{20} homolog. This is the first report of the presence of phytosphingosine in animal sources. C_{20}-phytosphingosine has been found in yeast (*42,43*). From the co-existence of sphingosines and phytosphingosines of the same chain length in various sources, Karlsson (*41*) suggested that sphingosine may be derived from dihydrosphingosine with phytosphingosine as intermediate (equation 1). He postulated further that the last step

$$\text{dihydrosphingosine} \xrightarrow{+O_2} \text{phytosphingosine} \xrightarrow{-H_2O} \text{sphingosine} \qquad (1)$$

does not exist in plants. This hypothesis leaves unexplained the failure to observe phytosphingosine in the central nervous system; if it were an immediate precursor of sphingosine, it should be present in appreciable amounts as is dihydrosphingosine. The distribution data presently available indicate that dihydrosphingosine may be the precursor of both sphingosine and phytosphingosine, because it is found with either or both of these bases in the different sources examined. Dehydrogenation or hydroxylation of dihydrosphingosine (equation 2) would yield sphingosine or phytosphingosine, respectively. No evidence has appeared for the occurrence of

$$\text{sphingosine} \xleftarrow{-2H} \text{dihydrosphingosine} \xrightarrow{+O_2} \text{phytosphingosine} \qquad (2)$$

sphingosine, except for a sphingosine isomer (*44–46*), in plant sources. The possible presence of sphingosine in the yeast, *Hansenula ciferri*, was suggested by Greene et al. (*19*), who observed about 2% tetradecanoic acid after periodate–permanganate oxidation of the base mixture.

Michalec (*47*) separated the DNP derivatives of the unsaturated long-chain bases from their saturated analogs by TLC on aluminum oxide with chloroform–methanol 100:1. The compounds were located by their yellow color or with ultraviolet light. Reverse-phase chromatography in the second dimension with methanol–tetralin–water 90:10:10, upper phase, accomplished the resolution of dihydrosphingosine; C_{16}-, C_{20}-dihydrosphingosines; sphingosine; C_{16}-, C_{20}-sphingosines; C_{20}-phytosphingosine; and *O*-methylsphingosine. Phytosphingosine did not migrate in the chloroform–methanol system of the first dimension. After removal of individual zones, the DNP bases were quantitatively recovered by elution with methanol. In contrast, bases sprayed with Ninhydrin failed to give quantitative recoveries (*30*).

D. Instability of Bases

The observations that *O*-methyl ethers of sphingosine and anhydro-phytosphingosine (*9–13*) were formed during hydrolysis of the intact lipid indicated the lability of the parent compounds to acid. When sphingosine was warmed on a steam bath (*48*), it progressively ceased to react with periodate; after 4–5 hr only a third of the original periodate activity remained, although the total nitrogen content was unchanged. Fujino and Zabin (*24*) noted the decomposition of ^{14}C-sphingosine preparations after TLC by the loss in radioactivity. The specific activities of sphingosine were higher when calculated on a colorimetric, rather than on a gravimetric, basis; this was attributed to changes in the amino group of sphingosine produced by drying, standing, and storage (*15*). During the preparation of DNP derivatives of the base mixture obtained after acid hydrolysis of sphingolipids, Karlsson (*49*) observed several side products after TLC; two appeared to be due to allylic rearrangements and two to dehydration of sphingosine and a related isomer. The role of oxygen, water, and light as causative agents of sphingosine decay has not been established; however, it appears that these agents may be involved because of the more rapid breakdown of the base on silica gel than in solution with the formation of two degradation products (*30*). Related long-chain bases with some of their reactive sites blocked did not deteriorate when stored. It would seem, therefore, that chromatography of *N*-substituted bases, such as DNP compounds (*34,41,47,49*), possesses obvious advantages. Of probably greater magnitude are the losses of minor and major components by degradation, rearrangement, and condensation incurred during hydrolysis of the sphingolipid to obtain the free base. Carter et al. (*50*) partially solved this problem by introduction of a mild degradative procedure for removing the hexose component from cerebrosides. Presently required is an equally gentle method, possibly an enzymatic one, for cleaving the amide bond to liberate the base from linkage with fatty acid.

IV. Gas Chromatography

A. Methods

A microanalytical procedure for the estimation of sphingosine and related long-chain compounds was first described by Sweeley and

Moscatelli (*51*). It depended on a method used earlier by Carter et al. (*3,7*) for determination of the structure of sphingosine and other long-chain bases by periodate oxidation. The procedure consisted of chromatographing the mixture of bases, isolated from acid hydrolysates of sphingolipids, on silicic acid to remove interfering traces of fatty acid esters. The purified bases (approximately 0.1 mM) were oxidized with sodium periodate or periodic acid (0.2 mM) in aqueous alcohol (usually 85–95%), and the long-chain aldehydes [Fig. 1 (p)], removed with methylene chloride or other suitable organic solvent, were analyzed by GC on a diethylene glycol adipate polyester column (15%, 60/80 mesh on Chromosorb W). For analysis, about 100 μg of aldehydes was used. The composition of a long-chain base mixture was calculated from the area under each component peak; sphingosine content was obtained by adding the area of the peak representing O-methylsphingosine to that of the peak representing sphingosine. The slope was obtained from a homologous series of authentic fatty aldehydes by a semilogarithmic plot of retention time versus carbon chain length. Hydrogenation was used to distinguish unsaturated aldehydes and to determine their chain length.

Since the description of the periodate procedure for the analysis of long-chain bases, several modifications have appeared (Fig. 3). These have been (a) determination of the aldehydes as dimethyl acetal derivatives (*46,54*); (b) oxidation of the base mixture to fatty acids with permanganate (*35,40*), CrO_3 (*40*), or a periodate–permanganate reagent (*19*); (c) oxidation of the aldehydes, obtained by lead tetraacetate treatment of the bases (*55,62*), with permanganate to fatty acids (*55*); (d) analysis of the unsaturated and saturated alcohols after reduction of the aldehyde carbonyl group (*40*); and (e) GC of the bases as their trimethylsilyl derivatives (*33*).

B. Analysis of Aldehydes Derived from Bases

In a survey of various sources for their long-chain base content, Sweeley and Moscatelli (*51*) found, in accord with previous work, phytosphingosine (*52*) and its C_{20} homolog (*42*) in crude yeast phosphatides. Their analysis of soybean phosphatides disclosed for the first time the presence of small quantities of dihydrosphingosine formerly believed to occur in animal sources only. The triacetyl derivative of dihydrosphingosine was isolated later by Stodola et al. (*53*) as a minor constituent from the lipids of *Hansenula ciferri*.

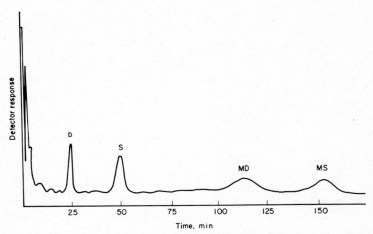

FIG. 3. Gas chromatography of long-chain aldehydes. Purified D-*erythro-trans*-sphingosine, D-*erythro*-dihydrosphingosine, *O*-methylsphingosine II, and *O*-methyldihydrosphingosine II, 5.0 mg of each, were degraded with 65.0 mg of lead tetraacetate [recrystallized from glacial acetic acid (*63*)] in 9.0 ml of dry benzene and 2 ml of glacial acetic acid at 55–60°C for 1 hr. The reaction mixture was diluted with 3.0 ml of water, and the long-chain aldehydes were removed by two extractions with 15.0-ml portions of *n*-heptane. The *n*-heptane solution was washed with water, filtered, concentrated to a sirup under reduced pressure, and dried over phosphorus pentoxide. After the addition of 3.0 ml of *n*-heptane, a 5 λ sample was injected onto a U-shaped stainless-steel column (⅛-in. internal diameter and 6 ft long) containing 15% diethylene glycol adipate polyester on Chromosorb W, 60/80 mesh. The column temperature was 190°C and the argon flow rate was 65 ml/min; detector (strontium 90) and vaporizer temperatures were 220° and 240°C, respectively. D, *n*-hexadecanal; S, *n*-hexadec-2-enal; MD, 2-methoxy-*n*-heptadecanal and 4-methoxy-*n*-heptadecanal; MS, 2-methoxy-*n*-heptadec-3-enal and 4-methoxy-*n*-heptadec-2-enal; The aldehydes D, S, MD, and MS were derived from D-*erythro*-dihydrosphingosine, D-*erythro-trans*-sphingosine, *O*-methyldihydrosphingosine II, and *O*-methylsphingosine II, respectively.

Human plasma sphingolipids (*51*) contained sphingosine as the major base, smaller amounts of dihydrosphingosine, and a new long-chain base tentatively identified as dehydrosphingosine.

The base pattern after hydrolysis of wheat flour cerebrosides was examined by Carter et al. (*44*) using the same procedure (*51*). Identification of the aldehydes disclosed the presence of dihydrosphingosine, phytosphingosine, and dehydrophytosphingosine. A new long-chain base was found; it was characterized as an isomer of sphingosine in which the double bond may be located in the 8,9 position similarly to that of dehydrophytosphingosine [Fig. 1(d)]. Similarly, Sastry and Kates (*45,46*) showed that the major base from runner bean leaves was a C_{18}-dehydrophytosphingosine that may be identical with the base from wheat flour cerebrosides (*44*). Smaller amounts of phytosphingosine, dihydrosphingosine, sphingosine isomer, and three unidentified phytosphingosine components were present. In these studies, the aldehydes were converted to dimethyl acetals (*54*) prior to chromatography on Apiezon L and butanediol succinate polyester.

n-Hexadecanal, 3.5%; *n*-hexadecen-2-al-1, 50%; and *n*-octadecen-2-al-1, 46.5%; representing dihydrosphingosine, sphingosine, and C_{20}-sphingosine, respectively, were obtained by Stanacev and Chargaff (*40*) after periodate oxidation of the bases from ox brain gangliosides. The structure of the bases was supported by (1) analysis of the unsaturated and saturated alcohols formed by reduction of the carbonyl group and by hydrogenation and (2) identification of the products from oxidation of the mixture of bases by OsO_4–HIO_4, $KMnO_4$, and CrO_3. Similarly, Sambasivarao and McCluer (*35*) found in human brain gangliosides approximately equal quantities of sphingosine, 41%, and C_{20}-sphingosine, 43%, 3% of dihydrosphingosine, and 13% of unidentified material consisting of two peaks; similar results were obtained with calf brain gangliosides. The position of the double bond was confirmed by detection of myristic and palmitic acids following permanganate oxidation of the base mixture; C_{20}-dihydrosphingosine, identified by the presence of stearic acid, could not be seen by the periodate procedure because *n*-octadecanal overlapped with *n*-hexadecen-2-al-1. One of the two unidentified peaks corresponded to dehydrosphingosine reported previously (*51*). In no instance was C_{20}-sphingosine or C_{20}-dihydrosphingosine observed in nonganglioside sphingolipids.

Carter and Hendrickson (*7*) followed the purification and determined the content of phytosphingosine and dehydrophytosphingosine from the seeds of various plants by analysis of the aldehydes obtained after oxidation of the bases. They showed that corn phosphatide contained mainly phytosphingosine and flax phosphatide predominantly dehydrophytosphingosine; other sources had intermediate values. The ratio of saturated to unsaturated base was flax, 15:85; soybean, 20:80; peanut, 50:50; and corn, 90:10.

The identification of the acids and aldehydes obtained after $KMnO_4$ and lead tetraacetate oxidation of the bases from human plasma sphingomyelin disclosed the presence of C_{16}- and C_{17}-dihydrosphingosines and C_{16}-sphingosine (*55*). The presence of sphingosine with double bonds in 4,5 and 12,13 positions was suggested by behavior of the aldehyde before and after hydrogenation and by chromatographic localization of the DNP derivative; it comprised about 10–15% of the total sphingosine.

The long-chain bases from the nonganglioside lipids of the corpus callosum and pons-oblongata of the human brain were examined by Moscatelli and Mayes (*56*). Sphingosine accounted for 94–99% and dihydrosphingosine for 1–5% of the total. Little difference was found between the two areas. Similar results were obtained for human brain cerebrosides by Radin and Akahori (*57*). C_{16}- and C_{14}-dihydrosphingosines were found in small amounts (*56*) along with possible traces of C_{16}-sphingosine. In confirmation of earlier findings (*35*), no C_{20}-sphingosine or C_{20}-dihydrosphingosine was found. The relative retention times for *n*-hexadecen-2-al-1 and *n*-octadecanal were sufficiently different in this system to permit differentiation between sphingosine and C_{20}-dihydrosphingosine (separation factor, 1.14). A column packed with 8% diethylene glycol succinate polyester on acid and solvent washed, silanized, 115/150 mesh Gas Chrom P was used. The column, however, did not discriminate between positional isomers of *O*-methylsphingosine [Fig. 1(e)–(h)]. These workers observed also that prolonged hydrogenation of the aldehydes over palladium–charcoal in absolute ethanol resulted in the formation of alcohols.

In a modification of the method of Sweeley and Moscatelli (*51*), Greene et al. (*19*) oxidized the base mixture from yeast with the periodate–permanganate reagent of von Rudloff (*58*) to fatty acids, CO_2, and NH_3. After methylation with diazomethane, the fatty acid esters were analyzed by GC on a column packed with 10% diethylene glycol succinate polyester on Chromosorb W. Sphingosine, phytosphingosine, and dihydrosphingosine would yield tetradecanoic, pentadecanoic, and hexadecanoic acids, respectively.

C. Direct Analysis of Bases

The method of Sweeley et al. (*59*) for the GC of sugars as their trimethylsilyl derivatives was extended to long-chain bases (*33*). A mixture of the dry bases (approximately 100 mg) is treated with an excess of hexamethyldisilazane–trimethylchlorosilane 2:1, 3.0 ml, in 10 ml of dry pyridine. After several minutes at room temperature, a sample is withdrawn from the reaction mixture and injected directly into a gas chromatograph. A U-shaped glass column packed with 2.5% SE-30 on acid-washed, silanized, 100/120 mesh Gas Chrom S was employed. The compounds separated were sphingosine, dihydrosphingosine, *O*-methylsphingosine, *N*-acetylsphingosine, and *O*-methyl-*N*-acetylsphingosine. From a sample of human plasma sphingomyelins hydrolyzed with aqueous methanolic HCl (1 *N* HCl and 10 *M* H_2O in methanol), the solvent found to give the best yields of base with a minimum of side products, the GC record of the trimethylsilyl bases showed sphingosine as the major product, small amounts of dihydrosphingosine and *O*-methylsphingosine, and an unknown peak believed to be C_{16}-sphingosine. The earlier observations (*55*) of the presence of C_{16}- and C_{17}-dihydrosphingosines in the same source remain to be explained. With the use of the trimethylsilyl procedure, sphingosine was identified in glycolipids of the lens of the human eye (*60*), and the purity was determined of several long-chain polyols, including hexadecane-1,2-diol, hexadecane-1,2,3-triol, 2,3-dihydroxypalmitic acid, and 2,3-epoxypalmitic acid (*61*).

The advantage of the present method over the periodate procedure, as stated by the authors, is that bases labeled with isotope may be separated and their specific activities determined. However, sufficient amounts of base must be available for degradation to determine the labeled positions. The quantitation of dihydrosphingosine, which was subject to error in the periodate procedure because of the formation of an insoluble periodate-base salt (*2,3*), would be more reliable now. A final advantage may be that bases that have changed their structure after a variety of manipulations and that are not susceptible to periodate oxidation can be studied. It would seem that both methods may be used together for structural determinations.

V. Conclusion

It is evident that the development of new analytical methods has made it possible to probe the fine composition of lipids. It may be that an array of bases of varying chain length will emerge as fractionation and isolation procedures improve. The general base pattern as it appears at present is C_{14}-, C_{16}-, C_{17}- ?, C_{18}-, C_{20}-dihydrosphingosines; C_{16}-, C_{18}-, C_{20}-sphingosines; C_{18}-, C_{20}-phytosphingosines; and C_{18}-dehydrophytosphingosine. Isomers of sphingosine in which the double bond is in a different position or in which two double bonds are present have been reported, along with three unidentified phytosphingosine components. A dehydrosphingosine of unknown structure has been postulated. Thus far, there has been no evidence for chain branching or additional functional groups. The observation of a vinyl ether linkage involving the secondary hydroxyl group of dihydrosphingosine in the intact lipid provides another problem for the biochemist. Although the metabolic significance of the minor base constituents is unknown at present, their importance to the organism may prove to be analogous to that of the trace elements.

Acknowledgment

This investigation was supported in part by Public Health Service Research Grant No. 03191-05 from the National Institute of Neurological Disease and Blindness.

The author wishes to express his gratitude to Dr. Warren M. Sperry and to Dr. David Rittenberg for advice and for reading this manuscript and to Mr. Richard L. Stiller for his competent assistance.

REFERENCES

1. H. E. Carter, W. J. Haines, W. E. Ledyard, and W. P. Norris, *J. Biol. Chem.*, **169,** 77 (1947).

2. H. E. Carter, W. P. Norris, F. J. Glick, G. E. Phillips, and R. Harris, *J. Biol. Chem.*, **170,** 269 (1947).

3. H. E. Carter, F. J. Glick, W. P. Norris, and G. E. Phillips, *J. Biol. Chem.*, **170,** 285 (1947).

4. H. E. Carter and C. G. Humiston, *J. Biol. Chem.*, **191,** 727 (1951).

5. H. E. Carter, D. Shapiro, and J. B. Harrison, *J. Am. Chem. Soc.*, **75,** 1007 (1953).

6. H. E. Carter and D. Shapiro, *J. Am. Chem. Soc.*, **75,** 5131 (1953).

7. H. E. Carter and H. S. Hendrickson, *Biochemistry*, **2,** 389 (1963).

8. H. E. Carter, P. Johnson, and E. J. Weber, *Ann. Rev. Biochem.*, **34,** 109 (1965).

9. H. E. Carter, O. Nalbandov, and P. A. Tavormina, *J. Biol. Chem.*, **192,** 197 (1951).

10. B. Weiss, *Biochemistry*, **3,** 1288 (1964).

11. F. Reindel, A. Weickmann, S. Picard, K. Luber, and P. Turula, *Ann. Chem.,* **544,** 116 (1940).

12. H. E. Carter, W. D. Celmer, W. E. M. Lands, K. L. Mueller, and H. H. Tomizawa, *J. Biol. Chem.,* **206,** 613 (1954).

13. P. W. O'Connell and S. H. Tsein, *Arch. Biochem. Biophys.,* **80,** 289 (1959).

14. D. B. Sprinson and A. Coulon, *J. Biol. Chem.,* **207,** 585 (1954).

15. B. Weiss, *J. Biol. Chem.,* **238,** 1953 (1963).

16. I. Zabin and J. F. Mead, *J. Biol. Chem.,* **211,** 87 (1954).

17. R. O. Brady and G. J. Koval, *J. Biol. Chem.,* **233,** 26 (1958).

18. R. O. Brady, J. V. Formica, and G. J. Koval, *J. Biol. Chem.,* **233,** 1072 (1958).

19. M. L. Greene, T. Kaneshiro, and J. H. Law, *Biochim. Biophys. Acta,* **98,** 582 (1965).

20. R. O. Brady and R. M. Burton, *J. Neurochem.,* **1,** 18 (1956).

21. T. Sakagami, *J. Biochem. (Tokyo),* **45,** 313 (1958).

22. E. Robins, O. H. Lowry, K. M. Eydt, and R. E. McCaman, *J. Biol. Chem.,* **220,** 661 (1956).

23. C. J. Lauter and E. G. Trams, *J. Lipid Res.,* **3,** 136 (1962).

24. Y. Fujino and I. Zabin, *J. Biol. Chem.,* **237,** 2069 (1962).

25. H. E. Carter and Y. Fujino, *J. Biol. Chem.,* **221,** 879 (1956).

26. N. K. Kochetkov, I. G. Zhukova, and I. S. Glukhoded, *Dokl. Akad. Nauk SSSR,* **147,** 376 (1962).

27. N. K. Kochetkov, I. G. Zhukova, and I. S. Glukhoded, *Biochim. Biophys. Acta,* **70,** 716 (1963).

28. N. K. Kochetkov, I. G. Zhukova, and I. S. Glukhoded, *Biokhimiya,* **29,** 570 (1964).

29. K. Sambasivarao and R. H. McCluer, *J. Lipid Res.,* **4,** 106 (1963).

30. B. Weiss and R. L. Stiller, *J. Lipid Res.,* **6,** 159 (1965).

31. R. H. Mazur, B. W. Ellis, and P. S. Cammarata, *J. Biol. Chem.,* **237,** 1619 (1962).

32. S. Gatt, *J. Biol. Chem.,* **238,** PC3131 (1963).

33. R. C. Gaver and C. C. Sweeley, *J. Am. Oil Chemists' Soc.,* **42,** 1 (1965).

34. K. A. Karlsson, *Acta Chem. Scand.,* **18,** 565 (1964).

35. K. Sambasivarao and R. H. McCluer, *J. Lipid Res.,* **5,** 103 (1964).

36. B. Majhofer-Orescanin and M. Prostenik, *Croat. Chem. Acta,* **34,** 161 (1962).

37. M. Prostenik and B. Majhofer-Orescanin, *Naturwiss.,* **47,** 399 (1960).

38. B. Majhofer-Orescanin and M. Prostenik, *Croat. Chem. Acta,* **33,** 219 (1961).

39. E. Klenk and W. Gielen, *Z. Physiol. Chem.,* **326,** 158 (1961).

40. N. Z. Stanacev and E. Chargaff, *Biochim. Biophys. Acta,* **59,** 733 (1962).

41. K. A. Karlsson, *Acta Chem. Scand.,* **18,** 2397 (1964).

42. M. Prostenik and N. Z. Stanacev, *Ber.,* **91,** 961 (1958).

43. N. Z. Stanacev and M. Kates, *Can. J. Biochem. Physiol.,* **41,** 1330 (1963).

44. H. E. Carter, R. A. Hendry, S. Nojima, N. Z. Stanacev, and K. Ohno, *J. Biol. Chem.,* **236,** 1912 (1961).

45. P. S. Sastry and M. Kates, *Biochim. Biophys. Acta,* **84,** 231 (1964).

46. P. S. Sastry and M. Kates, *Biochemistry,* **3,** 1271 (1964).

47. C. Michalec, *Biochim. Biophys. Acta,* **106,** 197 (1965).

48. J. Olley, unpublished results, quoted from J. A. Lovern, in *Biochemical Problems of Lipids* (G. Popjak and E. LeBreton, eds.), Butterworth, London, 1955, p. 99.

49. K. A. Karlsson, *Acta Chem. Scand.*, **17**, 903 (1963).

50. H. E. Carter, J. A. Rothfus, and R. Gigg, *J. Lipid Res.*, **2**, 228 (1961).

51. C. C. Sweeley and E. Moscatelli, *J. Lipid Res.*, **1**, 40 (1960).

52. H. E. Carter, W. D. Celmer, W. E. M. Lands, K. L. Mueller, and H. H. Tomizawa, *J. Biol. Chem.*, **206**, 613 (1954).

53. F. H. Stodola, L. J. Wickerham, C. R. Scholfield, and H. J. Dutton, *Arch. Biochem. Biophys.*, **98**, 176 (1962).

54. G. M. Gray, *J. Chromatog.*, **4**, 52 (1960).

55. K. A. Karlsson, *Acta Chem. Scand.*, **18**, 2395 (1964).

56. E. A. Moscatelli and J. R. Mayes, *Biochemistry*, **4**, 1386 (1965).

57. N. S. Radin and Y. Akahori, *J. Lipid Res.*, **2**, 335 (1961).

58. E. Von Rudloff, *Can. J. Chem.*, **34**, 1413 (1956).

59. C. C. Sweeley, R. Bentley, M. Makita, and W. W. Wells, *J. Am. Chem. Soc.*, **85**, 2497 (1963).

60. G. L. Feldman, L. S. Feldman, and G. Rouser, *J. Am. Oil Chemists' Soc.*, **42**, 742 (1965).

61. B. Weiss and R. L. Stiller, unpublished results, 1966.

62. B. Weiss, *Biochemistry*, **4**, 1576 (1965).

63. L. F. Fieser, *Experimental Organic Chemistry*, Heath, Boston, 1957.

GAS CHROMATOGRAPHY
OF INOSITOL AND GLYCEROL

12

Richard N. Roberts

ELECTRONICS LABORATORY
GENERAL ELECTRIC COMPANY
SYRACUSE, NEW YORK

I. Inositol

The most precise, rapid, and sensitive assay for myoinositol[†] is quantitative gas–liquid chromatography. The analysis is dependent upon the formation of a volatile derivative, since free inositol, like most carbohydrates, is nonvolatile and thus is not easily chromatographed. Most derivatives that reduce the polarity of all six hydroxyl functions will serve; however, only one has been found to be entirely suitable to date. Thus, hexa-*O*-trimethylsilylinositol, the trimethylsilyl ether formed by reaction with hexamethyldisilizane, has been the choice for the gas chromatographic quantitation of inositol by several laboratories (*9,24,31,38*).

Other derivatives of inositol that may be gas-chromatographed, such as the acetoxy (*15,22,23*), methoxy (*3*), hexa-*O*-dimethylsilyl ether, and hexatrifluoroacetoxy, do not fulfill many of the essential requirements of a suitable derivative.

[†] Myoinositol is hereafter called simply inositol.

Trimethylsilyl ether derivatives of most carbohydrates, and all the cyclitols, may be rapidly and quantitatively prepared on a microscale, and in addition, the reaction mixture may be used directly for chromatographic analysis. Silylation of all free hydroxyl groups occurs, and the yield of trimethylsilyl derivative is quantitative (*35*). The ether derivative has good thermal stability and is not easily hydrolyzed in the presence of excess reagent (*34*). Volatility is increased and the polarity is decreased to a greater extent than any other derivative investigated, enabling short retention times to be realized. Perhaps the most important property of hexa-*O*-trimethyl-silylinositol is that it may readily be separated from all other naturally occurring materials by gas–liquid chromatography (*9,24,27,30,35,37*). Even isomers and oxidation products of inositol may be resolved (Fig. 1 and Reference *24*). As an illustration of the utility of silylated inositol, the excellent procedure of Brower, et al. (*4*) may be cited. These authors use inositol as an internal standard for the gas chromatographic analysis of sugars.

Hexa-*O*-trimethylsilylinositol is prepared by the addition, to the anhydrous sample to be analyzed, of a solution of dry pyridine, hexamethyldisilizane, and trimethylchlorosilane 10:2:1, v/v/v. If the reagent solution is premixed, it should be no more than 3 hr old (*37*). The volume of the pyridine reaction mixture should be such that the total carbohydrate concentration is 1 to 10 mg/ml to ensure reagent excess. Knowledge of the total carbohydrate concentration is important, since the reagents will react with all free hydroxyl groups present. After the reaction is complete, the mixture is brought up to final volume with dry pyridine and an aliquot for gas chromatography is taken.

The pyridine must be rigorously dried before use even though the highest grade available is purchased. The procedure of choice is distillation from barium oxide and storage over potassium hydroxide pellets. However, the distillation technique may be omitted if the pyridine is treated with fully activated molecular sieve 5A.

The trimethylchlorosilane should be carefully hydrolyzed before use with single drops of water and mild mixing until rapid evolution of hydrochloric acid subsides. The partially hydrolzed reagent is then distilled and the cut boiling at 57.7°C is collected and used (*34*). The hydrolysis products, which would otherwise produce extra reagent peaks during the gas chromatographic analysis, remain in the distilla-

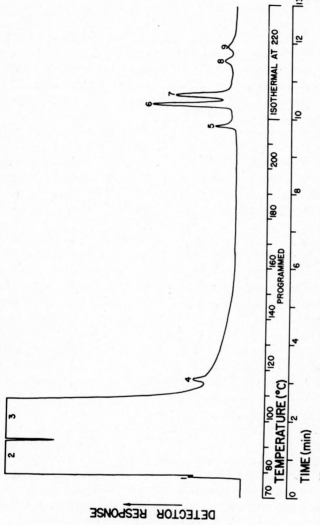

FIG. 1. Chromatography of trimethylsilyl derivatives separated on diethylene glycol succinate (*27*). 1, Air; 2, TMCS + HMDS; 3, pyridine; 4, pyridine impurity; 5, myoinosose; 6, 2-*O*,*C*-methylenemyoinositol; 7, myoinositol; 8, 2-*O*-hydroxymethylmyoinositol; 9, *N*-acetylglucosamine. Chromatographic conditions: 6 ft × ¼ in. o.d. 10% DEGS on Chromosorb W-HMDS, 80/100 mesh; helium flow at 55 ml/min; filament current, 150 ma; injection port, 290°C; detector temperature, 340°C; column temperature programmed from 70° to 220°C at 15°C/min.

tion pot. This reagent purification, although seemingly overcautious, is an essential part of the procedure and need, of course, only be done once for every batch of trimethylchlorosilane purchased. Impurities in the trimethylchlorosilane have been observed to cause low recoveries of carbohydrates. One of the main impurities was thought to be dimethylchlorosilane, which forms derivatives involving two hydroxyl groups and one dimethylchlorosilane molecule (*34*). The dimethylchlorosilane may be detected by a gas chromatographic analysis of the trimethylchlorosilane (*28*), but this is not necessary when the impurities are removed as described above.

Trimethylchlorosilane is a catalyst for the reaction and need not be present in equimolar amounts. Pyridine, although not an efficient solvent for inositol, is essential to the reaction, since it acts as a proton acceptor (*7*) and thus aids in driving the reaction to completion. During the reaction, ammonium and pyridinium chlorides form a white precipitate that settles out upon standing and does not interfere with sample transfer to the microliter syringe. If preferred, the reaction mixture may be centrifuged to ensure the use of only supernatant in the analysis.

Solvents other than pyridine have been investigated in an attempt to find one superior for inositol. These include dimethyl formamide, dioxane, diethylamine in pyridine, quinoline, dimethylsulfoxide, formamide, and *N*-methyl-2-pyrolidone. None of these are as suitable as pyridine, because they either cause chromatographic interference or inhibit the quantitative formation of the derivative (*27,30*). However, Flint, et al. (*9*) have reported that 25% dimethylsulfoxide in pyridine is superior to pyridine alone. Strongly basic solvents and those containing hydroxyl functions must be avoided, since they will destroy the reagent.

At room temperature, the formation of the derivative is complete in one hour [Fig. 2, (*27*)]. The reaction time may be greatly decreased by heating under reflux. The reaction rate is apparently controlled by the solubility of inositol in the pyridine reagent mixture, since derivatization of sugars, which are pyridine-soluble, has been found to be independent of time (*35*). The problem of mutarotation of sugars that is amplified by heating the basic solution (*35*) does not exist with inositol. It has been suggested by Freedman (*7*), and others, that the mechanism for the reaction involves formation of protonated nitrogen to give an ammonium intermediate of hexamethyldisilazane that then undergoes nucleophilic displacement by the OR⁻ group of the carbohydrate.

Caution should be observed at all times with these compounds. Pyridine and volatile silica derivatives are potential health hazards. The effluent from the gas chromatograph should be efficiently vented,

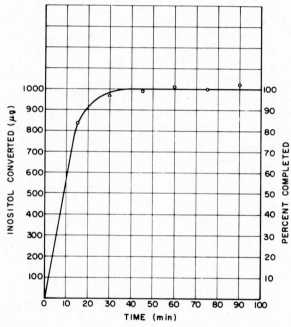

FIG. 2. Rate of inositol silylation at room temperature (27).

as should all distillation vapors and other handling procedures that are potential vapor sources.

Pure hexa-*O*-trimethylsilylinositol may be collected by preparatory gas chromatography using the same instrument parameters as for its quantitation. The chromatographic effluent is readily condensed in the simplest of trapping tubes without elaborate cooling or electrostatic precipitation precautions. It has even, for example, been collected by holding a 0.1-ml pipette against the collection port during the peak elution (30). The derivative may be seen to collect in the tube in a

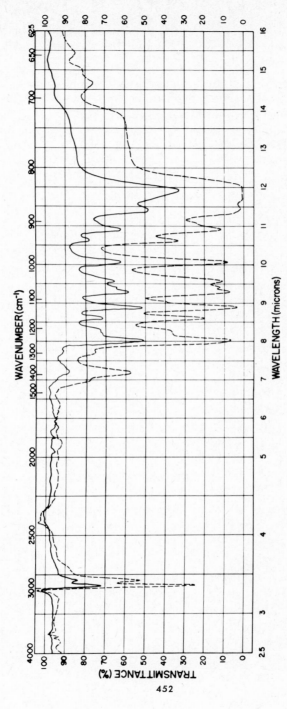

FIG. 3. Infrared spectra of gas-chromatographically pure hexa-*O*-trimethylsilylmyoinositol.

crystalline form. The infrared spectrum of pure hexa-*O*-trimethyl-silylinositol purified in this manner is shown in Fig. 3.

The preparation of the derivative in bulk quantities may be accomplished by scaling up the quantitative procedure outlined above and then, after 30 min of reflux, distilling off the pyridine and excess reagents and leaving behind essentially pure hexa-*O*-trimethylsilylinositol. Final purification may be accomplished by recrystallization. Free myoinositol may be recovered from its derivatized form by refluxing with anhydrous hydrochloric acid in chloroform. The inositol appears as a precipitate.

The silyl ether of inositol, even when stored in the original reaction mixture, has a considerable shelf life and has been observed to retain its quantitation for periods exceeding 4 months (*31*). In the dry, pure state the derivative has been stored for several years with no evidence of decomposition (*30*).

Hexa-*O*-trimethylsilylinositol may be chromatographed on a variety of columns coated with carbowax or silicone liquid phases, but care must be exercised in choice of both support and liquid phase when quantitation is desired. Columns prepared from nontreated support, such as acid-washed Chromosorb W, have been seen to cause partial loss of the inositol (*27,30*), particularly when first used (Fig. 4). Such columns may be conditioned in the instrument by heavy loading with chromatographically pure hexa-*O*-trimethylsilylinositol such that some degree of quantitation may be achieved, but it is strongly recommended that silanized support, such as chromosorb W-HMDS or Q, be used at the outset. In situ silanization of the column packing by injection of trimethylchlorosilane, dimethyldichlorosilane, or hexamethyldisilazane is not recommended. Commercially silanized support, or commercially packed columns using this support, are superior for analytical work. Preparatory gas chromatography may be carried out on nonsilanized supports, since minor losses are usually not important as long as the required resolution is realized.

Ethylene glycol succinate (DEGS) as a liquid phase promotes better resolution than the nonpolar silicone polymers SE-S2, SE-30, QF-1, and XE-60, but it is limited by adsorption effects (*32*). Richey, et al. (*29*), have reported an average error of 11% in the recovery of sugar mixtures from this liquid phase. Others (*27,30,31*) have used DEGS as a liquid phase and found no adverse effects on quantitation (Fig. 5). Carbowax 20M provides superior resolution to the silicone fluids with practically no adsorption effects (*32*). In general, the smallest-mesh silanized support consistent with adequate carrier gas flow and coated with the lowest per cent liquid phase should be chosen. Loading of columns with large amounts of liquid phase may promote lack of quantitation even though larger sample sizes may be then applied.

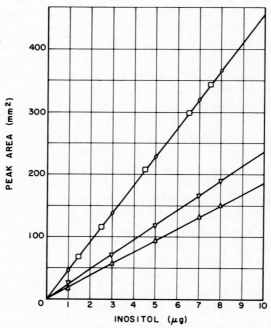

FIG. 4. Quantitation of hexa-*O*-trimethylsilylmyoinositol on four solid supports (*27*). ○, HMDS-treated Chromosorb W, 80/100 mesh, coated with 20% DEGS; □, HMDS-treated Chromosorb W, 60/80 mesh, coated with 10% DEGS; ▽, acid-washed Anakrom A, 60/80 mesh, coated with 20% DEGS; △, untreated Chromosorb W, 60/80 mesh, coated with 20% DEGS. (Chromatographic conditions same as Fig. 1.)

Many detectors may be applied to the gas chromatographic quantitation of inositol, but some are preferable to others. The best choice is the flame ionization detector because of its sensitivity and stability; but when high sensitivity is not necessary, the thermal-conductivity detectors are completely adequate. No advantage is realized with the electron-capture detector, since hexa-*O*-trimethylsilylinositol has a low affinity for electrons. In addition, if temperature programming is utilized to effect more complete resolution and decrease analysis time, the electron-capture detector may not be used. With a column such as 5% QF-1 on acid-base-washed silanized support and a flame ionization detector, quantitation is routine at the nanogram level (*24*), which is usually sufficient.

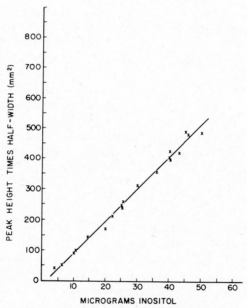

FIG. 5. Linear relationship between the mass of myoinositol hexa-*O*-trimethyl-silyl ether injected and the area of the resulting peak (*30*). Chromatographic conditions same as Fig. 1.

The actual instrument parameters necessary for inositol quantitation depend to a large extent upon the complexity of the mixture present. With simple mixtures, containing few carbohydrates, isothermal operation of a 10% DEGS column at 220°C or an SE-30 column at 300°C is usually adequate. With more complex mixtures temperature programming is necessary. The DEGS column would be programmed from 70° to 220°C at a rate of 15°C/min and held at the upper limit until the run was complete (Fig. 1).

Measurement of peak areas may be done simply with a millimeter scale. The best technique is peak height multiplied by width at half height, since peak areas are reproducible to better than 5%. A superior method of peak measurement involves the use of a digital integrator, which will record peak areas and retention times with good precision.

Free inositol has good chemical stability under quite severe conditions. Total cellular inositol of microorganisms has been determined after long-term hydrolysis in 6 N hydrochloric acid (*31*). With this technique of whole-cell acid hydrolysis, destruction of many cellular constituents occurs, but this does not interfere with inositol quantitation. The insoluble residue may be removed, after hydrolysis, by filtration or centrifugation with activated charcoal, but it has been observed that this residue does not impair inositol quantitation within detection limits of the thermal-conductivity unit (*30,31*). The only adverse effect noted was a slight reduction in useful column life.

All methods for the estimation of total inositol in biological systems must involve a technique of preliminary hydrolysis for liberation of the bound forms. Hydrolysis of phosphoinositides, the most common bound form, requires a strong acid. It is apparent from the studies of Nagy (*27*) that the most suitable hydrolytic procedure is 48-hr sealed-tube hydrolysis in aqueous 6 N hydrochloric acid at a temperature of 120°C (Fig. 6). The sample should be free of organic solvents before sealing the tube, since these apparently decrease the rate of free inositol production. Refluxing the cells or isolated phosphatides under atmospheric pressure with the same acid will not result in release of all of the bound inositol.

Any inositol present as phosphate ester, as a result of incomplete hydrolysis, will not be quantitated by this gas chromatographic procedure. Inositol phosphates may be gas-chromatographed only after methylation of the phosphate function before silylation of the remaining inositol hydroxyl groups. Caution is advised in accepting total inositol values from reflux hydrolysis procedures merely because replicate determinations check well. It has been shown (*31*) that consistent and reproducible values may be obtained although they represent low total inositol values. Therefore, the rigorous sealed-tube hydrolytic procedure should be used routinely when total inositol quantitation is important.

The inositol quantitation procedure can be summarized as follows:

1. Hydrolyze sample for 48 hr at 120°C with 5 volumes of 6 N hydrochloric acid in a sealed tube.

2. After stirring with acid-washed activated charcoal, filter and evaporate to dryness, first in a flash evaporator and then in a vacuum desiccator over sodium hydroxide pellets.

3. For each 10 mg of carbohydrate material add 1 ml anhydrous pyridine, 0.2 ml hexamethyldisilizane, and 0.1 ml trimethylchlorosilane.

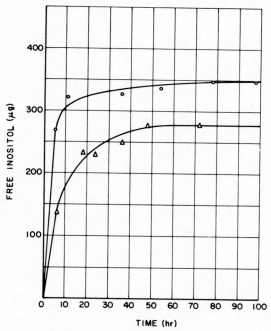

FIG. 6. Release of free inositol from phosphatidyl inositol as a function of acidic conditions and time (*27*). ○, sealed-tube hydrolysis of phosphatidyl inositol in 6 N aqueous HCl. △, sealed-tube hydrolysis in 3.15 N methanolic HCl.

4. Heat and stir the reaction mixture in a small glass-stoppered volumetric flask until the silylation is complete (approximately 15 min in boiling pyridine).

5. Dilute the contents of volumetric flask to the mark with dry pyridine, mix, and allow the white precipitate to settle.

6. Withdraw aliquot from the clear top portion of the flesk and apply to the chromatographic instrument.

II. Glycerol

The gas chromatographic analysis of free glycerol presents no particular technical problems. Free glycerol may be analyzed on a column of 10% Carbowax 20M–terephthalic acid on 70/80 mesh Chromosorb W(HMDS) (5). When temperature programmed from 115° to 250°C in 7 min this column has the ability to separate glycerol from other closely related glycols cleanly and without any peak tailing. This system will also chromatograph free fatty acids.

Another system, consisting of a 6 ft $\times$ ¼ in. o.d. column packed with 5% Carbowax 1500 on Haloport F and programmed at 9°C/min from 50° to 200°C, has the capability of glycerol analysis with good recovery (8). This column will separate glycerol from ethylene glycol, propylene glycol, diethylene glycol, 1,3-butylene glycol, dipropylene glycol, trimethylene glycol, and 1,2,6-hexanetriol (8). It is not suitable for free fatty acids. Other gas chromatographic procedures for free glycerol are those of Clifford (6), Murray and Williams (26), and Bauman (1).

In addition to the free form, glycerol may be chromatographed as a derivative. Glycerol triacetate (triacetin) is the most useful (14,19, 20,33), but isopropylidene glycerol (12,25,36) and tri-O-trimethylsilyl glycerol (30) have also been utilized. The latter form is obtained when the inositol assay procedure is carried out on a mixture containing glycerol. In this system the silyl derivative of glycerol chromatographs well. However, it emerges very near the pyridine solvent peak and, further, considerable glycerol is lost during sample preparation. For these reasons, the gas chromatographic quantitation of glycerol concurrently with inositol may not be conducted by the silinization technique described above.

The use of the trifluoroacetyl derivative of glycerol for gas chromatographic analysis of very low concentrations using the electron-capture detector has been postulated and is currently being investigated (30).

The difficulty in gas chromatographic quantitation of glycerol present in lipids arises not from the instrumental portion of the procedure, but rather at the sample preparation stage. Lack of quantitative recovery of glycerol from lipids is mainly due to incomplete lipid hydrolysis, but it may also be caused by volatilization and solubilization losses (10,21). Strong-acid hydrolysis of lipids, such as used for production of free inositol, may not be employed when glycerol recovery is desired because of extensive glycerol destruction (*11,13*).

In order to circumvent destruction of glycerol by strong acids, acetolysis of lipids was introduced by Bevan, et al. (*2*) in 1953. Hanahan and co-workers (*12,36*) employed this method in their laboratory with phosphatides. After acetolysis the lipid was saponified and isopropylidene glycerol was prepared for gas chromatographic analysis.

More recent work from the laboratory of Horrocks, Cornwell, and Holla (*16–18*) at Ohio State University has indicated that different methods of sample preparation are necessary for quantitative analysis of glycerol present in different types of lipids. The earliest report from this laboratory on the use of gas chromatography for lipid glycerol quantitation was published in 1962 (*19*). Glycerides were reduced with lithium aluminum hydride to yield the lithium aluminum alcoholate of fatty alcohols and glycerol. These were next acetylated with acetic anhydride. This method permits the simultaneous determination of fatty acids and glycerol in neutral lipids but not phosphatides.

A modification of this hydrogenolysis–acetylation technique that results in more reproducible results was introduced by Holla (*16,18*) and is outlined below. The glyceride (30 to 100 mg) and internal standard are weighed into a round-bottom flask and dissolved in 20 ml dry ether. Lithium aluminum hydride (200 mg) is suspended in 30 ml ether and added to the glyceride solution in 1-ml increments until boiling stops. One volume excess of the hydride is then added, and the solution is refluxed for 1½ hr. Acetic anhydride is then added dropwise to decompose the excess hydride, then 25 ml acetic anhydride and 30 ml xylene. The ether is removed by evaporation, and the solution is refluxed for 6 hr. It is next filtered, flash-evaporated to dryness, and dissolved in ether for gas–liquid chromatography. If early peaks are noted, more xylene is added and the evaporation is repeated.

This procedure is entirely satisfactory for neutral saturated triglycerides. Methyl eicosanoate is used as the internal standard, or, when eicosanic acid is present in the triglyceride, methyl erucate is used. Flash evaporation does not result in the loss of the glycerol derivative or C_{10}–C_{20} fatty alcohols. Octylacetate may be partially volatilized. Consistent and reproducible data are reported for this technique (*16*).

If one encounters eicosanoic acid and compounds that elute with triacetin during this gas chromatographic analysis, an alternative sample-preparation technique must be used for glycerol quantitation. This saponification–acetylation (*16*) technique is described below. The glyceride (30 to 100 mg) and hexadecanyl acetate (internal standard) are weighed out as before and dissolved in 30 ml absolute methanol. Sodium methoxide (25 mg sodium/10 ml methanol) is added, and the saponification is carried out by refluxing for 2 hr. The methanol is then removed by aspiration and 5 ml of water is added before the reflux is continued for one more hour. Then 35 ml each of acetic anhydride and xylene are added and treatment is continued as in the hydrogenolysis–acetylation technique above. The free fatty acids produced by this procedure are absorbed on the column and thus destroyed.

Gas chromatographic conditions were not extensively investigated by these workers. However, they found a 10–13% ethylene glycol succinate polyester liquid phase on a support of 60/80 mesh Gas Chrom P to give adequate separation of all peaks of interest (*16*). The column length was 10 ft of 0.25-in-i.d. stainless-steel tubing, and helium at 60 to 100 ml/min was the carrier gas. The column temperature was varied between 170° and 200°C depending upon the separation desired. In some cases, temperature programming was found to be necessary in order to reduce analysis time. Thermal conductivity detection was used, and peak areas were corrected for molar response relative to hexadecanyl acetate. All columns were conditioned 35–40 hr at 205°C under a flow rate of 30 ml/min before use.

The above acetolysis procedures do not result in quantitative recovery of glycerol from phosphatides. An acetolysis–saponification–acetylation procedure was developed for this purpose (*16,18*) and is outlined here. The phosphatide (50–100 mg) and internal standard (30 mg hexadecanylacetate) are weighed and dissolved in 10 ml of a 4:1, v/v, mixture of acetic acid and acetic anhydride. Reflux is then carried out for 10 hr, after which the excess reagents are removed with water aspiration and then high-vacuum evaporation. The flask is then flushed with nitrogen, and the residue is dissolved in 20 ml methanol. Sodium methoxide (50 mg sodium/10 ml methanol) is then added and refluxed for 2 hr. The methanol is flash-evaporated and 5 ml water is added. After a 2½-hr reflux, 35 ml each of acetic anhydride and xylene are added and again refluxed (6 hr). After final preparation as before, the acetates are analyzed by gas chromatography.

For lecithins and other more difficultly hydrolyzed phosphatides this procedure must be yet further modified in order to promote complete glycerol recovery (*16*). The original acetolysis solution is replaced by one containing trifluoroacetic acid and acetic anhydride and then carried out as above except the amount of sodium is increased and a sodium hydroxide solution is used in place of the 5 ml water. The presence of the stronger trifluoroacetic acid results in more complete acetolysis of the phosphatide for excellent glycerol recovery, but unsaturated fatty acids are destroyed. This modified acetolysis technique works well for glycerol analysis of cephalins and lecithins (*16,18*).

It is evident from the above review that the quantitation of lipid glycerol presents a considerable number of problems, particularly in the choice of hydrolytic procedure. It is felt that the best procedure is to use the technique that results in quantitative glycerol recovery from even the most difficult lipids. Since this procedure results in their loss, fatty acids must be analyzed on a separate aliquot of the lipid preparation. This "modified acetolysis–saponification–acetylation" procedure (*16,18*) is described below in detail:

1. The lipid (50–100 mg) and internal standard (30 mg, hexadecanylacetate) are weighed into a round-bottom flask and dissolved in 10 ml of a 1:4, v/v, mixture of trifluoroacetic acid and acetic anhydride.

2. After refluxing for 10 hr the excess reagents are removed first by water aspiration and then by high vacuum.

3. The flask is then flushed with nitrogen, and the residue is dissolved in 20 ml of methanol. Sodium methoxide (70 mg sodium/10 ml methanol) is added and the acetolysis is carried out by refluxing for 2 hr.

4. The methanol is removed by flash evaporation and 5 ml of 0.4 N NaOH is added.

5. After a 2½ hr reflux period, 35 ml each of acetic anhydride and xylene are added and reflux is continued for an additional 6 hr.

6. The filtered solution is then evaporated to dryness and dissolved in a small volume of ether for gas chromatographic analysis.

Instrumental parameters are those of Holla (*16*) described above. For thermal conductivity detection these large sample sizes are necessary. If the flame ionization detector is used, the sample size and reagent volumes may be reduced by at least one order of magnitude. When only a few milligrams of lipid is available, the more sensitive gas chromatographic detectors are essential. Most lipids contain 10%

or less glycerine by weight, and the initial sample size as well as the final volume of ether chosen should be adjusted to give a final concentration of glycerol of between 10 μg/ml and 1 mg/ml for flame ionization detection. For thermal conductivity detection the final concentration of glycerol should be approximately 100 times the above.

REFERENCES

1. F. Baumann and J. M. Gill, *Aerograph Research Notes,* Spring 1966, p. 6.
2. T. H. Bevan, D. A. Brown, G. I. Gregory, and T. Malkin, *J. Chem. Soc.,* **1953**, 127.
3. C. T. Bishop, in *Methods of Biochemical Analysis,* Vol. 10 (D. Glick, ed.), Wiley (Interscience), New York, 1962, p. 1.
4. H. E. Brower, J. E. Jeffery, and M. W. Folsom, *Anal. Chem.,* **38,** 362 (1966).
5. B. Byers, *Aerograph Research Notes,* Spring 1964, p. 2.
6. J. Clifford, *Analyst,* **85,** 475 (1960).
7. R. W. Freedman and P. P. Croitoru, *Anal. Chem.,* **36,** 1389 (1964).
8. R. L. Friedman and W. J. Raab, *Anal. Chem.,* **35,** 67 (1963).
9. D. R. Flint, Ten-Ching Lee, and C. G. Huggins, *Federation Proc.,* **24,** 662 (1965).
10. K. E. Guyer, W. A. Hoffman, L. A. Horrocks, and D. G. Cornwell, *J. Lipid Res.,* **4,** 385 (1963).
11. D. J. Hanahan, *Lipide Chemistry,* Wiley, New York, 1959, p. 188.
12. D. J. Hanahan, J. Ekholm, and C. M. Jackson, *Biochemistry,* **2,** 630 (1963).
13. D. J. Hanahan and J. N. Olley, *J. Biol. Chem.,* **231,** 813 (1958).
14. L. Hartman, *J. Chromatog.,* **16,** 223 (1964).
15. J. A. Hause, J. A. Hubicki, and G. G. Hazen, *Anal. Chem.,* **34,** 1567 (1962).
16. K. S. Holla, *Dissertation Abstr.,* **25**(12), 6930 (1965).
17. K. S. Holla and D. G. Cornwell, *J. Lipid Res.,* **6,** 322 (1965).
18. K. S. Holla, L. A. Horrocks, and D. G. Cornwell, *J. Lipid Res.,* **5,** 263 (1964).
19. L. A. Horrocks and D. G. Cornwell, *J. Lipid Res.,* **3,** 165 (1962).
20. E. Jellum and P. Bjornstad, *J. Lipid Res.,* **5,** 314 (1964).
21. P. Karrer and E. Jucker, *Helv. Chim. Acta,* **35,** 1586 (1952).
22. H. W. Kircher, *Methods in Carbohydrate Chemistry,* Vol. 1 (R. L. Whistler and M. L. Wolfrom, eds.), Academic Press, New York, 1962, pp. 13–20.
23. Z. S. Krzeminski and S. J. Angyal, *J. Chem. Soc.,* **1962,** 3251.
24. Y. C. Lee and C. E. Ballou, *J. Chromatog.,* **18,** 147 (1965).
25. M. E. Mason, M. E. Eager, and G. R. Waller, *Anal. Chem.,* **36,** 587 (1964).
26. W. J. Murray and A. F. Williams, *Analyst,* **86,** 849 (1961).
27. S. Nagy, Ph.D. thesis, Rutgers Univ., New Bruswick, N.J., 1965.
28. T. Oiwa, M. Sato, Y. Miyakawa, and I. Miyazaki, *Nippon Kagaku Zasshi,* **84,** 409 (1963).
29. J. M. Richey, H. G. Richey, Jr., and R. Shraer, *Anal. Biochem.,* **9,** 272 (1964).
30. R. N. Roberts, unpublished observations, 1966.

31. R. N. Roberts, J. A. Johnston, and B. W. Fuhr, *Anal. Biochem.,* **10,** 282 (1965).

32. J. S. Sawardeker and J. H. Sloneker, *Anal. Chem.,* **37,** 945 (1965).

33. J. S. Sawardeker, J. H. Sloneker, and A. Jeanes, *Anal. Chem.,* **37,** 1602 (1965).

34. R. R. Suchanec, *Anal. Chem.,* **37,** 1361 (1965).

35. C. C. Sweeley, R. Bentley, M. Makita, and W. W. Wells, *J. Am. Chem. Soc.,* **85,** 2497 (1963).

36. G. A. Thompson, Jr. and D. J. Hanahan, *J. Biol. Chem.,* **238,** 2628 (1963).

37. W. W. Wells, R. Bentley, and C. C. Sweeley, in *Biomedical Application of Gas Chromatography* (H. Szymanski, ed.), Plenum Press, New York, 1964.

38. W. W. Wells, T. A. Pittman, and H. J. Wells, *Anal. Biochem.,* **10,** 450 (1965).

13

GAS CHROMATOGRAPHIC ESTIMATION OF CARBOHYDRATES IN GLYCOLIPIDS

Charles C. Sweeley and Dennis E. Vance

DEPARTMENT OF BIOCHEMISTRY AND NUTRITION
GRADUATE SCHOOL OF PUBLIC HEALTH
UNIVERSITY OF PITTSBURGH
PITTSBURGH, PENNSYLVANIA

In 1958 Radin (*1*) reviewed various methods for the qualitative and quantitative analysis of the carbohydrate moiety in glycosyl ceramides, including descriptions of some general procedures for the extraction and purification of these substances. Some of the methods that were widely accepted at that time were relatively inadequate for one reason or another. There were problems, for example, associated with many of the conditions used for the hydrolysis of glycolipids, so that it was often difficult to achieve accurate results in the estimation of total carbohydrate and the qualitative identification of individual sugars. Although several colorimetric procedures were available for the measurement of total hexose in simple glycolipids such as cerebroside, the complex glycolipids were not easily analyzed by these methods because of differences in the color yields obtained with various sugars. More selective methods, involving paper chromatography or enzymatic determinations with glucose oxidase, for example, were generally time-consuming and laborious and usually demanded relatively large samples of material.

Considering the rapidly expanding list of simple and complex glycolipids known to occur naturally, the need is even greater today for rapid and reliable analytical methods for the wide variety of carbohydrates that are likely to be encountered as constituents of the glycolipids. It is hoped that at least some of these analyses can be performed by gas-liquid partition chromatography (GLC).

The principles and several applications of GLC were put forth in a remarkable display by James and Martin, beginning in 1952 (*2*), and there was little delay in developing gas chromatographic techniques for analyses of many types of easily volatilized molecules of biological interest. Few investigators attempted to relate this new form of chromatography to the analysis of nonvolatile or even to high-boiling compounds until several years later, however. Carbohydrates were considered in a class with steroids, amino acids, nucleosides, and other highly polar compounds as substances for which GLC techniques were inappropriate. This view was not to persist for long; following improvements in column packings and detector sensitivities, many new applications were described.

One of the earliest reports on the gas chromatography of carbohydrates was a classical paper by Bishop and his co-workers in 1958, in which were described conditions for GLC of a variety of *O*-methyl derivatives of pentoses and hexoses (*3*). Their encouraging results served to open the entire area of GLC of carbohydrates, taking advantage of the nonpolar and relatively volatile nature of the poly-*O*-methyl ethers. More recently, other volatile forms of carbohydrates have been examined, such as *O*-trimethylsilyl (TMSi) ethers (*4*) and *O*-acetyl esters (*5*), and the retention behavior of a wide variety of carbohydrates has been recorded. Extensive reviews by Bishop (*6*), Kircher (*7*), and Wells et al. (*8,9*) compare the separations that have been achieved with these various derivatives.

Though not yet widely exploited for analyses of carbohydrate components of glycolipids, gas chromatography appears to be the most rapid and accurate method yet described for both qualitative and quantitative estimations of the various components in neutral glycosphingolipids and gangliosides (*10,11*). Several advantages are the very high sensitivity that can be obtained with GLC and the fact that the method can be totally selective for any one of the carbohydrate components in a mixture, or it can be used to determine all of the components. GLC has also been used to determine fatty acids and long-chain bases of glycosyl ceramides as well; it is possible, therefore, that the analysis of carbohydrates and these other components can be combined into a single determination.

This review has been directed primarily to three analytical problems: (1) the quantitative liberation of component carbohydrates from a glycolipid, (2) the recognition of individual carbohydrates obtained from a given glycolipid, and (3) the accurate determination of the relative proportion of each of the individual carbohydrates in a mixture from complex glycolipids. Much of the developmental work reported here pertains specifically to glycosyl ceramides that occur in human organs and fluids, substances such as cerebroside, ceramide dihexoside, globoside, and the gangliosides. It is hoped, however, that these methods may also be used, perhaps with minor modifications, for the determination of carbohydrates in other types of glycolipids such as glycosyl glycerides, inositides, phytoglycolipid, and microbial glycolipids.

I. Recovery of Carbohydrates from Glycolipids

Thudichum (*12*) used several degradative procedures in his original structural studies on galactosyl ceramide from human brain. He found that this cerebroside could be hydrolyzed in aqueous sulfuric acid, aqueous barium hydroxide, or methanolic sulfuric acid, and recovered crystalline galactose after hydrolysis in aqueous 2% sulfuric acid at 130°C for 24 hr. Sphingosine and a mixture of fatty acids were obtained in the same manner. Following barium hydroxide hydrolysis, galactosylsphingosine was isolated by Thudichum. A great many minor modifications of these original procedures have since been described, but little attention has been directed to a quantitative evaluation of various procedures until relatively recently.

In one of the earliest quantitative studies of conditions for the hydrolysis of sphingolipids, Robins et al. (*13*) employed a colorimetric method with fluorodinitrobenzene to follow the liberation of long-chain base from various sphingolipids. Using aqueous hydrochloric acid and several conditions of concentration, temperature, and reaction time, they found that hydrolyses were incomplete with concentrations of hydrochloric acid less than about 3 N, whereas those carried out in 6 N HCl at 100°C for 60–90 min were satisfactory for high average recoveries of sphingosine and related bases. No information was given about the recoveries of hexose from cerebrosides under such strong acid conditions. Rosenberg and Chargaff (*14*) examined conditions for the hydrolysis of glucosyl ceramide, hoping to find optimal conditions for recovery of glucose. Using anthrone reagent to follow the liberation of glucose, they found that aqueous HCl at concentrations less than 2 N gave incomplete hydrolysis, whereas glucose was partially destroyed in 4 N HCl. Quantitative yields of glucose were obtained in 90 min with 3 N HCl at 100°C in a sealed tube, but some of the liberated glucose was apparently destroyed with longer periods of heating.

Unfortunately, quantitative recoveries of carbohydrates after hydrolyses in aqueous acids have not been achieved routinely in all laboratories. One of the difficulties might be related to the relatively low solubilities of most of the glycolipids in aqueous acid. This problem has been discussed by Radin (*1*), who proposed the use of a mixed solvent consisting of chloroform, ethyl alcohol, and concentrated HCl (*15*). All of the glycolipids are presumably soluble in this mixed reagent, and complete hydrolysis should be obtained much more readily as compared with aqueous hydrolyses. It is likely that carbohydrates are converted to a considerable extent to ethyl glycosides under these conditions.

In studies on the carbohydrate composition of several glycosyl ceramides from pig lung, Gallai-Hatchard and Gray compared recoveries of hexose by several procedures (*16*). Whereas hydrolyses of these substances were carried out originally (*17*) in aqueous 3 N HCl at 100°C for 2 to 3 hr, the authors later found inconsistent results by this method. Alternative procedures utilizing sulfuric acid, rather than HCl, or the mixed solvent of Radin et al. (*15*) did not, in these authors' experience, produce significantly better results. In fact, the authors concluded that the only method that gave them consistently quantitative recoveries of hexose from glycosyl ceramides was refluxing with anhydrous methanolic hydrogen chloride. We had selected similar conditions of methanolysis for the isolation of carbohydrates from glycosyl ceramides (*10*), but for different reasons, as described later.

When Blix isolated one of the sialic acids from bovine submaxillary mucoprotein in 1936 (*18*), it was observed that the substance was very unstable in dilute mineral acid. The entire class of *N*-acylated

neuraminic acids from glycoproteins, gangliosides, and other glycolipids are now known to be destroyed rather readily on heating for
prolonged periods in dilute acid. Early colorimetric procedures for
the determination of neuraminic acid (or *N*-acyl-neuraminic acid in
the intact glycolipid) therefore involved hydrolyses in approximately
0.05 *N* aqueous acid. Typical conditions for the preparation of *N*-

(1)

acetylneuraminic acid have been discussed in detail by Gottschalk
(*19*). It was also known that methanolysis of gangliosides yielded a
more stable derivative of neuraminic acid. Klenk (*20*) obtained the
methyl ester of methoxyneuraminic acid by treatment of ganglioside
with 5% methanolic HCl for 3 hr at 105°C. We have found that this
derivative completely survives heating in 0.5 *N* dry methanolic HCl
at 75°C for at least 24 hr (*10*).

Methanolysis of ganglioside in the presence of BF$_3$ as a catalyst
was shown recently by Rosenberg and Stern (*21*) to give quantitative

yields of fatty esters in about 3 hr at 100°C in sealed tubes. This might also be an interesting alternative method for degrading the oligosaccharide portion of gangliosides prior to GLC, but no studies were made of this fraction of the methanolysate. Methyl glycosides are presumed to be the only products of the reaction.

An interesting method has been described for the liberation of intact oligosaccharide moieties from glycosyl ceramides (*22,23*). Wiegandt and Baschang (*23*) obtained the oligosaccharides from several gangliosides by ozonolysis of the double bond of sphingosine, followed by hydrolysis in aqueous sodium carbonate. Yields by this sequence of reactions were moderately good; of course, any lipid consisting of dihydrosphingosine resisted degradation. In a more recent procedure described by Hakomori (*24*), acetylated glycosyl ceramides were oxidized with a mixture of osmium tetroxide and sodium metaperiodate, and intact oligosaccharides were liberated from the oxidized lipid by exposure to sodium methoxide. Since hematoside was converted to *N*-glycolylneuraminyl lactose by this procedure, as shown here, it is apparent that sialic acids might generally survive these mild conditions without hydrolysis of the glycosidic bond joining the keto group of neuraminic acid to other sugars. It is conceivable that the oligosaccharides liberated by these techniques might one day be

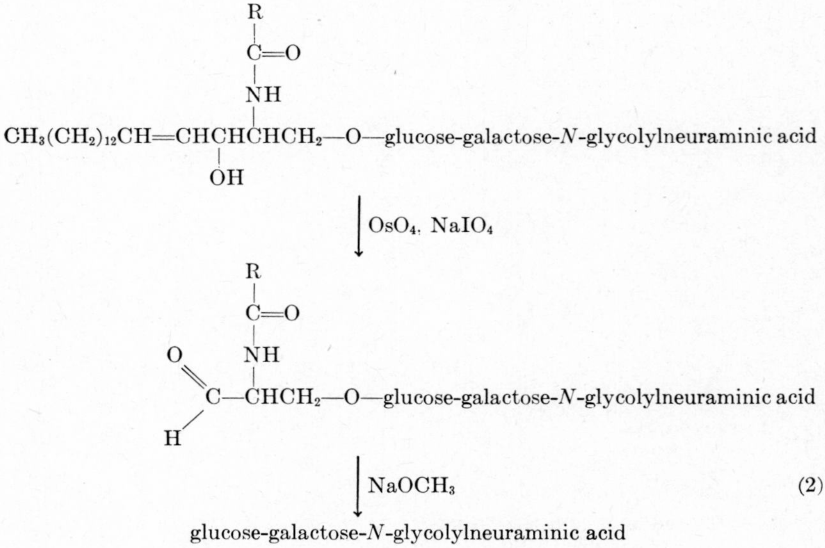

$$
\begin{array}{c}
R \\
| \\
C{=}O \\
| \\
NH \\
|
\end{array}
$$

$CH_3(CH_2)_{12}CH{=}CHCHCHCH_2{-}O{-}$glucose-galactose-*N*-glycolylneuraminic acid
$\qquad\qquad\qquad\qquad\ \ |$
$\qquad\qquad\qquad\qquad OH$

$$\Big\downarrow OsO_4, \ NaIO_4$$

$$
\begin{array}{c}
R \\
| \\
C{=}O \\
| \\
NH \\
|
\end{array}
$$

$\overset{O}{\underset{H}{\diagup}}C{-}CHCH_2{-}O{-}$glucose-galactose-*N*-glycolylneuraminic acid

$$\Big\downarrow NaOCH_3 \qquad\qquad\qquad\qquad (2)$$

glucose-galactose-*N*-glycolylneuraminic acid

subjected to GLC, but appropriate conditions have not yet been described to our knowledge.

Although there may be occasions when it is necessary to hydrolyze a glycolipid in aqueous mineral acid, such conditions are not likely to be generally useful for the liberation of carbohydrates in a GLC method of analysis. Several distinct disadvantages of aqueous acid are the marked instability of neuraminic acid, the relatively low solubilities of glycolipids, and the difficulty in obtaining an optimal time of hydrolysis when hexoses are completely liberated but survive further degradation. Anhydrous methanolysis, on the other hand, appears to be ideal for the routine isolation of high yields of the carbohydrates from glycolipids. The products are equilibrated anomeric mixtures of

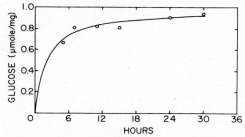

FIG. 1. Methanolysis of trihexosyl ceramide (0.5 mg) in 3 ml of 1.5 N methanolic HCl at 82°; glucose determined by GLC of the TMSi methyl glycosides with TMSi mannitol standard.

methyl glycosides, which are reasonably resistant to further modification by the reagent and which are well suited for conversion directly into volatile derivatives for GLC (*10*).

Complete methanolysis of glycosyl ceramides is a slow process with low concentrations of acid, as compared with observed rates of aqueous hydrolysis. As shown in Fig. 1, for example, it takes 24 hr to obtain a quantitative yield of glucose, as α,β-methyl glucopyranoside, from galactosyl-(1 → 4)-galactosyl-(1 → 4)-glucosyl-(1 → 1)-ceramide when the glycolipid is heated in a sealed† tube with dry 1.5 N methanolic HCl at 82°C. Although the formation of methyl glucopyranoside from this ceramide trihexoside required methanolysis of two glycosidic bonds, it is probable that the glycosidic bond of glucose (or galac-

† Screw-capped vials with Teflon inserts.

tose) to sphingosine is more resistant to cleavage than any of the other glycosidic bonds. The rather low rate of liberation of glucose by methanolysis, in Fig. 1, is therefore presumed to reflect primarily the rate of cleavage of glucosyl ceramide, an intermediate that has been shown to accumulate during very mild methanolysis of this lipid (*25*). It is interesting to note that nearly identical curves were obtained in an earlier study of the rate of methanolysis of galactosyl ceramide, using anthrone reagent to follow the liberation of sugar (*26*).

The concentration of HCl in dry methanolyses probably can be varied within certain limits, but incomplete liberation of carbohydrates will be observed with low concentrations, and selective destruction

TABLE 1

Various Conditions for Methanolysis of Glycosyl Ceramides[a]

Glycolipid	Time, hr	HCl conc., N	Temp., °C	Gal/glu[b]
Cytolipin H	18	0.5	72	1.16
(gal-glu)	24	0.5	82	1.00
Ceramide trihexoside	18	0.5	72	2.18
(gal-gal-glu)	24	0.5	82	2.10
	24	1.0	82	1.97
Globoside	18	0.5	72	2.48
(N-Ac-galam-gal-gal-glu)	24	1.0	82	2.08
	24	1.5	82	2.10

[a] The products were measured as TMSi derivatives by GLC.
[b] Ratios of observed areas on GLC chromatograms.

and/or other secondary alterations may occur at high concentrations. Predicted ratios of galactose to glucose can be obtained with 0.5–1 N HCl and 24 hr of heating, as shown in Table 1. In these studies, gal/glu ratios greater than predicted probably indicate incomplete methanolysis of the glycosidic bond to ceramide, and values lower than predicted are interpreted as indication of secondary changes in the methyl glycosides (or, relatively, some destruction of galactose). With greater than 1.5 N HCl, gal/glu ratios as low as 1.5 have been observed with ceramide trihexoside when GLC was used to determine the yields of products. This observation was attributed (*10*) to the formation of some unknown galactose derivative, whose proportion was increased with the strength of acid and whose GLC retention time as

TMSi derivative was unfortunately coincident with that of one of the glucose peaks. Total peak area for the galactose was therefore less with higher concentrations of HCl, and the glucose area was proportionately increased, leading to falsely low gal/glu ratios.

It is possible that some free galactose is formed during methanolysis, and that the yields of this form are dependent on the HCl concentration. Kishimoto and Radin (*27*) recently reported that significant amounts of methyl chloride and water are formed in mixtures of dry methanol and hydrogen chloride, as shown below. Despite the most rigorous efforts to prepare absolutely dry mixtures, some water is

$$CH_2OH + HCl \rightleftharpoons CH_3Cl + H_2O \tag{3}$$

always present in the methanolysis reagent, therefore, and its concentration will depend on the amount of HCl in the reagent. As the HCl concentration is increased, there may actually be sufficient water to cause increasing partial hydrolysis of the methyl glycosides to free hexoses. That this might account for the interfering galactose peak in GLC after methanolysis with strong HCl is suggested by the fact that TMSi α-galactose and TMSi α-methyl glucopyranoside have nearly identical retention times (*10*).

The reaction of methanol with hydrogen chloride is probably slow at room temperature, and HCl is only slowly depleted as the reagent stands. An equilibrium is presumably attained quickly when the reagent is heated, however, and the effective concentration of HCl during methanolysis might actually be low. Radin found, for example, that only 3.7% of the acid remained after methanolic HCl was heated at 100°C for 5 hr, whereas only half of the titratable acid was lost in 1.5 months at room temperature (*27*). This relatively low actual concentration of HCl after heating for a short time might explain the limited capacity for cleaving glycolipids completely with this reagent. We have found that, at most, 1 mg of glycosyl ceramide can be degraded in 3 ml of 0.5 to 0.75 *N* methanolic HCl at about 70–80°C. An advantage of the rapidly decreasing concentration of HCl during methanolysis is that the products are probably much more stable than they would be in the original reagent.

Various parameters in the methanolysis reaction are not easily balanced for optimal results. The HCl concentration should be high to promote complete liberation of the sugars in a minimum time, but it should be moderately low to avoid secondary changes in the methyl glycosides. It appears that a suitable HCl level for one lipid may not be as satisfactory with another. For example, it appears that globoside (with an *N*-acetyl galactosamine unit) requires higher HCl concentration than other glycosyl ceramides (Table 1). The temperature of methanolysis undoubtedly influences the rate to a certain extent, but higher temperatures also tend to give a more extensive loss of HCl, by conversion to methyl chloride, and there may actually be an optimal temperature above and below which the rate is lower. As the effective HCl concentration decreases with increasing temperature, the capacity of the reagent also could be expected to decrease. No final decisions about these questions have been made yet, but methanolysis of glycosyl ceramides in this laboratory are currently conducted in 0.75 *N* methanolic HCl at 80°C for 24 hr. Calculated yields of glucose, and appropriate galactose/glucose ratios, have been obtained consistently with cerebroside, ceramide dihexoside, ceramide trihexoside, and globoside by this procedure (*28*).

II. Preparation of TMSi Derivatives of Carbohydrates

The poly(*O*-TMSi) derivative of a carbohydrate is easy to prepare, and quantitative yields are usually obtained at room temperature within a few minutes (*4*). The reagent generally used for trimethylsilylation consist of a mixture of hexamethyldisilazane and trimethylchlorosilane in dry pyridine. This reagent can be added to a dry residue of carbohydrate or to a solution of one or more sugars in solvents such as pyridine, acetonitrile, and dimethylformamide. All free hydroxyl groups are converted to TMSi ethers under these conditions, by a reaction sequence that is not well understood and may be exceedingly

$$3ROH + (CH_3)_3SiNHSi(CH_3)_3 + (CH_3)_3SiCl \xrightarrow{\text{pyridine}} 3ROSi(CH_3)_3 + NH_4Cl \downarrow \quad (4)$$

complex. The role of trimethylchlorosilane has been assumed to be primarily a catalytic one, even though it is probably consumed in the over-all reaction as shown in reaction (*4*). Horii et al. (*29*) and Wells (*30*) proposed a sequence of reactions in which the actual active species for trimethylsilylation is *N*-TMSi-pyridinium chloride. As shown in the first reaction, Horii et al. believe this intermediate might be formed initially by a reaction of pyridine and trimethylchlorosilane. They propose, then, that it is generated subsequently from

$$\text{(5)}$$

$$\text{(6)}$$

$$\text{(7)}$$

pyridine hydrochloride and hexamethyldisilazane in a reaction driven by the precipitation of ammonium chloride. One difficulty with this mechanism is that no TMSi glucose can be observed by GLC ten minutes after mixing glucose, hexamethyldisilazane, and pyridine hydrochloride in pyridine (*31*). It is possible, therefore, that pyridine hydrochloride does not enter the reaction as shown in (7). The exact nature of the process of trimethylsilylation in pyridine still has aspects that are not clear, and further studies of the mechanism are needed.

The composition of an anomeric mixture remains essentially unchanged during trimethylsilylation in pyridine (*4*), presumably because the rate of formation of the TMSi ether of the anomeric OH group is much faster than the rate of mutarotation in dry pyridine at room temperature.† Thus, single anomeric forms of a sugar give almost exclusively the TMSi derivative of that same anomer, whereas the proportions of individual anomers in aqueous equilibrium do not change to a different equilibrium during trimethylsilylation. This is not necessarily the case when the pyridine solution is warmed, however, and changes in the composition of anomeric mixtures may occur (*4,9*). If it is important to maintain a certain composition of anomeric forms, the temperature of the reaction mixture should be maintained at or below room temperature. Normally this is not an important

† This is supported by recent findings that the rate of trimethylsilylation is relatively fastest with the anomeric OH group, followed by progressively lower rates with α-glucose as follows: 6OH > 2OH > 3OH = 4OH (*31*).

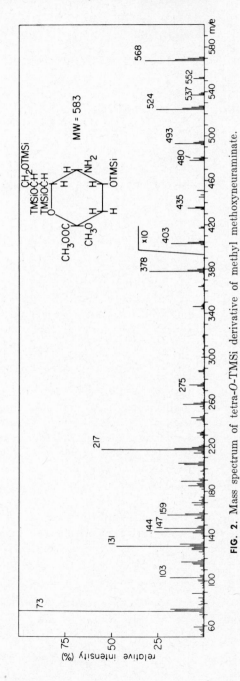

FIG. 2. Mass spectrum of tetra-*O*-TMSi derivative of methyl methoxyneuraminate.

consideration in the preparation of TMSi derivatives of methyl-glycosides isolated from glycolipids, but it should be remembered that warming the reaction will lead to erratic ratios of the various GLC peaks for individual anomers.

Volatile TMSi derivatives are also formed from 2-amino-2-deoxy-hexoses and related substances (*4,10,32–34*). The nature of the reaction of hexamethyldisilazane with amino hexoses had not been considered, but it was assumed that the usual reaction in pyridine would yield a penta-TMSi derivative with four *O*-TMSi groups and an *N*-TMSi group from the amine. This view was shown recently to be incorrect, however, when a mass spectrum of the TMSi derivative of galactos-amine was examined (*35*). The findings of a molecular ion at m/e 467 and other ions at M-15 (m/e 452), M-90 (m/e 377), and M-90-15 (m/e 362) were clearly not compatible with a penta-TMSi derivative, but rather provided strong evidence for a tetra-TMSi form of galac-tosamine (mol. wt. 467). The strongest peak in the mass spectrum was at m/e 131. Since this is probably due to a fragmentation process lead-ing to $[CH(NH_2)CHOTMSi]^+$, it is probable that the reaction of galactosamine with hexamethyldisilazane and trimethylchlorosilane in pyridine leads to the formation of the tetra-*O*-TMSi derivative shown below. Other hexosamines are presumed to react in a similar fashion. In this regard, Karlsson recently reported that sphingosine gives a di-*O*-TMSi derivative with an unreacted amino group (*36*).

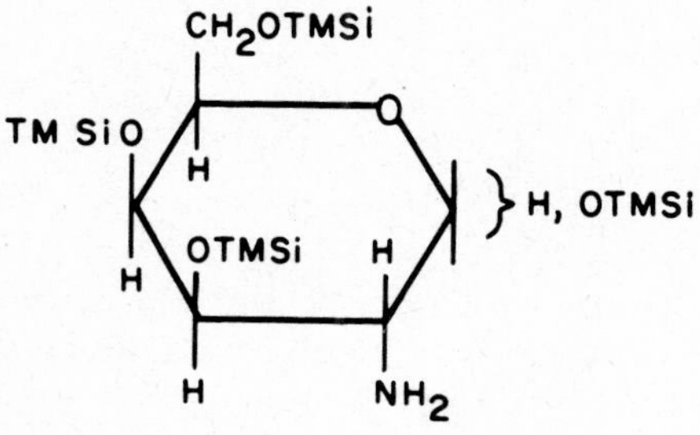

A completely analogous TMSi derivative has been found to be formed from the methyl ester of methoxyneuraminic acid. A mass spectrum of the TMSi derivative, recorded at 70 e.v. with an LKB 9000 for combined gas chromatography–mass spectrometry (*37*), is shown in Fig. 2. Though a molecular ion was not observed in this case, the presence of ions at m/e 568 (M-15)$^+$, m/e 552 (M-31)$^+$, m/e 524

(M-59)$^+$, m/e 493 (M-90)$^+$, and m/e 403 (M-2 $\times$ 90)$^+$ provided strong evidence for a molecular ion at m/e 583, consistent with the molecular weight of a tetra-TMSi derivative. There was a strong peak at m/e 131 and, like TMSi galactosamine, it was assigned to [CH(NH$_2$) CHOTMSi]$^+$. Although the mass spectral data do not completely exclude other possibilities, it is proposed that the product of trimethylsilylation of methyl methoxyneuraminate is that shown in Fig. 2.

It is possible that another donor of TMSi groups, bis(trimethylsilyl) acetamide (*38*), might give a different derivative with both *N*-TMSi and *O*-TMSi groups in reactions with hexosamines and derivatives of neuraminic acid, but this possibility has not yet been examined to our knowledge.

It appears that the over-all rate of trimethylsilylation is dependent on the concentrations of hexamethyldisilazane and carbohydrate in pyridine, and the reaction is therefore at least bimolecular in its kinetics (*31*). For quantitative conversions of carbohydrates to TMSi derivatives, the relative proportions of the silanes and ROH are not critical so long as the equivalent concentration of hexamethyldisilazane is in considerable excess. When the concentration of hexamethyldisilazane is limiting, however, a variety of partial TMSi derivatives are formed (*39*); the nature of these derivatives depends somewhat on the relative reactivities of various OH groups in the molecule (*40*). Many of these partial derivatives are easily separated on packed GLC columns (*33,34,40*), and not only is quantitation of a given sugar impossible but the investigator may be confused about qualitative identifications when partial TMSi forms are present. Unless partial TMSi derivatives are particularly sought, therefore, it is expedient to conduct the trimethylsilylation reaction with a relatively large excess of the TMSi donors. Partial TMSi derivatives are also observed when insufficient time has been allowed for the reaction to be completed (*39*). To be certain with methyl glycosides, especially when a hexitol is used as internal standard, the reaction should be allowed to react for at least 15 min at room temperature.

Although the TMSi derivatives can be hydrolyzed relatively easily in water or aqueous alcohol (*41*), they can be kept for long periods if water and other hydroxyl solvents are carefully excluded from the reaction mixture. Test solutions of TMSi glucose have been stored for several months in this laboratory with little or no decrease in the concentration of the derivative, and TMSi derivatives of glucose and methyl glucopyranosides have actually been distilled and stored as neat oils (*42*). Comparisons were made of the relative stabilities of the TMSi derivatives of the methyl glycosides of galactose to glucose and mannitol to glucose. As shown in Table 2, there was little effect of

TABLE 2

RELATIVE STABILITIES OF TMSi METHYL GLYCOSIDES AND TMSi MANNITOL

Time after preparation	Galactose/glucose	Mannitol/glucose
2 hr	0.96	1.27
1 day	0.97	1.28
2 days	0.98	1.32
4 days	1.01	1.38
7 days	1.00	1.27
14 days	0.99	1.29

time on either galactose/glucose ratios or the concentration of the TMSi glucopyranosides relative to that of TMSi mannitol. Although these data do not exclude the possibility that all of the TMSi derivatives may be decreasing slightly with time, it is clear that the relative concentrations do not change. Our data do not, in fact, support the contention that the GLC analyses of TMSi derivatives must be made at some precise time after the reaction is started. This observation may not be valid in the case of mixtures that contain hexosamines, however. There have been observations of changes in concentrations of these derivatives with time.

Following methanolysis of a glycolipid, methanol and HCl must be removed before conversion of the methyl glycosides to TMSi derivatives. This step can be accomplished by distillation of the solvent in vacuo or by evaporation with a stream of inert gas. Residual HCl usually requires a second evaporation with added solvent in both cases. Our quantitative data on recoveries of carbohydrates from authentic glycolipids have not been satisfactory when either of these methods is used, however. Erratic recoveries of galactose are usually observed for some reason that we have not yet found, and the gal/glu ratios are affected by the lower galactose concentrations. When the HCl is removed from the methanolysate by percolation of the cooled solution through a small column of a weak anion-exchange resin (OH⁻), difficulties in the recovery of calculated yields of galactose are not encountered. After the methyl glycosides are removed from the column with methanol, solvent can be removed by any convenient method. The problem associated with galactose loss by the other two methods must be associated with contact of the methyl glycosides with strong HCl during the last stages of the evaporation.

III. Gas Chromatographic Separation of TMSi Derivatives

The TMSi derivatives of various anomeric forms of free glucose and galactose, isolated from a fully equilibrated aqueous solution of the sugars, are reasonably well separated on a short packed column containing a nonpolar liquid phase such as SE-30. A typical separation of these derivatives is illustrated in Fig. 3; the retention time of the

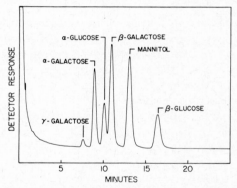

FIG. 3. GLC separation of anomeric forms of TMSi glucose and TMSi galactose on 2% SE-30, 160°C.

internal standard, hexa-O-TMSi-mannitol, is approximately 1.3 times that of TMSi-α-glucose, and its peak is therefore conveniently located about halfway between those of β-galactose and β-glucose. The ratio of galactose to glucose in this mixture was 2:1, so that the resulting chromatogram would show the relative peak sizes that would be observed with a mixture of free hexoses from one of the naturally occurring glycosyl ceramides (gal-gal-glu-ceramide). Aside from the problems associated with aqueous hydrolyses of glycolipids, discussed earlier, there are difficulties in the GLC of the free hexoses. The pyranose forms of free α-glucose and β-galactose are difficult to separate completely as TMSi derivatives on nonpolar columns, and overlapping to the degree shown in Fig. 3 is commonly observed. These two anomeric forms of TMSi-glucose and TMSi-galactose are well separated on a polar ethylene succinate polyester (4), but the two α anomers have nearly the same retention times on this liquid phase. Sawardeker and Sloneker (43) achieved very good separations of all the anomeric forms of glucose, galactose, and mannose, as the TMSi derivatives, with another polar liquid phase (Carbowax 20M), but the column was 12 ft long and the analysis therefore required a much longer time (over one hour). When time is not a critical factor and a good SE-30 column is available for the analyses, equally complete separations can be achieved in about 40 min (9). In addition to the overlapping of some of the hexose peaks, the order of elution of peaks may cause confusion on occasion. On the SE-30 column, for example, TMSi-α-glucose has a retention time between the retention times of TMSi-α- and TMSi-β-galactose, so that glucose and galactose peaks alternate in the elution sequence. The same relative retention behavior was observed with Carbowax 20M (43). Despite these difficulties in GLC of TMSi hexoses, a great many applications of the technique have already been described, including analyses of samples that contained glucose and galactose (34,44–49).

To avoid the obvious shortcomings of a GLC method in which the various anomeric forms give separate peaks on the chromatogram, Sawardeker et al. (50) recommended the use of alditol acetates, which were well separated on a chemically combined copolymer of a cyano-silicone and poly(ethylene succinate) (ECNSS-M). As Gunner et al. (51) had indicated in their early work on GLC of O-acetyl derivatives, working with the alditols has the great advantage of single peaks for each of the sugars. The method described by Sawardeker et al. (50) has been evaluated in considerable detail in recent reports by Crowell and Burnett (52) and Sjöström et al. (53). Similarly, the TMSi derivatives of N-acetyl hexosaminitols was recommended by Horowitz and Delman (54) for the GLC analysis of a mixture containing reducing sugars, sugar alcohols, and amino sugar alcohols. Methods that employ other derivatives of the components of glycolipids, such as alditol acetates, may eventually prove to be useful, but simple and quantitative procedures are not currently available for the several steps in the conversion of methyl glycosides to these derivatives.

In contrast to the difficulties encountered in achieving complete resolution of the TMSi derivatives of free hexoses, it is relatively easy to separate the TMSi derivatives of some methyl glycoside mixtures. This is fortunately the case with the anomeric mixture obtained from glucose and galactose, and very rapid analyses can be made for these two components in glycolipids by GLC. The three chromatograms reproduced in Fig. 4 were among the first that were obtained in this laboratory, and while the separations have since been improved considerably (*10*), these chromatograms show the ease with which different gal/glu ratios can be differentiated. The middle chromatogram was obtained with a mixture from the methanolysis of an unknown glycolipid from the kidney of a patient with Fabry's disease (*25*), and it actually provided the first evidence that the lipid might be a ceramide trihexoside with a gal/glu ratio of 2 (*55,58*). The top chromatogram was obtained with the methyl glycosides from an authentic 2:1 mixture of galactose and glucose; the bottom chromatogram was of the TMSi derivatives of methyl glycosides from lactose.

A better resolution of the five peaks from TMSi methyl glucosides and TMSi methyl galactosides is shown in Fig. 5. TMSi mannitol was added to the mixture as an internal standard to aid in quantitative analyses of the constituents. The sample in this case was lactosyl ceramide isolated from 50 ml of packed human erythrocytes. In addition to the good resolution of the peaks, and the fact that the analysis required less than 20 min, Fig. 5 illustrates several important points. Since the average concentration of lactosyl ceramide in human erythrocytes was shown recently by Vance and Sweeley (*28*) to be about 0.6 μmole per 50 ml, GLC has adequate sensitivity for analyses of biological samples at least this small in quantity. In fact, it has been possible to obtain proper hexose ratios and nearly theoretical total yields of glucose and galactose from 0.1 μmole of ceramide trihexoside isolated from human plasma (*28*). A second important feature of the GLC analysis is the relative freedom from interfering substances of biological or manipulative origin. After the glycolipid has been purified by column and thin-layer chromatography, small amounts of material have given remarkably clean GLC chromatograms, such as that shown in Fig. 5.

The methyl esters of fatty acids, produced in the methanolysis of glycolipids, are a source of interfering material, however, and must be removed before GLC. As shown in Fig. 6, the retention time of methyl palmitate is nearly coincidental with that of α-methyl TMSi gluco-

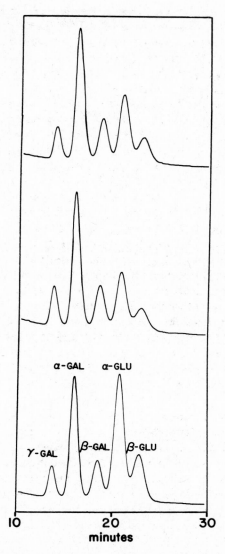

FIG. 4. Comparison of areas of TMSi methyl glycosides from synthetic mixture of glucose and galactose (1:2) (top), from kidney glycolipid (middle), and from lactose (bottom); methanolysis in 0.5 N methanolic HCl at 75°C for 24 hr; 3% SE-52 at 140°C.

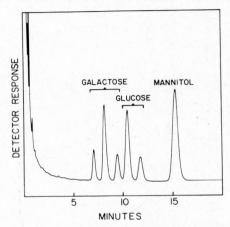

FIG. 5. GLC of TMSi methyl glycosides from ceramide dihexoside of erythrocytes and added TMSi mannitol; methanolysis with $0.5\,N$ methanolic HCl at 75°C for 18 hr; GLC on 2% SE-30 at 160°C.

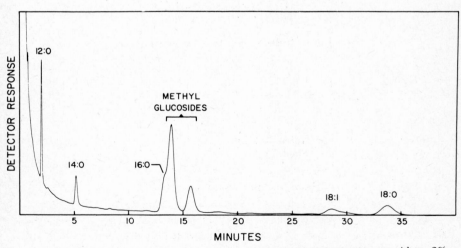

FIG. 6. GLC of mixed methyl esters and TMSi methyl glucopyranosides; 2% SE-30 at 160°C.

pyranoside, and methyl heptadecanoate would probably cochromatograph with TMSi mannitol. The methyl esters are easily separated from the methyl glycosides, just after methanolysis, by several extractions with hexane or petroleum ether (*10*), and this step should always be included in the method.

When glycolipids contain galactosamine or glucosamine as a constituent in the oligosaccharide moiety, several problems usually result. The GLC chromatogram is always more complicated in the area of the peaks for TMSi methyl glycosides when these substances are present. As shown in Fig. 7, for example, the tetra-*O*-TMSi derivative

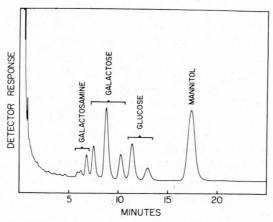

FIG. 7. GLC of TMSi methyl glycosides from plasma ceramide tetrahexoside containing galactosamine, galactose, and glucose (1:2:1); methanolysis and GLC as described in Fig. 5.

of galactosamine (methyl glycoside) has several anomeric forms that have retention times that are only slightly less than those of the TMSi methyl galactosides. Also, additional peaks are often observed at a point somewhat later in the chromatogram. These were shown to be TMSi methyl glycosides of *N*-acetylgalactosamine that survived methanolysis. Similar interference in the glucose area would probably be encountered with samples that contained glucosamine or its *N*-acyl derivatives. Quantitative analyses of the hexosamine and the corresponding hexose are both difficult to achieve in these cases, although some success was claimed with an alternative procedure in which the sample was treated with acetic anhydride in methanol after methanolysis to convert free hexosamines completely to N-acetyl derivative (*10*). The TMSi derivatives were then prepared and analyzed in the usual way. Perry has described a very good procedure for the preparation of N-acetyl derivatives of hexosamines and has discussed GLC of the TMSi-N-acetyl derivatives in some detail (*32*). Richey et al. (*34*) have also presented some data on the TMSi derivatives of acetamido sugars.

Once a particular hexosamine has been shown to be a constituent of a given glycolipid and its proportion relative to the hexoses has been determined, it is probably much better to remove the hexosamine from the mixture before GLC in routine analyses of that glycolipid. This step was incorporated into a procedure for the quantitative analysis of glucose and galactose in gangliosides described by Penick and McCluer (*11*). The entire diluted methanolysis mixture was percolated through a short column of Dowex-50-X_4 (H^+) that had been carefully preextracted with methanol before use. Although they did not discuss subsequent analysis of the hexosamine, it is implicit in their method that this fraction could be recovered for separate analysis by GLC. In fact, in a procedure that has been described by Kärkkäinen et al. (*33*) an ion-exchange step was used to separate the hexosamines from neutral sugars prior to GLC of the TMSi hexosamines. It was possible, in this way, to determine both glucosamine and galactosamine in aqueous hydrolysates of mucopolysaccharides without interference by the hexoses in the gas chromatograms.

The chromatogram shown in Fig. 7 is of the carbohydrate constituents of one of the N-acetyl-galactosaminyl-digalactosyl-glucosyl ceramides [globoside (*56*)], and despite the complex area of peaks for galactose and galactosamine derivatives, the ratio of galactose to glucose could be determined from this chromatogram. The same sample was then analyzed after the hexosamine had been removed by ion-exchange chromatography. The resulting mixture, as shown in Fig. 8, contained no free galactosamine, and the rate of galactose to glucose was the same (2:1) as determined by the direct procedure. Subsequent analyses of globoside in human plasma and erythrocytes were therefore made in this laboratory by the modified procedure, in which the galactosamine was removed, since the relative galactosamine content would not be expected to change in serial samples from one individual or even samples from different persons. The glycolipid was actually identified by its location on thin-layer plates, not by GLC, whereas the amount of that glycolipid was determined by the yield of TMSi methyl glucosides observed by GLC.

In a study of the glycolipids of pig lung, Gallai-Hatchard and Gray (*16*) discuss the possibility that within a given class, such as cerebrosides and ceramide dihexosides, the glycolipids from various sources may not always be chemically identical. Although lactosyl ceramide is the dihexoside that is nearly always found in mammalian material, digalactosyl ceramide has been shown to occur in brain (*57*)

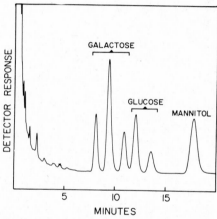

FIG. 8. GLC of TMSi methyl glycosides from plasma ceramide tetrahexoside containing galactosamine, galactose, and glucose (1:2:1) after removal of galactosamine products; methanolysis and GLC as described in Fig. 5.

and kidney (*58*), and both galactosyl and glucosyl ceramides are common in the cerebroside class. In human plasma, glucosyl ceramide is the predominant type, but Svennerholm and Svennerholm (*59*) found a small amount of galactosyl ceramide in large-scale isolations of the individual glycolipids from blood serum. This observation has been confirmed with 50-ml samples of plasma from a number of individuals (*28*). A typical GLC record showing the relative concentrations of galactose and glucose from plasma cerebroside is shown in Fig. 9. The galactosyl ceramide has been as high as 10% of the total mixture, but it is usually about 5% of the glucosyl ceramide concentration.

The determination of sialic acids by GLC has received relatively less attention than is deserved, because the product from methanolysis is stable and it is readily converted to the TMSi derivative, which is determined rapidly by GLC in a nonpolar column (*10*). When a highly pure sample of sialolactose was subjected to methanolysis and the products were converted to TMSi derivatives, the neuraminic acid derivative, now known to be the tetra-TMSi *O*-methyl ester shown in Fig. 2, was eluted from SE-30 with a retention time about 4.2 times that of TMSi α-methyl glucopyranoside (*10*), as shown again in Fig.

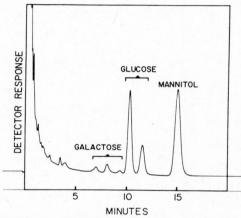

FIG. 9. GLC of TMSi methyl glycosides from human plasma cerebroside fraction; GLC with internal standard (TMSi mannitol) and methanolysis as described in Fig. 5.

10. The peak was nearly symmetrical with the nonpolar column, and the recovery through methanolysis and conversion to TMSi derivative was high, since the relative ratio of the areas for sialic acid and glucose was 0.96.

If mild aqueous hydrolysis were used to liberate the *N*-acyl neuraminic acids from various sources, an advantage would be that the *N*-acetyl, *N*-glycolyl, and other forms would probably separate as their TMSi derivatives. The retention time for the methyl ester of *N*-acetyl-*O*-TMSi-neuraminic acid, reported by Bolton et al. (*60*), was 3.2 relative to that for methyl α-galactopyranoside on SE-30, but it is not clear whether this retention factor was for an isothermal or a programmed analysis.

A summary is given in Table 3 of the relative retention behavior of some TMSi methyl glycosides that are found in methanolysates of glycolipids and related substances. These data were obtained with a typical packed column (6 ft × 3 mm i.d.) of 2% SE-30. The two TMSi methyl glycosides of mannose would be eluted from the column just before those of galactose (*60*), giving complex chromatograms when galactose, mannose, and galactosamine are present in the same sample.

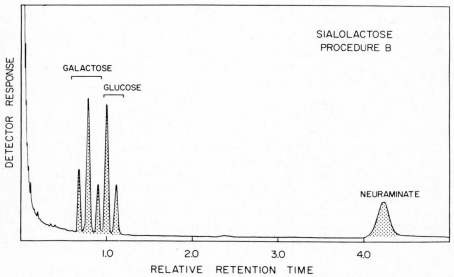

FIG. 10. GLC of TMSi methyl glycosides from sialolactose on 2.5% SE-30 at 160°C. [Reproduced from *Analytical Chemistry*, **36**, 1461 (1964); copyrighted by the American Chemical Society and reprinted by permission of the copyright owner.]

A few words must be said about other derivatives that may be preferred in special instances or that are preformed under the conditions of an experiment. In the determination of the positions of glycosidic bonds in glycolipids, for example, it is customary to react the glycolipid with a permethylating agent. The products obtained after subsequent methanolysis can be determined directly by GLC (*3,6,61*) in such cases. In very unusual cases, one of the products of permethylation and methanolysis might have more than two free OH

TABLE 3

RELATIVE RETENTION BEHAVIOR OF TMSI METHYL GLYCOSIDES[a]

Compound[b]	Retention time[c]
Rhamnose	0.23
Fucose	0.23
	0.26
	0.28
Galactosamine	0.56
	0.58
	0.63
Galactose	0.68
	0.78
	0.91
Glucose	1.00
	1.13
Mannitol	1.49
Myo-inositol	2.86
Methyl methoxyneuraminate	4.10

[a] Mannitol and myo-inositol were not heated in methanolic HCl. Their TMSi forms were prepared directly in the usual manner.

[b] The methyl glycosides were prepared by heating the sugar in 0.50 N methanolic HCl at 72°C for 18 hr. The TMSi derivatives of galactosamine and methoxyneuraminate have free —NH₂ groups (see text).

[c] The retention times are relative to the time for α-methyl-2,3,4,6-tetra-O-TMSi-glucopyranoside, with an observed retention time of 12 ± 2 min on 2% SE-30 at 160°C.

groups remaining, and it might be necessary in such cases to convert the product to a more volatile TMSi derivative.

When the Smith degradation is used in studies of the structure of an oligosaccharide portion of a glycolipid (*62*), the products are mixtures of alkane polyols such as ethylene glycol, glycerol, and erythritol. These products are more conveniently separated as O-acetyl derivatives on a cyanosilicone (XE-60) or JXR silicone column than as the TMSi derivatives (*63,64*).

Complex GLC chromatograms were obtained by Perry and Hulyalkar (*65*) in attempted analyses of acidic glycoses by methanolysis and formation of TMSi derivatives of the products. The hexuronic acids themselves gave several peaks when treated with methanolic HCl before forming TMSi derivatives, presumably from mixtures of various lactones and the methyl esters. To avoid these problems, Perry and Hulyalkar have recommended a procedure in which the hexuronic acids in oligosaccharide chains are converted by reductive hydrolysis to acid-stable aldonic acids, which were then analyzed as the tetra-*O*-TMSi-aldono-1,4-lactones. The GLC column contained 10% neopentylsebacate polyester. Reference retention times of the derivatives of a number of hexuronic acids were given, and Perry and Hulyalkar pointed out that the pentoses and hexoses in the carbohydrate, converted to TMSi pentitols and hexitols by this procedure, did not interfere with the GLC determination of the hexuronic acid derivatives.

IV. Quantitative Relationships

Direct comparisons of the areas observed on GLC chromatograms will *sometimes* provide sufficiently accurate molar ratios of the individual carbohydrates, and corrections for yields of TMSi derivative and for differences in detector response are not necessary. This applies mainly to oligosaccharide chains composed of neutral hexoses only. In the glycosyl ceramides, for example, the observed total area of the three TMSi methyl galactoside peaks can be compared directly with the total area of the two TMSi methyl glucopyranosides to calculate molar ratios in the lipids, assuming that a flame ionization detector has been used for GLC. In Table 4 are shown three gal/glu ratios from areas of the peaks on GLC of the TMSi methyl glycosides. The average ratio was 0.99, indicating (1) no differences in the two sugars at all or (2) completely balancing effects in the preparation of methyl glycoside, conversion of the glycoside to TMSi derivative, and relative

TABLE 4

CONVERSION FACTORS FOR GLC ESTIMATION OF GLUCOSE AND GALACTOSE IN METHANOLYSIS MIXTURES

	Galactose/glucose	Mannitol/glucose
Preparation 1	1.000	1.220
Preparation 2	0.973	1.267
Preparation 3	1.003	1.249
Average	0.99	1.25
Average deviation	0.016	0.017

responses of these compounds in the detector. The sample of D-glucose for this study was from the National Bureau of Standards (standard sample 41) and was used as supplied. A commercial sample of D-galactose was recrystallized from hot aqueous alcohol as described by Wolfrom and Thompson (66). The areas of the various anomeric TMSi methyl glycosides were determined by planimetry. This result is in agreement with the report by Richey et al. (34) that the TMSi derivatives of free glucose and galactose give equal responses in the argon ionization detector and with the report by Sjöström et al. (53) in which are listed almost identical relative responses by five different alditol acetates (flame detector). Considerably different values were reported by Penick and McCluer (11), who found a ratio of 1.11 for the areas of the glucose over those of galactose, and by Bolton et al. (67), who found a value of about 1.45. These discrepancies cannot be accounted for at the present time.

For the direct determination of the amounts of individual sugars recovered from a given glycolipid, an internal standard should be used with GLC methods. Various substances have been used as internal standards, depending on the composition and retention times of the mixture of sugars under investigation. Mannitol is an excellent standard for glycosyl ceramides. The retention time of TMSi mannitol is different from the times of the carbohydrate components likely to be encountered in methanolysis mixtures. If a known quantity of recrystallized mannitol is added to a cooled methanolysis mixture before extraction of the fatty esters and removal of the HCl by ion-exchange chromatography, the area that is observed with a flame ionization detector for TMSi mannitol is greater than that for an equal amount of glucose or galactose carried through methanolysis and the subsequent steps to TMSi methyl glycosides, as shown in Table 4. Using the ratio of 1.25 observed for TMSi mannitol to total TMSi methyl glucoside, the yield of glucose is calculated from GLC data by the equation

$$\mu\text{moles glucose} = \frac{\text{area of glucose peaks}}{\text{area of mannitol peak}}$$
$$\times\ 1.25 \times \mu\text{moles mannitol added}$$

At the present time, the determination of glucose in a methanolysis sample can be made by this GLC method with an accuracy of about 2%. Greater errors than this have been observed in employing the procedure for the determination of individual glycosyl ceramides in blood samples (*28*). When an entire sample of plasma or erythrocytes is analyzed in duplicate, each of the glycosyl ceramides can be determined with an over-all accuracy of about 10%. Hopefully, the precision with which these analyses can be performed will improve with further study of the procedure.

It is in the quantitative determination of the carbohydrates in glycolipids that the TMSi derivative is likely to be superior to other types of derivatives such as *O*-methyl ethers and acetates. Unless the derivative is a highly nonpolar substance, difficulties with adsorptive phenomena are generally encountered when very small quantities of substance are injected into the GLC system. It is in this particular area of micromolar quantities that most of the analyses of glycolipids from biological tissues and fluids will be carried out in the years ahead. The results that have been reported so far certainly encourage further effort in extending the application of GLC for investigations of glycolipids.

Acknowledgment

This investigation was supported in part by U.S. Public Health Research Grants AM-04307 and GM-13423 from the National Institute of Arthritis and Metabolic Diseases and the National Institute of General Medical Science.

REFERENCES

1. N. S. Radin, in *Methods of Biochemical Analysis,* Vol. 6 (D. Glick, ed.), Wiley (Interscience), New York, 1958, Chap. 7.

2. A. T. James and A. J. P Martin, *Biochem. J.,* **50**, 679 (1952).

3. A. G. McInnes, D. H. Ball, F. P. Cooper, and C. T. Bishop, *J. Chromatog.,* **6**, 556 (1958).

4. C. C. Sweeley, R. Bentley, M. Makita, and W. W. Wells, *J. Am. Chem. Soc.,* **85**, 2497 (1963).

5. C. T. Bishop and F. P. Cooper, *Can. J. Chem.,* **38**, 388 (1960).

6. C. T. Bishop, in *Advances in Carbohydrate Chemistry,* Vol. 19 (M. L. Wolfrom and R. S. Tipson, eds.), Academic Press, New York, 1964.

7. H. W. Kircher, in *Methods in Carbohydrate Chemistry,* Vol. 1 (R. L. Whistler and M. L. Wolfrom, eds.), Academic Press, New York, 1962, p. 13.

8. W. W. Wells, C. C. Sweeley, and R. Bentley, in *Biomedical Applications of Gas Chromatography* (H. A. Szymanski, ed.), Plenum Press, New York, 1964, Chap. 5.

9. C. C. Sweeley, W. W. Wells, and R. Bentley, in *Methods of Enzymology,* Vol. VIII A (E. F. Neufeld and V. Ginsburg, eds.), Academic Press, New York, 1966, Sect. 1–7.
10. C. C. Sweeley and B. Walker, *Anal. Chem.,* **36,** 1461 (1964).
11. R. J. Penick and R. H. McCluer, *Biochim. Biophys. Acta,* **116,** 288 (1966).
12. J. L. W. Thudichum, *The Chemical Constitution of the Brain,* Archon Books, Hamden, Conn., 1962.
13. E. Robins, O. H. Lowry, K. M. Eydt, and R. E. McCaman, *J. Biol. Chem.,* **220,** 661 (1956).
14. A. Rosenberg and E. Chargaff, *J. Biol. Chem.,* **233,** 1323 (1958).
15. N. S. Radin, J. R. Brown, and F. B. Lavin, *J. Biol. Chem.,* **219,** 977 (1956).
16. J. J. Gallai-Hatchard and G. M. Gray, *Biochim. Biophys. Acta,* **116,** 532 (1966).
17. G. M. Gray, *Biochem. J.,* **94,** 91 (1965).
18. G. Blix, *Z. Physiol. Chem.,* **240,** 43 (1936).
19. A. Gottschalk, *The Chemistry and Biology of Sialic Acids and Related Substances,* Cambridge, New York, 1960, p. 40.
20. E. Klenk, *Z. Physiol. Chem.,* **237,** 76 (1942).
21. A. Rosenberg and N. Stern, *J. Lipid Res.,* **7,** 122 (1966).
22. R. Kuhn and H. Wiegandt, *Z. Naturforsch.,* **19b,** 256 (1964).
23. H. Wiegandt and G. Baschang, *Z. Naturforsch.,* **20b,** 164 (1965).
24. S. Hakomori, *J. Lipid Res.,* **7,** 789 (1966).
25. C. C. Sweeley and B. Klionsky, *J. Biol. Chem.,* **238,** PC3148 (1963).
26. N. M. Oldham and C. C. Sweeley, unpublished work, 1961.
27. Y. Kishimoto and N. S. Radin, *J. Lipid Res.,* **6,** 435 (1965).
28. D. E. Vance and C. C. Sweeley, *Federation Proc.,* **26,** 277 (1967).
29. Z. Horii, M. Makita, I. Takeda, Y. Tamura, and Y. Ohnishi, *Chem. Pharm. Bull. (Tokyo),* **13,** 636 (1965).
30. W. W. Wells, in press.
31. S. M. Kim, R. Bentley, and C. C. Sweeley, unpublished work, 1966.
32. M. B. Perry, *Can. J. Biochem.,* **42,** 451 (1964).
33. J. Kärkkäinen, A. Lehtonen, and T. Nikkari, *J. Chromatog.,* **20,** 457 (1965).
34. J. M. Richey, H. G. Richey, and R. Schraer, *Anal. Biochem.,* **9,** 272 (1964).
35. J. Kärkkäinen, E. Haahti, and C. C. Sweeley, unpublished work, 1966.
36. K. A. Karlsson, *Acta Chem. Scand.,* **19,** 2425 (1965).
37. R. Ryhage, *Anal. Chem.,* **36,** 759 (1964).
38. J. F. Klebe, H. Finkbeiner, and D. M. White, *J. Am. Chem. Soc.,* **88,** 3390 (1966).
39. C. C. Sweeley, *Bull. Soc. Chim. Biol.,* **47,** 1477 (1965).
40. S. M. Kim, R. Bentley, and C. C. Sweeley, *Carbohydrate Res.,* in press.
41. E. J. Hedgley and W. G. Overend, *Chem. Ind. (London),* **1960,** 378.
42. F. A. Henglein, G. Abelsnes, H. Heneka, Kl. Lienhard, Pr. Nakhre, and K. Scheinost, *Makromol. Chem.,* **24,** 1 (1957).
43. J. S. Sawardeker and J. H. Sloneker, *Anal. Chem.,* **37,** 945 (1965).
44. R. J. Alexander and J. T. Garbutt, *Anal. Chem.,* **37,** 303 (1965).
45. H. E. Brower, J. E. Jeffery, and M. W. Folsom, *Anal. Chem.,* **38,** 362 (1966).

46. W. W. Wells, T. Chin, and B. Weber, *Clin. Chim. Acta,* **10**, 352 (1964).
47. P. O. Bethge, C. Holmström, and S. Julin, *Svensk Papperstid.,* **69**, 60 (1966).
48. M. D. G. Oates and J. Schrager, *Biochem. J.,* **97**, 697 (1965).
49. G. Wulff, *J. Chromatog.,* **18**, 285 (1965).
50. J. S. Sawardeker, J. H. Sloneker, and A. Jeanes, *Anal. Chem.,* **37**, 1602 (1965).
51. S. W. Gunner, J. K. N. Jones, and M. B. Perry, *Can. J. Chem.,* **39**, 1892 (1961).
52. E. P. Crowell and B. B. Burnett, *Anal. Chem.,* **39**, 121 (1967).
53. E. Sjöström, P. Haglund, and J. Janson, *Svensk Papperstid.,* **69**, 381 (1966).
54. M. I. Horowitz and M. R. Delman, *J. Chromatog.,* **21**, 300 (1966).
55. C. C. Sweeley and B. Klionsky, *Abstracts, Sixth International Congress of Biochemistry,* New York, 1964.
56. T. Yamakawa, S. Yokoyama, and N. Handa, *J. Biochem. (Tokyo),* **53**, 28 (1963).
57. S. Gatt and E. R. Berman, *J. Neurochem.,* **10**, 43 (1963).
58. C. C. Sweeley and B. Klionsky, in *The Metabolic Basis of Inherited Disease* (J. B. Stanbury, J. B. Wyngaarden, and D. S. Fredrickson, eds.), McGraw-Hill, New York, 1966, Chap. 29.
59. E. Svennerholm and L. Svennerholm, *Biochim. Biophys. Acta,* **70**, 432 (1963)
60. C. H. Bolton, J. R. Clamp, and L. Hough, *Biochem. J.,* **96**, 5C (1965).
61. T. Yamakawa, S. Nishimura, and M. Kamimura, *Japan. J. Exptl. Med.,* **35**, 201 (1965).
62. M. Abdel-Akher, J. K. Hamilton, R. Montgomery, and F. Smith, *J. Am. Chem. Soc.,* **74**, 4970 (1952).
63. C. C. Sweeley, unpublished work, 1964.
64. R. Ledeen and K. Salsman, *Biochemistry,* **4**, 2225 (1965).
65. M. B. Perry and R. K. Hulyalkar, *Can. J. Biochem.,* **43**, 573 (1965).
66. M. L. Wolfrom and A. Thompson, in *Methods in Carbohydrate Chemistry,* Vol. 1 (R. L. Whistler and M. L. Wolfrom, eds.), Academic Press, New York, 1962, p. 121.
67. C. H. Bolton, J. R. Clamp, G. Dawson, and L. Hough, *Carbohydrate Res.,* **1**, 333 (1965).

14

CHROMATOGRAPHIC ANALYSIS OF NITROGEN BASES DERIVED FROM LIPIDS

John M. McKibbin

DEPARTMENT OF BIOCHEMISTRY
UNIVERSITY OF ALABAMA MEDICAL CENTER
BIRMINGHAM, ALABAMA

Chromatographic separations of nitrogen bases derived from lipids have served two areas of lipid research: the analysis of lipid mixtures containing phosphatide and glycolipid and studies of the metabolism of the individual bases. The need for the former has been largely superseded by present methods for chromatographic resolution of the intact lipids and by specific methods for determination of the bases in hydrolysate mixtures. However, the complete resolution of lipid mixtures, including the identification of minor components, probably has not been achieved for any tissue, and methods of resolution of the hydrolysis components, including the nitrogen bases, will undoubtedly prove useful in reaching this goal. This chapter concerns the chromatographic separations of base mixtures that have been useful in metabolic and analytical studies. The resolution of the sphingosines and eicosisphingosines is considered in Chapter 11.

I. Resolution of Serine, Ethanolamine, N-Methyl Ethanolamines, and Choline

A. Ion-Exchange Chromatography

The biosynthesis of lecithins by methylation of phosphatidyl ethanolamine (PE), *N*-monomethyl PE, and *N*-dimethyl PE has been studied in animal tissues and in the select microorganisms that synthesize lecithin. This subject area has been reviewed recently by van Deenen and DeHaas (*1*). Effective separations of the four bases are required for study of the action of PE *N*-methyl-*trans*-ferases, using isotopically labeled methyl groups. Pilgeram et al. (*2*) hydrolyzed the lipid extracts with acid and placed the water-soluble fraction of the hydrolysate on a Dowex-50 column. Elution with 1.5 M HCl achieved separations of serine, ethanolamine, and choline. Most authors have used this system for separation of the methylated ethanolamines. Wolf and Nyc (*3*) and Crocken and Nyc (*4*) described a Dowex-50 H⁺ column 15 × 450 mm for use with *Neurospora crassa* lipid hydrolysates. The water-soluble fraction was placed on the column and eluted with 500 ml 1.5 N HCl. Serine, ethanolamine, monomethyl ethanolamine, dimethyl ethanolamine, and choline were eluted in that order as essentially homogeneous fractions. Gibson, et al. (*5*) used a Dowex-50-X8 H⁺ column 50 × 1 cm for fractionation of these bases from rat liver lipids. The column was eluted in the same fashion with comparable results. Bremer and Greenberg (*6*) and Bremer et al. (*7*) achieved comparable separations with Dowex-50 H⁺ columns eluted with 1.5 N HCl and observed that methionine elutes at about the same point as choline. This mixture was resolved with paper chromatography (*7,8*).

B. Paper Chromatography

Chargaff et al. (*9*) and Levine and Chargaff (*10,11*) first reported techniques for the preparation and resolution of several water-soluble lipid bases and amino acids. The author has found no system reported for the complete resolution of all of the bases with paper chromatography. However, several solvent systems are available for partial resolution, and these are given with R_f values for the several bases, presumably applied to the paper as chloride or hydrochloride salts, in Table 1.

The primary amines are usually detected with Ninhydrin or with naphthoquinone sulfonate. The latter has been particularly useful in this laboratory because it gives a relatively constant and similar molar color yield for amines and amino acids and gives a chromatogram with stable color. The chromatograms are sprayed with a solution containing 0.5% sodium 1,4-naphthoquinone sulfonate and 2% sodium tetraborate in 50% aqueous ethanol. They are then developed by heating in an oven at 90–100°C for 5 min. The amines appear as dark brown to purple spots against a light brown background. Choline can be detected by spraying with Dragendorff's reagent or by spraying with phosphomolybdic acid followed by reduction with stannous chloride (*9*).

II. Resolution of Hexosamines

A. Hexosamine Derivatives

The resolution of hexosamine mixtures in lipid hydrolysates, principally glucosamine and galactosamine, is of current significance in the analysis and characterization of glycolipids. The use of gas–liquid chromatography for this purpose is considered in Chapter 13. Several new methods have been devised for direct analysis of hexosamine mixtures or for separation of hexosamine derivatives rather than the more laborious separation of the hexosamines. Earlier chromatographic methods with dinitrophenyl (DNP) derivatives, with DNP hexosaminitols, or with pentoses formed by ninhdrin treatment have been reviewed by Lederer and Lederer (*14*). Cessi and Serafina-Cessi (*15*) treated hexosamine mixtures with acetyl acetone under anhydrous conditions. Glucosamine gave a nonvolatile pyrrole derivative and galactosamine a volatile pyrrole that was determined with Ehrlich's reagent.

Good and Bessman (*16*) substituted borate for carbonate buffers in both the Morgan-Elson and Elson-Morgan reactions. The glucosamine chromogen was about twice the galactosamine in the former and about one-quarter in the latter, permitting direct analysis of mixtures.

TABLE 1

R_f Values for Lipid Bases

Solvent system	Ref.	Serine	Glucosamine	Ethanolamine	N-methyl ethanolamine	N-dimethyl ethanolamine	Choline
n–Butanol–diethylene glycol–H₂O 4:1:1	11	0.19		0.38	0.47	0.49	0.38
	12	0.20		0.37			0.51
n–Butanol–morpholine 3:1 (sat. with water)	11	0.18		0.48	0.42		0.36
n–Butanol–morpholine 3:1	9	0.11		0.51			0.30
n–Butanol–dioxane 4:1 (sat. with water); upper phase used	11				0.26	0.23	0.25
n–Butanol–dioxane 4:1	9	0.14		0.27			0.26
n–Butanol–pyridine 4:1 (sat. with water); upper phase used	11				0.24	0.18	0.16
n–Butanol–pyridine 4:1	9	0.05		0.25			0.24
n–Butanol–ethylene glycol–water 4:1:1	13	0.32	0.34	0.45			0.46
n–Butanol–acetic acid–water 2:1:2	13	0.42		0.53			0.59
n–Butanol–acetic acid–water 4:1:5; upper phase used	13	0.17	0.09	0.18			0.17
n–Butanol–phenol–80% formic acid–water 50:50:3:10, v/w/v/v	8			0.16	0.38	0.63	0.78

Johnston (*17*) used yeast hexokinase for the specific phosphorylation of glucosamine followed by separation of the cationic galactosamine from 6-phospho glucosamine on a Dowex-50-X5 column. Luderitz et al. (*18*) used a yeast enzyme preparation containing both hexokinase and an acetylating system that converted glucosamine to *N*-acetyl glucosamine-6-phosphate. The incubation mixture was then treated with borate followed by the Morgan-Elson reagent. Because of the specificity of either or both the phosphorylation and *N*-acetylation reactions, neither galactosamine nor mannosamine interfered.

Sempere et al. (*19*) used the galactose oxidase of *Polyporus circinatus* (*20*) for the specific oxidation of galactosamine and *N*-acetyl galactosamine in the presence of glucosamine. Scott (*21,22*) prepared the phenylisothiocyanate derivatives of these hexosamines and separated them by electrophoresis in the presence of molybdate or by ascending paper chromatography on molybdate-treated paper using isopropanol–H_2O 4:1, v/v, as the developing solvent. Quantitative determinations of the hexosamine derivatives were carried out by elution of the paper with 10% aqueous barium acetate and measurement of absorption at 240 mμ. The molar color yields of the derivatives of glucosamine and galactosamine were found to be identical.

B. Ion-Exchange Chromatography

Hexosamines in glycolipid hydrolysates are predominantly in the free form rather than the *N*-acyl form, since conditions drastic enough to hydrolyze glycosidic bonds of amino sugars are more than sufficient to cleave the *N*-acyl bonds. Gardell (*23*) first separated galactosamine from glucosamine on Dowex-50 H$^+$ columns. The hydrolysate containing 60–800 μg of each hexosamine was taken up in 0.3 N HCl and placed on a 0.6 × 40 cm column (cross-linking not specified). The columns were eluted with 0.3 N HCl and a flow rate of 1.5 to 2.0 ml/hr. Glucosamine eluted completely in the volume range 68–75 ml; galactosamine in 83- to 90-ml range. Higher concentrations of HCl gave lower effluent volumes and poorer separation. This procedure was used successfully by Brante (*24*) with hydrolysates of liver polysaccharides in gargoylism and by Svennerholm (*25*) with lipid hydrolysates from nervous tissue.

Pearson (*26*) used 0.1 M pH 5.0 citrate buffer for elution of Dowex-50 columns in the separation of glucosamine from galactosamine in the presence of large amounts of amino acids. In the absence of hydroxylysine, Amberlite CG-120 was equally effective when eluted with either 0.3 N HCl or 0.35 M pH 5.26 citrate buffer. Column chromatography of the hexosamines is conveniently monitored with the Ninhydrin reaction in the absence of other amines or amino acids or with the Elson-Morgan reaction.

C. Paper Chromatography

The similar chromatographic properties of glucosamine and galactosamine make effective separations on paper unusually difficult. A number of solvent systems useful in the resolution of other sugar mixtures do not achieve separation of glucosamine, galactosamine, or their hydrochlorides (27). However, several solvent systems offer enough difference in R_f to give good separations on long descending chromatograms. Masamune and Yosizawa (28) presented data on seven different solvent systems useful in separation of sugars and amino sugars by descending chromatography. In most of these the hexosamine sulfates were better separated than the hydrochlorides. Caldwell (29) preferred system II, *n*-butanol–pyridine–water 5:3:2, using a development time of 40 hr and allowing the solvent to drip from the serrated end of 50-cm paper strips. This system was especially useful for qualitative work with glycolipid hydrolysates (30). Water-saturated collidine has also provided useful separation of these hexosamines (31). Hornung (32) used descending chromatography with the system ethyl acetate–pyridine–water 12:5:4 and a filter pad fastened to the bottom of the chromatograph paper. This continued solvent flow or, in effect, increased the length of the chromatogram. Glucosamine had a higher R_f than galactosamine, and good separations were obtained. All of the systems required 24 to 48 hr for good separation. Detection systems for the chromatograms include Ninhydrin, the Elson-Morgan reagent, and the naphthoquinone-4-sulfonate reagent.

Rosenthal et al. (33) used quaternary amine salts for identification of sugars and amino sugars on paper chromatograms. After development, the chromatograms were sprayed with solutions of the amine salts, dried, heated 3 to 5 min at 100°C, and examined with ultraviolet light of 3600 Å. The chromatograms appeared faintly fluorescent, and the sugar areas showed quenching with slight coloration due, presumably, to complex formation. Proflavine hydrochloride was the most sensitive of eight quaternary amine salts tried, requiring 10 to 50 μg of sugar to give visually detectable color. Five solvent systems giving separation of glucosamine from galactosamine were found (Table 2).

TABLE 2

SEPARATION OF GALACTOSAMINE FROM GLUCOSAMINE BY
PAPER CHROMATOGRAPHY (*33*)

	R_f		Color	
Solvent system	Galactos-amine	Glucos-amine	Galactos-amine	Glucos-amine
Pyridine–ethyl acetate–acetic acid–water 5:5:1:3	0.26	0.18	Green	Green
Butanol–ethanol–water 4:1.1:1.9	0.06	0.16	Green	Green
t-Amyl alcohol–propanol–ethanol–water 4:1.3:0.5:2	0.18	0.12	Violet	Green
Butanol–acetic acid–water 5:1.4:2.9	0.08	0.18	Violet	Green
t-Amyl alcohol–formic acid–water 4:1:1.5	0.09	0.16	Violet	Green

Conversion of free amino sugars to *N*-acetyl amino sugars offers some advantages in separation. The amino sugars may be converted to the *N*-acetyl derivative by the method of Roseman and Ludowieg (*34*). After development, the *N*-acetyl hexosamines may be detected by spraying with a mixture of 95% ethanol–0.05 M Na$_2$B$_4$O$_7$ 1:1 and heating in an oven at 100°C for 10 min, followed by spraying with a mixture of 2% *p*-dimethyl amino benzaldehyde in glacial acetic acid (10 ml), 30 ml *n*-butanol, and 0.4 ml conc. HCl (*35*). The chromatograms develop by drying at 20–35°C for ½–2 hr. The pigment may be eluted with ethyl acetate–glacial acetic acid–water 3:1:1 for colorimetric determination, inasmuch as the spectra are identical with the usual modified Elson-Morgan pigment (*35*).

The system *n*-butanol–pyridine–water 6:4:3 gave good separation of *N*-acetyl glucosamine, R_f 0.26, from *N*-acetyl galactosamine, R_f 0.12, on borate-impregnated paper (*36*). Good separations were also obtained on borate-treated paper with ethyl acetate–pyridine–water 2:1:2 (*37*). In 20 hr *N*-acetyl glucosamine moved 24.7 cm and *N*-acetyl galactosamine 20.2 cm. Chromatography of the *N*-acetyl hexosamines has an advantage in that these substances migrate in a single spot rather than in two spots, as seen frequently in the chromatography of hexosamine salts in acidic solvents.

D. Thin-Layer Chromatography

Very little has been reported thus far on hexosamine separations with this important time-saving analytical technique. Gunther and Schweiger (*38*) used cellulose powder plates (MN 300, Machery and Nagel, Duren, Germany) 0.25 mm thickness with three solvent systems, which gave significant separation of hexosamines (presumably the salts). *N*-acetyl derivatives moved much faster than the hexosamine salts but were not as well separated (Table 3). The amino

TABLE 3

THIN-LAYER CHROMATOGRAPHY OF HEXOSAMINES AND *N*-ACETYL
DERIVATIVES WITH CELLULOSE POWDER (*38*)

	R_G values[a] for solvent systems[b]		
	I	II	IV
Glucosamine	1.00	1.00	1.00
Galactosamine	0.91	0.83	0.88
N-Acetyl glucosamine	1.28	1.62	1.82
N-Acetyl galactosamine	1.24	1.53	1.70

[a] R_f relative to glucosamine.
[b] Solvent systems: I, butanol–ethanol–isopropanol–ammonia–water 2:4:0.5:0.5:1.5; II, pyridine–ethyl acetate–acetic acid–water 5:5:1:3; IV, ethyl acetate–pyridine–tetrahydrofuran–water 7:3:2:2. Development time 2 to 3 hr.

sugars were detected with a thiobarbituric acid reagent or with Ninhydrin. By spraying the plates with an aqueous borate buffer (0.2 *M* boric acid, 0.05 *M* $Na_2B_4O_7$, 0.05 *M* NaCl) excellent separations of the *N*-acetyl derivatives were obtained with the system ethyl acetate–isopropanol–pyridine–water 7:3:2:2. Two-dimensional thin-layer chromatography was equally successful using cellulose MN-300 A, which gives good detection with the Elson-Morgan reagent. This system gives separations in 3 hr comparable with those obtained with paper chromatography in 24 hr.

Stahl and Kaltenbach (*39*) used Alusil plates (silica gel G with aluminum oxide G 1:1) "active" (heated) and "inactive" (not heated).

Active plates gave reasonably good separation of the hexosamines (R_f 0.40 and 0.48 for galactosamine and glucosamine, respectively) with *n*-propanol–ethyl acetate–water–25% ammonia 6:1:3:1. Separation on the "inactive" plates was not as effective, with R_f values of 0.37 and 0.42, respectively.

III. Resolution of Nitrogen Bases from Whole-Lipid Hydrolysates

The chromatographic resolution of all of the nitrogen bases from hydrolysates of whole lipid extracts has not been reported, although the resolution of groups of the major bases has been considered above. The nature and significance of the minor components found in hydrolysates of whole lipid extracts are of some interest, although there are few published accounts of studies directed at this problem. Several technical problems must be considered, the first being the nonlipid nitrogenous components found in lipid extracts, which are artifacts of extraction. It is difficult to prepare lipid extracts that contain all of the lipids of a tissue and a minimum of artifact. Most workers use mild extraction procedures designed to reduce chemical artifacts and accept an incomplete extraction. The use of large solvent volumes helps to reduce extraction artifacts by all methods in general use. Regardless of the method of extraction, the contaminants should be removed by some purification procedure.

Another problem is the destruction of bases and other artifacts produced by hydrolysis. Hydrolysis with aqueous or alcoholic hydrochloric or sulfuric acids under conditions drastic enough to release all of the bases may give low recoveries of ethanolamine, serine, or sphingosine and, especially, sialic acids. To reduce hydrolysis losses, McKibbin et al. (*40*) saponified lipid extracts from eight different dog tissues in aqueous $1 N$ NaOH at ambient temperature for 3 days. After acidification and chloroform extraction, 76 to 89% of the non-sphingosine nitrogen was in the acid aqueous phase as free base and about 5% more as phosphate ester. The desalted base hydrochloride preparation was fractionated on Dowex-50 Na$^+$. Five fractions were eluted: an acidic fraction with 0.01 N HCl, one with 0.05 M pH 5.0 sodium acetate buffer, one with 0.05 M pH 9.0 NaHCO$_3$–Na$_2$CO$_3$ buffer, one with 0.1 M Na$_2$CO$_3$, and finally, one with 4 N HCl. Free serine, ethanolamine, and choline were specifically eluted in the second, fourth, and fifth fractions, respectively. The small amounts of phosphoric acid esters of these bases were found only in the first fraction. Phosphoryl ethanolamine and phosphoryl serine were hydrolyzed with $2\,N$ HCl at 120°C and phosphoryl choline with $5.5\,N$ HCl at 130°C. Any destruction resulting from hydrochloric acid hydrolysis was thus confined to this small fraction. The minor constituents were found in the first three fractions, which contained only 17 to 30% of the original total nitrogen. Hexosamines and sialic acid were not determined in these fractions, since they were expected in the nonsaponifiable fraction.

The chloroform-soluble or nonsaponifiable fraction contained 10 to 30% of the original total lipid nitrogen. Hydrolysis with barium hydroxide released only 50 to 70% of this as choline and sphingosine, an unexpectedly low proportion of this fraction. The other major constituent was ethanolamine, and it exceeded choline in six of the eight lipid extracts analyzed. The nature of this ethanolamine-containing substance has not been established, but it is not ethanolamine glyceryl ether phosphatide (*41*). Wren and Holub (*42*) have suggested that it may be ethanolamine fatty acid amide formed by base-catalyzed aminolysis of esters. After analysis of all of the fractions it was found that 1.4 to 9.5% of the total lipid nitrogen in the several tissues could not be identified with the known lipid bases and ammonia. A large part of this, 0.9 to 5.1% of total nitrogen, was considered to be substituted amine, since it was principally in the pH 9 eluate and not hydrolyzed by acid to primary amine. No significant amount of this fraction has been identified. In most of the tissues the fraction was too large to have been a sialic acid and hexosamine artifact. It is still uncertain whether this nitrogen is truly of lipid base origin or, if so, whether the bases have not been altered in the course of preparation.

REFERENCES

1. L. L. van Deenen and G. H. DeHaas, *Ann. Rev. Biochem.,* **35,** 157 (1966).

2. L. O. Pilgeram, E. M. Gal, E. N. Sassenrath, and D. M. Greenberg, *J. Biol. Chem.,* **204,** 367 (1953).

3. B. Wolf and J. F. Nyc, *Biochim. Biophys. Acta,* **31,** 208 (1959).

4. B. J. Crocken and J. F. Nyc, *J. Biol. Chem.,* **239,** 1727 (1964).

5. K. D. Gibson, J. D. Wilson, and S. Undenfriend, *J. Biol. Chem.,* **236,** 673 (1961).

6. J. Bremer and D. M. Greenberg, *Biochim. Biophys. Acta,* **46,** 205 (1961).

7. J. Bremer, P. H. Figard, and D. M. Greenberg, *Biochim. Biophys. Acta,* **43,** 477 (1960).

8. T. Kaneshiro and J. H. Law, *J. Biol. Chem.,* **239,** 1705 (1964).
9. E. Chargaff, C. Levine, and C. Green, *J. Biol. Chem.,* **175,** 67 (1948).
10. C. Levine and E. Chargaff, *J. Biol. Chem.,* **192,** 465 (1951).
11. C. Levine and E. Chargaff, *J. Biol. Chem.,* **192,** 481 (1951).
12. M. M. Gertler, J. Kream, and O. Baturay, *J. Biol. Chem.,* **207,** 165 (1954).
13. J. M. McKibbin and R. Freebern, unpublished data.
14. E. Lederer and M. Lederer, in *Chromatography,* Elsevier, Amsterdam, 1957, Chap. 28.
15. C. Cessi and F. Serafina-Cessi, *Biochem. J.,* **88,** 132 (1963).
16. T. A. Good and S. P. Bessman, *Anal. Biochem.,* **9,** 253 (1964).
17. I. R. Johnston, *Biochem. J.,* **86,** 254 (1963).
18. O. Luderitz, A. R. Simmons, O. Westphal, and J. L. Strominger, *Anal. Biochem.,* **9,** 263 (1964).
19. J. M. Sempere, C. Gancedo, and C. Asensio, *Anal. Biochem.,* **12,** 509 (1965).
20. G. Avigad, D. Amaral, C. Asensio, and B. L. Horecker, *J. Biol. Chem.,* **237,** 2736 (1962).
21. J. E. Scott, *Biochem. J.,* **82,** 43P (1962).
22. J. E. Scott, *Biochem. J.,* **92,** 57P (1964).
23. S. Gardell, *Acta Chem. Scand.,* **7,** 207 (1953).
24. G. Brante in *Cerebral Lipidoses, A Symposium* (J. N. Cumings, ed.), Charles C Thomas, Springfield, Ill., 1957, p. 164.
25. L. Svennerholm, in *The Amino Sugars,* Vol. II (E. A. Balazs and R. W. Jeanloz, eds.), Academic Press, New York, 1965, Chap. 36.
26. C. H. Pearson, *Biochem. J.,* **88,** 540 (1963).
27. E. Heftmann, in *Chromatography* Reinhold, New York, 1961, p. 505.
28. H. Masamune and Z. Yosizawa, *Tohoku J. Exptl. Med.,* **59,** 1 (1953).
29. R. C. Caldwell, Ph.D. thesis, Univ. Alabama, Tuscaloosa, 1964.
30. W. R. Vance, C. P. Shook, III, and J. M. McKibbin, *Biochemistry,* **5,** 435 (1966).
31. S. M. Partridge and R. G. Westall, *Biochem. J.,* **42,** 238 (1948).
32. M. Hornung, *J. Bacteriol.,* **86,** 1345 (1963).
33. W. A. Rosenthal, S. Spaner, and K. D. Brown, *J. Chromatog.,* **13,** 152 (1964).
34. S. Roseman and J. Ludowieg, *J. Am. Chem. Soc.,* **76,** 301 (1954).
35. M. R. Salton, *Biochim. Biophys. Acta,* **34,** 308 (1959).
36. J. L. Strominger, *Biochim. Biophys. Acta,* **30,** 645 (1958).
37. R. G. Spiro, *J. Biol. Chem.,* **234,** 742 (1959).
38. H. Gunther and A. Schweiger, *J. Chromatog.,* **17,** 602 (1965).
39. E. Stahl and U. Kaltenbach, in *Thin Layer Chromatography* (E. Stahl, ed.), Academic Press, New York, 1965, p. 463.
40. J. M. McKibbin, S. Meltzer, and M. J. Spiro, *J. Lipid Res.,* **2,** 328 (1961).
41. S. Nakagawa and J. M. McKibbin, *Proc. Soc. Exptl. Biol., Med.,* **111,** 634 (1962).
42. J. J. Wren and D. S. Holub, *Biochem. J.,* **90,** 3P (1964).

AUTHOR INDEX

Numbers in parentheses are reference numbers and indicate that an author's work is referred to although his name is not cited in the text. Numbers in italics show the page on which the complete reference is listed.

M

SUBJECT INDEX

A

Acetolysis, of glycerol, 459–461

Acyl esters, GLC of, 254

Adsorption chromatography, of alkoxy lipids, 348–349
of neutral glycerides and fatty acids, 206, 207–223
on alumina, 221–222
on Florisil, 217, 218–221
on silicic acid, 207–218

Alcoholysis, of phosphatides and glycerides, analysis, 163–189
methods, 168–171

Aldehydes, long-chain base-derived, GLC of, 439–442
long-chain, GLC of, 401
acid derivatives, 405–406
alcohol-derivatives, 405
from biochemical sources, 407–409
in biochemical studies, 422–426
DMA derivatives, 403–405, 408, 412–422, 423–425
identification, 413–422
methods, 409–412
reference standards, 406–407
volatile derivatives, 403–406

Alkoxy acetaldehydes, GLC of, 351

Alkoxy lipids, adsorption chromatography of, 343–345
detection of, 343–345
fractionation of classes, 348–354
GLC and isolation of, 339–360
applications of, 355–357
identification of, 354–355
isolation, 345–348
occurrence and structure, 340–343

Alkyl diglycerides, isolation, 346–348

Alkyl glycerol-(1) ethers, alkoxy acetaldehydes of, 351
GLC of, 349–350
applications, 355–357
isopropylidene derivatives of, 349–350
"TFA" derivatives of, 351–352
"TMS" derivatives of, 352–354

Alkyl 2,3-*O*-isopropylidene glycerol-(1) ethers, GLC of, 349–350

Allyl ethers, GLC of, 255

Alumina column chromatography, of neutral glycerides and fatty acids, 221–222

Anacystis nidulans, lipids, TLC of, 91

Apiezon greases, use in GLC, 410 ff.

Argentation TLC, of glycerides and fatty acids, 194–197, 348

Autoradiography, of phosphatides and sulfolipids, 10–11

B

Bacteria, lipids, analysis, 26–31

Bases (*See also* Sphingosine)
aldehydes from GC of, 439
biochemistry of, 432
chemistry of, 430–431
C_{18} type, TLC of, 433–436
direct analysis, 442–443
instability of, 438
long-chain, TLC of, 436–437
nitrogen, *see* Nitrogen bases
TLC of, 432–438

Benzylidene monoglyceride derivatives, GLC of, 255

Bile, lipid analysis of, 156–157

Bitrifluoroacetyl alkyl glycerol ethers, GLC of, 351–352

Bitrimethylsilyl glycerol ethers, GLC of, 352